Rolf Brendel

Thin-Film Crystalline Silicon Solar Cells

Physics and Technology

Rolf Brendel

Thin-Film Crystalline Silicon Solar Cells

Physics and Technology

With a Foreword of A. Goetzberger

WILEY-VCH GmbH & Co. KGaA

Autor:
Dr. Rolf Brendel
Bayerisches Zentrum für Angewandte Energieforschung e. V.
ZAE Bayern,
e-mail: brendel@zae.uni-erlangen.de

Library of Congress Card No. applied for.

British Library Cataloguing-in-Publication Data:
catalogue record for this book is available from the British Library.

Bibliographic information published by Die Deutsche Bibliothek
Die Deutsche Bibliothek lists this publication in the Deutsche Nationalbibliografie; detailed bibliographic data is available in the Internet at <http://dnb.ddb.de>.

Printed in the Federal Republic of Germany

Printed on acid-free paper

Printing Strauss Offsetdruck GmbH, Mörlenbach

Bookbinding Großbuchbinderei J. Schäffer GmbH & Co. KG, Grünstadt

ISBN 3-527-40376-0

O send Your light and Your truth,
let them lead me.

Psalm 43:3

Foreword

This book by Rolf Brendel closes a gap in the literature on photovoltaics, in particular on silicon solar cells. While there are several books on the general aspects of this topic available, they are limited mostly to the theory and practice of bulk silicon solar cells. The present book emphasizes thin silicon solar cells and treats the subject in a very comprehensive manner. Dr. Brendel is exceptionally qualified to write such a book because he has contributed personally in important ways to this field.

The crystalline silicon solar cell in its conventional form dominates today, with about 90% of the world market. This dominance of the market is on one hand surprising, because silicon as an indirect semiconductor has a relatively low absorption coefficient for a large fraction of the wavelengths of the solar spectrum. On the other hand silicon photovoltaics benefits from the large know-how developed in the past for all kinds of silicon devices. In order to absorb enough of the infrared sunlight to achieve a high efficiency, silicon cells have to have a thickness of several hundred micrometers. In addition, the material has to be of extreme purity and good crystalline perfection. Therefore the potential for cost reduction of this technology is limited. The impressive cost reduction achieved so far results partly from increased production volume and corresponding improvements of technology, but also from the availability of cheap surplus semiconductor-grade feedstock material. Photovoltaics profits from the fact that off-spec silicon not suitable for the semiconductor industry can still be used for solar cells. This dependence on the raw material base of another industry has its limitations, which are now being reached. Shortages of silicon for the photovoltaic industry are occurring periodically, depending on the ups and downs of demand in the semiconductor device market.

A completely different approach are thin-film materials with a direct bandgap. These genuine thin-film materials are characterized by a very high absorption. Therefore they are used with a thickness in the micrometer range. The oldest such material is amorphous silicon, which is mainly used for consumer products. Other strong contenders are chalcogenides like CIS (copper indium diselenide) and cadmium telluride. All these materials have been under development for many years and are now still in the stage of pilot production. It is still doubtful if they will reach the ambitious cost goals planned for them. The main reason for their slow progress seems to be the fact that both materials and technology have to be developed from scratch.

An alternative are thin layers of crystalline silicon on foreign substrates. As mentioned above, the problem of low absorptivity of this material has to be overcome. This can be done by clever optical design, as was pointed out many years ago. The key is multiple reflections of the light within the thin film. These concepts, however, remained theoretical until recently when the bulk silicon technology began to reach its limits. Several approaches exist for the realization of the crystalline thin-film solar cell. The most straightforward is the deposition of silicon from the gas phase by chemical vapor deposition. High-temperature and low-temperature approaches are possible. The best results have been achieved with the transfer technique, which uses films transferred from the surface of monocrystalline wafers. This technique, to wich the author of the present book has contributed extensively, requires a very small amount of silicon because the substrate can be used many times over.

The book starts with describing the present state of the technology of crystalline silicon cells. Then a very complete introduction to the theory of thin solar cells is given. The thermodynamic and quantum mechanical limitations of efficiency are outlined and then the practical limitations of efficiency are introduced step by step. Several new concepts are introduced. The experimental part starts with an exhaustive overview over the techniques for the realization of crystalline thin-film solar cells. The chapters on layer transfer processes are particulary interesting because the author describes many of his own results. Brendel has contributed a new concept, the PSI cell, wich combines the transfer technique with optimal light trapping by a waffle structure. The appendices contain more detailed theoretical treatments of some important subjects.

This book can be highly recommended for all interested in a new chapter of silicon solar cells which is just opening up.

A. Goetzberger
Fraunhofer Institute for Solar Energy Systems

Preface

Photovoltaics with thick crystalline Si wafers is a mature technology that is currently entering large-scale production. For a widespread solar electric power generation, however, a substantial reduction of the fabrication cost is required. For this purpose thin-film technologies are being developed with only micron-thick semiconductor layers for light absorption. While thin-film modules from amorphous Si, $Cu(In,Ga)Se_2$, and CdTe are already being commercially produced on a small scale, the development of thin-film modules from crystalline Si is still in the laboratory phase. This phase is characterized by competition of many different approaches for depositing and fabricating the thin crystalline Si solar cells.

This book is adressed to the physicist and the engineer who are interested in finding their way through the many approaches that are currently under investigation. It is also hoped that the reader gains an insight into the fundamental physical loss mechanisms that occur in solar cells. These physically inevitable losses set upper efficiency limits for thin-film cells that are more restrictive than for thick cells. The book also covers advanced device characterization by quantum efficiency analysis. If possible, analytic treatments of the optical and the transport properties are preferred. Such models permit a time-efficient and transparent modeling of many – obviously not all – the effects observed in thin-film cells. I encourage the reader to apply and modify these models to solve his own research problems. A review of recent developments in the field of thin-film crystalline Si cells discloses a wealth of novel technological routes towards highly efficient and potentially easy-to-fabricate thin-film crystalline Si modules. However, it is still by no means clear that any of these routes is clever enough to compete with conventional crystalline Si wafer technology, which is a fast-moving target for all thin-film technologies. If this book could inspire one of its readers to introduce new concepts for fabricating and understanding thin-film solar cells, it was really worth the effort of writing it.

The foundation of this book is my research conducted at the Max-Planck-Institut für Festkörperforschung (MPI-FKF) in Stuttgart from 1992 to 1997 and at the Bavarian Center for Applied Energy Research (ZAE Bayern) in Erlangen from 1997 to 2001. The other source of the book is a course on the "Physics of crystalline Si solar cells" that I held for graduate Physics students at the University of Erlangen-Nuremberg. The exchange of ideas with the students and colleagues contributed valuable aspects to my current understanding of solar cells and made the research on thin-film photovoltaics an exciting delight.

I thank Prof. H. J. Queisser, director at the MPI-FkF, for actively supporting my post-doc research and for his continuous encouragement to leave the beaten tracks. Much of my thin-film Si solar cell work had not been possible without successful cooperation with Prof. J. H. Werner during my time at MPI-FkF, sharing of visions with Dr. R. Plieninger, open-minded exchange of ideas with Dr. U. Rau, tedious lifetime measurements carried out by Dr. M. Schöfthaler, Dr. M. Wolf's quantum efficiency analysis work, and the excellent technical support of Dipl.-Ing. B. Fischer, Dipl.-Ing. B. Winter, and G. Markewitz.

I thank Prof. M. Schulz, director at ZAE Bayern, for giving me the chance to head the department for Thermosensorics and Photovoltaics and for the scientific freedom to establish new photovoltaic research activities at ZAE Bayern. I thank Dipl.-Ing. R. Auer for leading the technological work with an apparently never ending idealism, and Dr. V. Gazuz, Dipl.-Ing. W. Kinzel, and Dipl.-Ing R. Horbelt for fully committing themselves to solar cell fabrication. Many thanks also go to our PhD students Dipl.-Phys. M. Bail for lifetime measurements, Dipl.-Phys. K. Feldrapp for cell analysis, Dipl.-Phys. G. Kuchler for ion-assisted deposition, Dipl.-Phys. G. Müller for porous Si multi-layer design, and Dipl.-Phys. D. Scholten for multi-dimensional device simulations. The administrative skills of A. Kidzun greatly helped me to devote more of my time to science.

I also thank our project partners, Dr. S. Oelting from ANTEC GmbH in Kelkheim, Dr. H. Artmann and Dr. W. Frey from Robert Bosch GmbH in Gerlingen, Dr. H. v. Campe and Dr. W. Hoffmann from RWE Solar GmbH in Alzenau, Dipl.-Ing. J. Krinke and Prof. H. P. Strunk from the Institute of Microcharacterisation at the University of Erlangen-Nuremberg, Dipl. Phys. H. Nagel, M. Steinhof, and Prof. R. Hezel from the Institut für Solarenergieforschung Hameln (ISFH) in Hameln, and Dr. G. Wagner from the Institut für Kristallzüchtung in Berlin. I thank Dr. W. Appel from the Institut für Mikroelektronik Stuttgart for his willingness to perform high-temperature Si depositions on unusual substrates like glass and porous Si.

I thank my wife Christiane, who tolerated my absence from home on many evenings, weekends, and holidays throughout the last few years and actively supported my work on this book with her love.

Erlangen, March 2001 R. Brendel

Contents

Symbols and Acronyms

Latin symbols

symbol	unit	
A		optical absorption
A_c	m^2	macroscopic cell area
A'		injection-dependent optical absorption
c	$m\ s^{-1}$	vacuum velocity of light
C		optical concentration factor
C_{max}		maximum optical concentration
C_n	$m^6\ s^{-1}$	Auger recombination coefficient for eeh processes
C_p	$m^6\ s^{-1}$	Auger recombination coefficient for ehh processes
D_{it}	$m^{-2}J^{-1}$	interface state density
D_n	$m^2\ s^{-1}$	electron diffusion coefficient
D_p	$m^2\ s^{-1}$	hole diffusion coefficient
E_C	J	edge of the conduction band
E_F	J	Fermi level at equlibrium
E_{Fn}	J	quasi-Fermi level of the electrons
E_{Fp}	J	quasi-Fermi level of the holes
E_g	J	semiconductor energy gap
EQE		external quantum efficiency
E_V	J	edge of the valance band
g	$m^{-3}\ s^{-1}$	carrier generation rate
G	A	photogeneration current
G	m	grain size
h	J s	Planck's constant
$\hbar$	J s	Planck's constant divided by 2π
$I_{AM1.5G}$	$W\ m^{-2}\ nm^{-1}$	energy flux density of global AM1.5 spectrum per wave-length interval
I_{mpp}	A	current at the maximum-power point
IQE		internal quantum efficiency
IQE^*		corrected internal quantum efficiency under forward bias
j_E	$s^{-1}\ m^{-2}$	energy flux density per solid angle and energy interval
j_h	$A\ m^{-2}$	hole current density
j_n	$A\ m^{-2}$	electron current density
j_{sc}	$A\ m^{-2}$	short-circuit current density
j_{sc}^*	$A\ m^{-2}$	maximum short-circuit current or photogeneration
k	$J\ K^{-1}$	Boltzmann's constant
l	m	path length of light in the cell
l		reduced minority carrier diffusion length L/G
$\bar{l}$	m	average path length of light in the cell
L	m	minority carrier diffusion length in the base of the cell

L_C	m	Collection length derived from quantum efficiency measurements under spatially homogeneous carrier generation
L{*f*}		Laplace transform of function *f*
L_J	m	diffusion length from the current-voltage curve $j(U)$
L_Q	m	diffusion length from *IQE* for strongly absorbed light
$L_{Q\infty}$	m	diffusion length from *IQE* for weakly absorbed light
L_α	m	optical absorption length
m		multiplicity: number of electron-hole pairs created per absorbed photon
n	m^{-3}	electron concentration
N_A	m^{-3}	acceptor concentration
N_D	m^{-3}	donor concentration
n_s		index of refraction of Si
n_{sur}	m^{-3}	surface or interface concentration of electrons
n_W	m^{-3}	electron concentration at the edge of the space charge region
N_γ	$m^{-2}\ s^{-1}$	photon flux density
$n_{\gamma c}$	$J^{-1}\ s^{-1}\ m^{-2}$	luminescence photon flux per photon energy interval and étendue
$n_{\gamma s}$	$J^{-1}\ s^{-1}\ m^{-2}$	photon flux of the sun per photon energy interval and étendue
p	m^{-3}	hole concentration
(p, q, r)		unit vector of the direction of propagation of a ray
P_{abs}	W	solar radiation power absorbed by the cell
P_{inc}	W	solar radiation power irradiating the cell
p_{sur}	m^{-3}	surface or interface concentration of holes
p_W	m^{-3}	hole concentration at the edge of the space charge region
q	C	elementary charge
Q_f	C	fixed charges in the dielectric layer
Q_G	C	charge on the metal gate or corona charges
Q_{it}	C	charge in interface states
Q_o	C	interface or surface charge at equilibrium
Q_{sc}	C	charge in semiconductor
$\boldsymbol{r}$	m	position vector
R		optical reflectance
R	$J\ s^{-1}\ m^{-2}$	radiance: power per area and projected solid angle
R_{Aug}	A	Auger recombination current
R_b		reflectance of the back reflector
R_{grb}	A	grain boundary recombination current
R_{rad}	A	radiative recombination current
R_s		surface reflectance
R_{SRH}	$m^{-3}\ s^{-1}$	Shockley-Read-Hall recombination rate
R_{sur}	A	surface recombination current
R_σ		ratio of electron to hole capture cross-section
S	$m\ s^{-1}$	surface recombination velocity
s_b		reduced back surface recombination velocity $S\ G/D_n$
S_b	$m\ s^{-1}$	back surface recombination velocity
S_{diff}	$m\ s^{-1}$	differential surface recombination velocity
S_{grb}	$m\ s^{-1}$	grain boundary recombination velocity

T		optical transmittance
T_c	K	temperature of the solar cell
T_d	K	deposition temperature
T_f		transmittance of the front surface of the cell
T_s	K	temperature of the sun
T_t		transmittance of front surface from inside the cell
U	V	voltage
U_{grb}	$s^{-1}\ m^{-2}$	grain boundary recombination rate
U_{mpp}	V	voltage at the maximum-power point
U_{oc}	V	open-circuit voltage
U_{rad}	$s^{-1}\ m^{-3}$	radiative recombination rate
U_{rec}	$s^{-1}\ m^{-3}$	recombination rate
U_{sur}	$s^{-1}\ m^{-2}$	surface recombination rate
U_t	V	thermal voltage
V_c	m^3	cell volume
W_{bas}	m	thickness of the cell's base
W_e	m	thickness of the cell's emitter
W_{eff}	m	effective film thickness: cell volume divided by macroscopic cell area
W_f	m	film thickness measured perpendicular to the collecting junction
W_{scr}	m	thickness of space charge region
W_{sub}	m	thickness of the cell's substrate
x		reduced *x*-coordinate X/G
X	m	*x*-coordinate
y		reduced *y*-coordinate Y/G
Y	m	*y*-coordinate
z		reduced *z*-coordinate Z/G
Z	m	*z*-coordinate

Greek symbols

symbol	unit	
α		Si facet angle relative to macroscopic cell surface
α_{eff}	m^{-1}	effective optical absorption coefficient
α_s	m^{-1}	optical absorption coefficient of Si
β		glass facet angle relative to macroscopic cell surface
γ		facet angle of the Si film relative to macroscopic cell surface
γ_E		photon of energy E
ε_o	$F\ m^{-1}$	dielectric constant of vacuum
ε_s	$F\ m^{-1}$	static relative dielectric constant of Si
$\mathcal{E}$	m^2	étendue: area times projected solid angle
η		efficiency: cell output power devided by incident radiation power
η_{abs}		efficiency: cell output power devided by absorbed radiation

		power
η_c		local carrier collection efficiency
η_{Car}		Carnot efficiency
η_{net}		efficiency: cell output power devided by net input radiation power
ϑ	rad	angle of light ray relative to cell normal
λ	m	wavelength of light
λ_g	m	wavelength of photons with bandgap energy E_g
Λ		Lambertian character of surface
μ_n	$m^2 V^{-1} s^{-1}$	electron mobility
μ_p	$m^2 V^{-1} s^{-1}$	hole mobility
τ	s	minority carrier lifetime in the base of the cell
Φ	$J C^{-1}$	electrical potential
Φ_o	J	neutrality energy level
Φ_λ	$m^{-2} s^{-1}$	flux of photons with wavelength λ
Ψ_{sur}	V	surface potential
ω	s^{-1}	frequency times 2 π
Ω	rad	solid angle

Latin acronyms

ac	alternating current
ABS	Alig-Bloom-Struck theoryfor impact yield
AM	air mass
AM1.5G	global solar spectrum of air mass 1.5
ARC	antireflection coating
a-Si	amorphous silicon
BSF	back surface field
CLEFT	cleavage of lateral epitaxial films for transfer
CPM	constant photocurrent technique for measurement of optical absorption index
c-Si	crystalline silicon
CV	capacitance voltage measurements
CVD	chemical vapor deposition
CZ	Czochralski
dc	direct curruent
EBIC	electron beam-induced current
ECR	electron cyclotron resonance
EQE	external quantum efficiency
GDMS	glow discharge mass spectroscopy
HF	high frequency
HTS	high-temperature substrate
IAD	ion-assisted deposition
IQE	internal quantum efficiency
ITO	indium tin oxide, a transparent conductor
LBIC	light beam-induced current
LCAO	linear combination of atomic orbitals

LPE	liquid-phase epitaxy
LTP	layer transfer process
LTS	low-temperature substrate
mc-Si	multi-crystalline Si
MNOS	metal oxide nitride semiconductor structure
MOS	metal oxide semiconductor structure
ONO	silicon oxide/silicon nitride/silicon oxide multi-layer stack
PCD	photoconductance decay
PC-plot	parameter confidence plot
PECVD	plasma-enhanced chemical vapor deposition
PERL	passivated emitter rear locally diffused
poly-Si	polycrystalline silicon
PSI	porous silicon
QMS	quasi-monocrystalline Si
QSSPC	quasi-steady-state photoconductance decay technique
rms	root mean square
SCR	space charge region
SEM	scanning electron microscope
SIMOX	separation by implantation of oxygen
SIMS	secondary ion mass spectroscopy
SiN_x	silicon nitride
SPC	solid-phase crystallization
SPS	sintered porous silicon
SRH	Shockley-Read-Hall recombination model
SRV	surface recombination velocity
SSP	silicon sheet from powder
STAR	surface texture with enhanced absorption and back reflector
TCA	$C_2H_3Cl_3$
TCO	transparent conducting oxide such as ZnO:Al or SnO_2:F
TEM	transmission electron microscopy
tpa	jump trials per surface atom
VEST	via hole etching for separation of thin films
VHF	very high frequency
VPE	vapor-phase epitaxy
XRD	X-ray diffraction
ZMR	zone melt recrystallization
1D	one-dimensional
2D	two-dimensional
3D	three-dimensional

Greek acronym

μc-Si	microcrystalline silicon with grain sizes < 1 μm

1 Introduction

Semiconducting photovoltaic cells convert solar radiation to electric power. A photon that enters the cell will contribute to the electric current if it is absorbed by exciting an electron from the valence band into the conduction band and if this electron recombines neither in the volume nor at the surfaces of the semiconductor. High optical absorption and little carrier recombination are therefore two prerequisites for an efficient power conversion. The physical mechanisms and the solar cell design which maximizes the power output are well understood.

1.1 Highest-efficiency crystalline Si solar cells

The current world record in power conversion efficiency with Si solar cells is 24.7% and was achieved with a solar cell that is shown schematically in Figure 1.1 [1]. The design features that are decisive for high photogeneration and low carrier recombination are listed in Table 1.1. A double-layer antireflection coating and the photolithographically defined front surface texture with regular inverted pyramids minimize the reflection loss at the front surface. The Si wafer is 400 μm thick to offer a long optical path length that permits close to complete absorption of all those photons that have an energy larger than the electronic bandgap of Si. A dielectric SiO_2 layer is inserted between the Si wafer and the Al back conductor to achieve a high optical reflectance at the back of the cell. This measure and the path length enhancement due to total internal reflection at the pyramids [2] enhance the response to long-wavelength light that is not fully absorbed by the first pass through the cell. The Si material has a high minority carrier lifetime in the range of milliseconds to minimize charge carrier recombination. Such long lifetimes are feasible with monocrystalline and highly purified Si, e.g. with a wafer from float-zone (FZ) silicon. In a long-lifetime material, all charge carriers can reach the surfaces

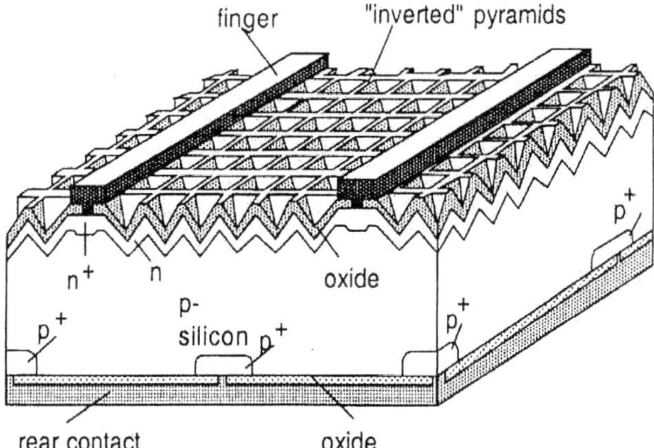

Figure 1.1. Schematic of a high-efficiency solar cell with the PERL (passivated emitter rear locally diffused) design. Figure reproduced from Ref. [3].

Table 1.1. Design features of high-efficiency cells that achieve efficiencies of 24% and technological measures to realize these features

Design features	Technological measures
For high absorption	
Antireflection coatings	ZnS/MgF_2 double-layer
Thick Si wafer	Sawing 400 μm thick wafers from ingot
Inverted pyramids on front	High-temperature oxidation, photolithography, and etching
Back reflector	Intermediate oxide layer perforated by photolithography
For low recombination	
High-lifetime Si	Monocrystalline and highly purified Si material
Passivated surfaces	High-temperature oxidation in TCA cleaned furnace, alneal on front and back
Small contacts	Photolithography to open oxide on front and back
High doping at contacts	Photolithography for additional local diffusion at front and back

because the diffusion length exceeds the wafer thickness. Surface passivation is thus an important topic for high-efficiency solar cells. Dangling bonds at the surface of the Si crystal are commonly saturated by growing SiO_2 layers in a clean furnace at a temperature around 1000°C. Special post-oxidation treatments, such as covering the SiO_2 with an Al layer and annealing it in forming gas, reduce the surface recombination. Electrical contact from the Al back to the Si wafer is made through holes in the SiO_2 layer. The distance of the holes is a compromise between the series resistance loss for a large separation and the recombination losses for a small separation [4]. The size of the photolithographically defined holes is small and in the range of 50 microns because metal/Si interfaces are locations of high recombination. In addition to keeping the contact area small another helpful measure to reduce the amount of interface recombination at the metal contacts is to place the contacts onto highly doped regions of the Si cell. In this way, the concentration of the minority carriers is reduced which makes it less likely for a majority carrier to find a partner to recombine with. Hence, the recombination rate decreases. Local doping at the contacts is realized technically by additional oxidation, photolithography, etching, and diffusion. A similar procedure is applied to the front surface of the cell where the metallization contact is finger-shaped. Again the contact of the narrow fingers with the locally highly doped emitter is made through a few micron-wide openings in the SiO_2 layer. Photolithography is used to control the position of doped regions and contacts on a micron scale.

The Si solar cell design described above is named PERL (passivated emitter rear locally diffused). The PERL design uses a fabrication sequence that was developed to achieve optimum performance. Cost issues were not considered. As a consequence, the PERL cells are by orders of magnitude too expensive for competitive electric power generation on a large scale. The PERL cell was, however, helpful to understand the transport and recombination processes in Si solar cells. Even subtle details of the electronic device performance are understood today [5] and can be modeled with numerical computer codes [6, 7].

1.2 Industrial crystalline Si solar cells

For cost reasons, commercial cells cannot use most of the above-mentioned high-efficiency features: double-layer antireflection coatings are replaced by single-layer antireflection coatings, and a surface texture is often omitted. Monocrystalline Si wafers are increasingly replaced by block-cast multicrystalline wafers. The 30% kerf loss when sawing the Si wafers from an ingot is avoided by directly pulling multicrystalline wafer ribbons [8]. Screen printing replaces the photolithographic definition of the fingers. Local diffusions are not applied. These simplifications of the fabrication sequence reduce the fabrication costs of industrial cells by a factor that is larger than the accompanying reduction of the cell efficiency. Industrial Si cells currently reach power conversion efficiencies ranging from 12 to 16%. The cost of photovoltaic energy generated with these cells is 0.5 to 1 €/kWh, depending on the fabrication technology and the location of power generation. In 2001, a cost reduction by a factor of 10 is still required to make photovoltaics competitive with fossil fuel or nuclear power generation. The high cost of photovoltaic energy is currently the main obstacle for a wider spread of this sustainable source of energy, in particular in the developing countries.

Innovative cell processing

With Si solar cells, a major cost reduction cannot be achieved by an efficiency enhancement alone. A further reduction of the fabrication cost is also necessary. In the laboratory innovative processes are under investigation that yield high-performance cells with processing sequences much simpler than that of the PERL cell. We give two examples. At the ISFH in Hameln a 20%-efficient 10×10 cm^2 cell was fabricated without using photolithography [9-11]. One of the innovative features is a front surface metallization that is applied by oblique and mask-free evaporation in a high-throughput vacuum chamber. This technique generates grid fingers of low electrical resistance and low front surface shadowing. Silicon nitride provides the surface passivation, thus avoiding high-temperature SiO_2 passivation. At the FhG-ISE in Freiburg 20%-efficient cells with point contacts similar to those at the back of the PERL cell were generated by laser ablation [12]. One of the photolithographic steps of the PERL cell is thus avoided. If included into a mass production line such innovative fabrication concepts are capable of reducing the fabrication cost and keeping the cell efficiency above 18%.

1.3 Thin-film crystalline Si cells

The reduction of the fabrication costs increases the relative significance of the material costs in industrial solar cells. In today's commercial modules the cost of the raw Si wafer is about half of the total cost. This fraction will further increase with the steadily growing global production volume of Si solar modules that brings about a shortage in Si supply. Hence there is a necessity to use thinner Si wafers in order to save Si material. Wafers thinner than 200 μm are, however, difficult to process without breakage.

Thin-film crystalline Si cells are an alternative to wafer cells. By definition, these cells have a thickness of less than 50 μm and are deposited onto a suitable substrate. The substrate enhances the mechanical strength and avoids the breakage of the thin films. The reduction of Si consumption constitutes the driving force for the world-wide rapidly increasing efforts to fabricate solar modules from thin films of crystalline Si.

1.4 Physical problems with thin-film crystalline Si cells

Aiming at a high power conversion efficiency from Si cells with a thickness of only a few microns raises questions on the physical limits of power conversion, on device fabrication, and on device characterization. The subsequent sections will introduce the questions treated in this work.

What is the maximum photogeneration?

For very thin Si films the optical absorption severely limits the device current. Figure 1.2 shows the optical absorption A on a path length of l = 1, 10, and 1000 μm in crystalline Si. For a path length of 1 to 10 μm the near-infrared fraction of the solar spectrum is hardly absorbed since silicon is an indirect semiconductor with a small absorption coefficient in this spectral region. The enhancement of the photogeneration is thus important for thin crystalline Si cells.

In most of the previous work on the efficiency limits of thick photovoltaic devices it was assumed that the optical absorption is unity for all photon energies exceeding the bandgap [14, 15] and that it is zero for sub-bandgap radiation. This assumption is inappropriate for thin-film crystalline Si cells as Figure 1.2 shows: The optical absorption of a 1 μm-thin film is not a step function. In this work we therefore use the literature values for the dielectric function of crystalline Si, including sub-bandgap absorption due to phonon-assisted electron-hole pair generation, to calculate the maximum photogeneration in thin-film cells.

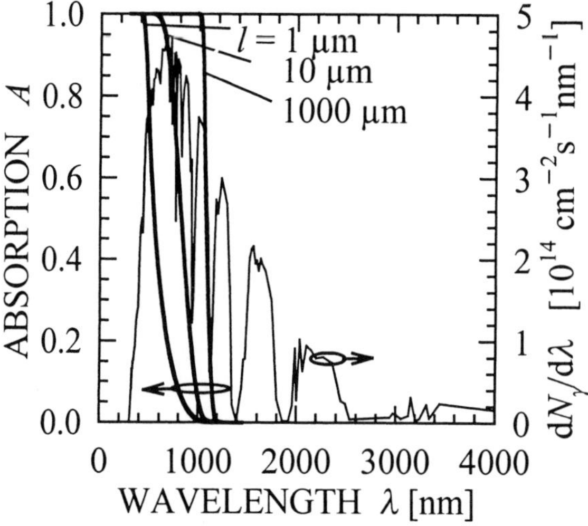

Figure 1.2. Optical absorption $A = 1 - \exp(-\alpha_s l)$ in crystalline Si (at 300 K) for an optical path length l. The terrestrial solar photon flux $dN_\gamma/d\lambda$ per wavelength interval [13] is shown as the lighter line .

What is the maximum path length enhancement?

A structured surface of the thin Si films enhances the optical path length since light is internally reflected frequently. The path length enhancement depends on the shape of the surface texture. In the framework of geometrical optics, which means that the size of the surface texture and the film thickness are large compared to the wavelength, the maximum path length enhancement was previously derived for solar cells having surfaces of zero reflectance [16]. In this work we give a generalization of this theorem to the more realistic case of cells with non-vanishing surface reflectance. One consequence of an upper limit of the average path length is an upper limit for the optical absorption and thus for the photogeneration in the cell. In this work cells that reach the maximum photogeneration under isotropic illumination are said to exhibit optimum light trapping.

We also investigate the question of whether the geometrical optics limit can possibly be surpassed in the framework of wave optics.

What is the relative significance of the various loss mechanisms?

Losses in power conversion efficiency are caused by non-absorption of solar radiation (e.g. sub-bandgap radiation), by a less than optimum light trapping scheme, by non-concentration of solar light, by the thermalization of hot carriers, by Auger recombination, by luminescence radiation [17], by surface recombination, and by grain boundary recombination. In a gedanken experiment we will construct a highly idealized photovoltaic device that has the efficiency of a Carnot machine operating between 5780 K, the temperature of the sun, and 300 K. We quantify the impact of the above-mentioned loss mechanisms on the device efficiency by adding one loss mechanism after the other. Our modeling thus becomes more realistic and even pessimistic, finally ending at a 4%-efficient small-grained thin-film cell that is dominated by grain boundary recombination. Some of the losses considered are avoidable by applying an appropriate technology. Our analysis will give us a hint which of the losses are most worth avoiding.

What model is suitable to study the significance of the various physical loss mechanisms?

Our intention to study photovoltaic cells that work at the Carnot efficiency, and to add intrinsic and extrinsic loss mechanisms step by step, demands a physical model that describes hypothetical devices that do not exist in reality. A thin-film cell that has the dielectric function of crystalline Si and excludes thermalization losses is one example. We model such cells by assuming that every absorbed photon generates as many electron-hole pairs as is energetically possible. The classical detailed balance model [14, 63] is not applicable to these carrier-multiplying devices since conversion efficiencies exceeding the Carnot limit would result. To resolve this problem we have to account for the sharing of the photon's chemical potential among the multiple generated electron-hole pairs [18, 66]. This modified detailed balance model is the basis for including all further loss mechanisms.

What is the maximum efficiency of ideal thin-film Si cells?

The inevitable intrinsic recombination processes such as radiative recombination and Auger recombination in Si are well-known effects. Efficiency limits resulting from these recombination losses have been calculated before [19, 20], demonstrating that efficiencies well above 25% are theoretically feasible with crystalline Si films thinner than 10 μm. In the previous investigations the authors assume Lambertian light trapping, which is not the optimum case. We calculate the efficiency limits considering optimum light trapping. We also use updated data for the Auger recombination rate [21], the sub-bandgap absorption in crystalline Si [22], and the intrinsic carrier concentration in Si [23].

How can efficiency limiting processes be revealed by quantum efficiency analysis?

The current-voltage curve of a solar cell does not elucidate where in the device the recombination losses occur. In contrast, the measurement of the quantum efficiency spectrum provides a depth resolution since the optical absorption length of monochromatic light in Si varies from the 10 nm into the centimeter range with varying wavelength. Quantum efficiency data also provide a lateral resolution if the light spot is scanned across the cell. Unfortunately, the standard technique (see Appendix C on p. 245) that is commonly used to analyze the quantum efficiency spectra of thick cells is not applicable to thin-film cells for several reasons:

Light trapping not included: The standard analysis assumes an exponentially decaying carrier generation profile and is restricted to those wavelengths that are sufficiently strongly absorbed to make light trapping effects irrelevant. Thus, in the case of thin-film cells, only a small fraction of the solar spectrum is covered by the standard model. Basore extended the quantum efficiency analysis to the opposite extreme of weakly absorbed light [24]. The intermediate spectral range of neither strongly nor weakly absorbed light is not covered by previous analytical models.

Current from the emitter not included: In thin solar cells, the optical absorption in the front emitter is a noticeable fraction of the light absorption in the cell. Photogeneration in the emitter is not accounted for by the standard evaluation technique.

Back surface fields not included: Thin-film Si solar cells are particularly sensitive to surface recombination because the minority carriers are generated close to a surface. Highly doped surface layers are frequently applied to reduce the recombination rate. The standard internal quantum efficiency (IQE) analysis does not account for these so-called back-surface field layers.

Grain boundaries not included: Most of the thin-film cells are polycrystalline with the preferential grain boundary orientation being perpendicular to the junction. The standard evaluation technique is based on a one-dimensional transport model and does not account for the three-dimensional transport that is induced by grain boundary recombination.

Quantum efficiency models that account for these thin-film features [25] will be described and applied in this work.

Is ray-tracing analysis appropriate for thin-film cells?

We determine the carrier generation rate in thin-film cells with textured surfaces, such as those shown in Figure 5.8 on p. 128, by Monte Carlo ray-tracing. The author's ray-tracing program SUNRAYS [26] is extensively used in the present study and by many Si solar cell groups. While many simulation studies can be found in the literature, direct comparisons of experimental and simulated reflectance spectra of textured Si solar cells are rare. In order to justify the ray-tracing approach we compare simulated reflectance spectra of textured thin-film cells with the spectra measured.

Using the program SUNRAYS, we investigate various texture-shapes that are applicable to thin-film Si cells [27, 28]. We search for texture shapes that have a large path length enhancement. This theoretical study will also reveal what the key design parameters of faceted light trapping textures are.

How can high back surface reflectance be achieved in thin-film cells?

A high back surface reflectance is important in thin cells. We analyze dielectric back reflectors, and reflectors detached from the textured back surface of the cell. We also discuss optical back reflectors from multi-layers of porous Si that we introduced [29] to improve the light confinement in thin-film cells on ceramic substrates.

How can recombination parameters be extracted from quantum efficiency spectra?

A general approach of extracting recombination parameters such as the base diffusion length L and the back surface recombination velocity S_b is to model the quantum efficiency theoretically and to vary the recombination parameters until a fit to the experimental data is achieved. For this purpose we developed an analytical model that accounts for light trapping, back surface fields and current contributions from the emitter [25]. We pay special attention to the problem, to whether or not these recombination parameters are determined uniquely, and to what information we can still extract if they are not unique. An analytical model is advantageous here, because fitting experimental data becomes feasible in a short time.

The electronic transport equations are comparatively easy to solve when the recombination rate is linear in the excess minority carrier concentration. This linearity does not hold in general, however. The oxidized surface of crystalline thin-film cells shows a surface recombination velocity that decreases by more than two orders of magnitude with increasing carrier concentration. In such cases the data analysis using a linear theory for the quantum efficiency spectra does not permit extraction of the actual surface recombination velocity. Our approach to solve this difficulty is to perform differential quantum efficiency measurements using weak chopped monochromatic light at various voltage biases. The small signal measurements permit a linearization of the transport equations. Measurements at various bias levels permit a determination of the actual recombination parameters.

In this work we exemplify the concept of differential recombination parameters that the author introduced to analyze recombination rates in non-linear cells [30-32].

How can carrier recombination be modeled analytically in polycrystalline cells?

Grain boundaries in thin-film cells make the transport problem three-dimensional. Hence fitting of experimental data becomes almost impossible with computer codes that use a finite element approach. The computation time would be too long even with today's computers. We therefore developed an analytical solution of the three-dimensional minority carrier diffusion equation [33] that accounts for the three locations of recombination: the volume of the grain, the grain boundary, and semiconductor surface. We solve the transport equations in Fourier space following the ansatz of Dugas [34]. We then apply our model to extract the grain boundary recombination velocity in polycrystalline thin-film cells.

Recombination at the intersection of a grain boundary with the p-n junction is in general stronger than the grain boundary recombination in the neutral base. In this work we publish an analytical solution describing carrier recombination in the depleted space charge region of polycrystalline Si films. Assuming an interface state density that does not depend on the energy, results in a grain boundary charges that is proportional to the position of the Fermi energy [35]. This linear relation permits a Fourier decomposition of the Poisson equation in three dimensions.

What can we learn from effective diffusion lengths extracted from quantum efficiency data?

Since the recombination occurs at various locations in polycrystalline thin-film cells it is common practice to define effective diffusion lengths. The standard IQE analysis [36] derives an effective diffusion length L_Q from the measured data. Alternatively, an effective diffusion length L_J could be derived from the diode saturation current density of a thin-film cell, provided base recombination dominates. The effective diffusion length L_J is the equivalent diffusion length of a thick monocrystalline base-dominated cell that has the same value of the dark saturation current as the polycrystalline thin-film cell. For monocrystalline thin-film cells with spatially homogeneous doping (no electric fields) and spatially homogeneous minority carrier properties it is known that the effective diffusion length L_Q equals the effective diffusion length L_J [24]. This is a useful relation since the quantum efficiency analysis permits extraction of the recombination rate under forward bias in the dark. The value of L_Q can be measured locally by scanning the illuminating light spot laterally across the cell. We can thus derive local dark-current-voltage curves from light beam-induced current mappings without the necessity to separate the cell into many individual diodes.

In this work we study the interrelation of the effective diffusion lengths L_Q and L_J for inhomogeneous semiconductors, e.g. for polycrystalline Si. The value of L_Q is derived from an experiment where carriers are collected at the junction, while the value of L_J follows, in principle, from an experiment where carriers are injected into the base. Since Donolato's reciprocity theorem of charge carrier collection [37] relates the local carrier collection probability with the excess minority carrier concentration under forward injection, the reciprocity theorem is an appropriate tool to study the relation of the effective diffusion lengths L_Q and L_J in spatially inhomogeneous material.

Does the reciprocity theorem of charge carrier collection hold for Fermi statistics?

Donolato's reciprocity theorem was extended by a series of generalizations such that it was finally proved for non-homogeneously doped semiconductors with an arbitrary spatial dependence of minority carrier lifetime τ, of the diffusion coefficient D_n, and of the equilibrium minority carrier concentration n_o [38-41]. Variations of the quantity n_o arise from changes of the electrical potential close to ohmic contacts, at grain boundaries, or at floating junctions, as well as from potential fluctuations due to heavy doping or from bandgap variations due to fluctuations of the material composition. The reciprocity theorem is, however, in all these previous versions restricted to free charge carriers obeying Boltzmann statistics [38, 41].

Electrons and holes are, however, fermions. Boltzmann statistics is therefore not applicable if the difference between the Fermi level and the energy of the charge carrier is less than a few times the thermal energy. Thus the reciprocity theorem, in its previous form, excludes degenerately doped junctions in thin-film cells. It also excludes deep states, e.g. in grain boundaries or at interfaces. In this work we give our generalization of the reciprocity theorem to Fermi statistics [42].

Shockley and co-workers [43] as well as Misiakos and Lindholm [38] mentioned that, in general, reciprocity relations known from different fields of statistical physics arise from the principle of detailed balance and from Onsager's principle of micro-reversibility [44]. In order to generalize the reciprocity theorem to Fermi statistics, we thus derive the theorem from the principle of detailed balance, which does not explicitly refer to a particular type of statistics.

How can we discriminate betenn surface and bulk recombination in thin-film cells?

Losses in commercial wafer cells are often dominated by recombination in the Si bulk. Surface recombination is negligible. In thin-film cells, all electrons and holes are always generated close to a surface and thus surface recombination easily dominates over bulk recombination. It is then difficult to differentiate the contributions of bulk and surface recombination. A non-destructive experimental technique to switch off the surface recombination would be a helpful tool for the experimentalist. Our approach, namely to electrostatically repel one type of charge carrier from the interface by corona charges [45], provides such a technique.

What are the limitations to current thin-film Si cells on foreign substrates?

Thin-film crystalline Si solar cells require a substrate for sufficient mechanical strength. Since this substrate must be available at low cost it is typically a foreign, that means a non-Si, substrate. Growing Si films on foreign substrates with a low defect density is a difficult task. Although significant progress has been made in fabricating thin crystalline Si cells on metal, glass, graphite, and ceramic substrates, in March 2001 the power conversion efficiencies of the currently best devices were still only in the range of 9 to 11% despite more than a decade of international research [46-49].

We review previous approaches to fabricating polycrystalline thin-film Si cells on foreign substrates. We apply our advanced quantum efficiency analysis to identify optical losses and grain boundary recombination as significant loss mechanisms in current thin-film cells on foreign substrates.

A high-throughput fabrication of thin-film cells is currently hindered by the fact that a low-cost foreign substrate that would permit the growth of high-quality Si layers at a high growth rate (>0.5 $\mu m\ min^{-1}$) does not yet exist.

How to overcome previous technological limitations?

Monocrystalline wafers from electronic-grade Si are an ideal substrate, since they permit the growth of high-quality monocrystalline Si layers at high temperature and at high rates. The only obstacle is the high cost of the Si wafer. These costs could, however, be reduced if the Si wafer was re-used frequently by separating the epitaxial Si film from the wafer. This is the concept of a layer transfer process (LTP) for photovoltaics that is shown schematically in Figure 1.3.

Layer transfer was initially introduced for GaAs solar cells by McClelland et al. [50]. A special surface layer is applied prior to the epitaxial growth of the device layer. After film growth, this surface conditioning layer permits the transfer of the device layer from a re-usable Si growth substrate to a low-cost device carrier. The device carrier does not have to withstand the high growth temperatures. It may thus be low-cost window glass or even plastics. Layer transfer circumvents a key problem of thin-film cells on foreign substrates: high-rate epitaxy at a high deposition temperature becomes possible despite using a low-cost device carrier.

The decisive feature of all layer transfer processes is the nature of the surface conditioning layer, since it determines the quality of the epitaxial film. Surface conditioning with porous Si was introduced by Yonehara et al. [51]. The so-called ELTRAN process yields planar Si films without light trapping. The ELTRAN process sacrifices the Si substrate and is therefore not applicable to thin-film crystalline Si photovoltaics. Tayanaka et al. introduced the sintered porous Si process that can re-use the substrate wafer and yields a planar Si film [52]. The author's so-called porous Si (PSI) process [53] uses porous Si on the surface of a textured monocrystalline Si substrate to fabricate surface

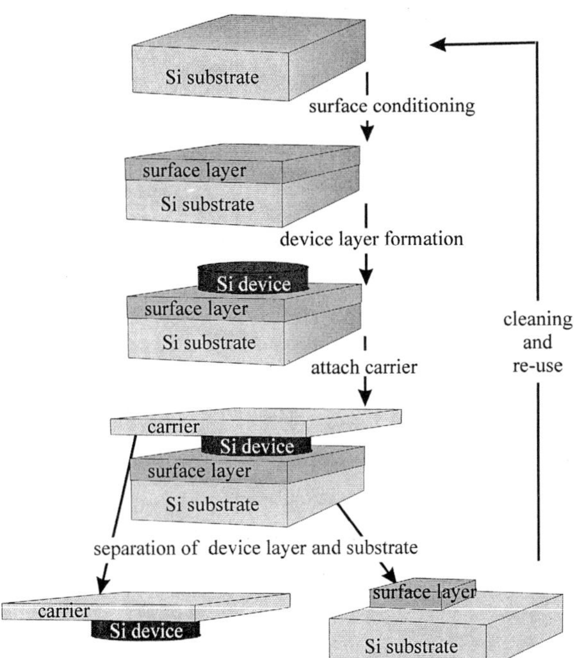

Figure 1.3. The layer transfer process (LTP) starts with a Si substrate that receives a surface conditioning. After film growth, a carrier is attached to the device layer to enhance the mechanical strength of this layer. A special surface layer permits detachment of the device layer. The Si substrate may be re-used for further layer fabrication.

textured films with efficient light trapping. The porous Si functions (i) as a seed for epitaxial growth and (ii) as a separation layer that enables the transfer. The costly Si substrate can thus be re-used since it is not sacrificed.

In contrast to conventional thin-film cells, cells from the PSI process have no grain boundaries, because homo-epitaxy of Si on porous Si yields monocrystalline material. By growing epitaxial films on periodically or randomly textured Si wafers with a porous surface layer, it becomes possible to fabricate very thin waffle-shaped films (see for example Figure 6.7 on p. 177). In this work we analyze their light trapping capability by optical reflection measurements and by ray-tracing analysis.

During the heating of the reactor and during the epitaxy the porous Si film changes its morphology due to the migration of Si atoms on the large inner surface of the porous Si. We observe these morphological changes by scanning electron microscopy. Numerical modeling of the reconstruction of porous Si during annealing reveals the driving force for the morphological changes. We investigate the electronic quality of the epitaxial films that grow on annealed porous Si by transmission electron microscopy and by analyzing the electronic transport in solar cells.

Photolithography and high-temperature oxidation, as for the PERL cell, is not acceptable for low-cost thin-film cells. Thus, a photolithography-free solar cell process needs to be developed to permit the safe handling of 2 to 15 μm-thin monocrystalline and textured cells. At ZAE Bayern, we developed such a process that has led to the fabrication of efficient thin-film cells and modules. In this book we apply our advanced quantum efficiency modeling to study the optics and the electronic transport in these novel devices.

We finally investigate what efficiencies are technically possible with waffle-shaped monocrystalline thin-film Si cells from the porous Si process.

2 Physical loss mechanisms

Shockley and Queisser calculated the efficiency of an ideal semiconductor solar cell that exhibits only radiative recombination [14]. These authors applied the principle of detailed balance between carrier recombination and carrier generation at thermal equilibrium. Thereby the calculated power conversion efficiencies are independent of empirical estimates of the material quality. The assumption of zero absorption below the fundamental energy gap E_g and unity absorption above the bandgap makes the efficiency a function of the bandgap. An optimum efficiency of 30% was calculated for an energy gap of $E_g = 1.1$ eV under illumination with a black body spectrum that has the temperature of the sun [14].

Using a similar approach, Würfel and Ruppel derived the same efficiency limit at an optimum energy gap of 1.3 eV [54] from a thermodynamic model that is based on the assumption of a chemical equilibrium between the electron-hole gas in the semiconductor and the luminescent photon radiation field. These authors used the concept of a chemical potential of luminescence photons [55]. Green [56] and Tiedje et al. [19] showed that Auger recombination is the dominating intrinsic recombination loss in Si cells.

All of the above work assumed that every absorbed photon creates a single electron-hole pair. Deb and Saha [57] provided a first semi-empirical estimate of solar cell efficiency enhancements to be expected from carrier multiplication by impact ionization. Impact ionization permits the generation of more than one electron-hole pair per absorbed photon; hence the quantum efficiency can exceed unity. They determined a maximum solar cell efficiency of 31% for a hypothetical material with $E_g = 0.8$ eV. For Si, quantum efficiencies exceeding unity were measured by Vavilov and Britsyn [58]. The measurement of quantum efficiencies above unity in high-efficiency crystalline Si cells by Kolodinski et al. [59] led to renewed interest in this effect [60-65]. Different assumptions on the optical absorption, the recombination rates, the relaxation, and the multiplication of the carriers resulted in various efficiency limits. Werner et al. calculated an efficiency limit of 43% [63] if the solar photons generate as many electron-hole pairs as is energetically allowed. The author developed a detailed balance model for carrier multiplying cells [66, 67, 18] that confirmed the efficiency limit of 43%. Luque and Marti proved that, in contrast to earlier approaches, the author's model for carrier multiplying cells complies to the second law of thermodynamics [68].

The model that we develop in this chapter combines various features presented in earlier work. Applying the concept of a chemical photon potential [54, 55], we describe carrier multiplication [18] to model an electron-hole gas that is de-coupled from the lattice. Using realistic models for the optical absorption makes the limiting efficiency dependent on the film thickness [19]. We investigate this thickness dependence in detail. We consider cells that are illuminated by the global air mass 1.5 spectrum [13], which is relevant for terrestrial photovoltaics, while most of the above-mentioned papers assume a black body type of illumination. We introduce laser action [69] into our efficiency model, which is necessary if the cell voltage exceeds the smallest band-to-band excitation energy accounted for in the model. We also use updated experimental data for the intrinsic carrier concentration of Si [70], for the optical constants of Si [22], and for the Auger recombination rates [21, 71].

Starting from a highly idealized device working at the Carnot efficiency of 94.8%, we add one loss mechanism after another to finally end up with a cell without light trapping and grain boundary recombination wich has an efficiency of only 4%. This procedure enables the estimation of the significance of the various loss mechanisms.

We confine ourselves to crystalline Si cells. The efficiency limits of multi-bandgap cells are analyzed in Refs. [15, 72-74].

2.1 Limitations to photogeneration

2.1.1 Solar spectrum

The standard terrestrial solar spectrum is the global air mass 1.5 (AM1.5G) spectrum with a photon flux density per wavelength interval $dN_\gamma/d\lambda$, as shown in Figure 2.1. The spectrally integrated energy flux density is 1000 W m^{-2}. The spectrum is similar to the emission spectrum of a black body at the temperature of T_s = 5780 K, which is the temperature of the sun's surface. The deviations from the black body spectrum stem from atmospheric molecules such as ozone, which absorbs ultraviolet light with wavelengths smaller than 300 nm, and H_2O, CO_2, and other molecules that have absorption lines in the near-infrared. Figure 2.1 also displays the current density

$$j_{sc}^{*}(\lambda)=\frac{q}{hc}\int_{0}^{\lambda}\lambda' I_{AM1.5G}(\lambda')d\lambda' \tag{2.1}$$

that a black solar cell would generate if all photons with a wavelength smaller than λ were converted into an electron-hole pair and if all these generated pairs were collected at the junction. The symbol $I_{AM1.5G}$ in Eq. (2.1) denotes the solar energy flux density (W m^{-2} nm^{-1}) that is tabulated in Ref. [13]. The maximum current density is j_{sc}^{*}(4045 nm) = 69.9 mA cm^{-2}, corresponding to a photon flux integral j_{sc}^{*}/q = 4.3×10^{17} cm^{-2} s^{-1}. Half of the solar photons have a wavelength smaller than 1000 nm and the other half has a wavelength larger than 1000 nm. Hence, λ = 1000 nm is a "typical" solar photon wavelength that corresponds to a photon energy of 1.24 eV. At this energy Si has an optical ab-

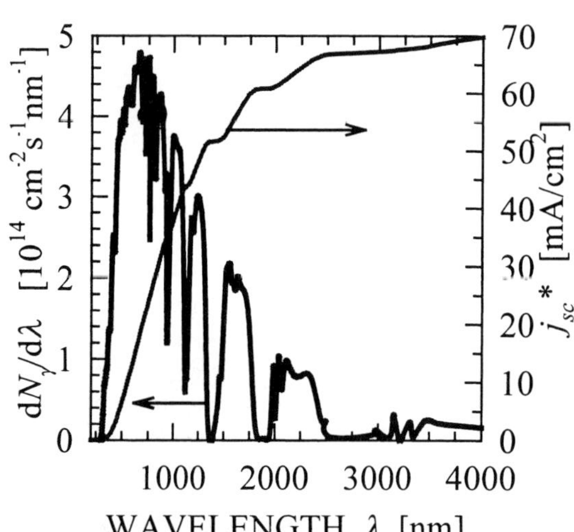

Figure 2.1. Solar photon flux density $dN_\gamma/d\lambda$ of an AM1.5G spectrum [13] and the integrated photon flux N_γ, according to Eq. (2.1).

sorption length of L_α = 156 μm. The optical absorption length is the inverse of the Si absorption coefficient α_s. A cell thickness of several hundred μm is thus required for complete optical absorption. That is one reason why conventional Si wafer cells have a thickness of 300 μm. The other reason is that a thickness of 300 μm permits safe handling of Si wafers.

For the efficiency calculations we require the solar photon flux per photon energy E and per étendue. The étendue is the product of the cell area and the illuminating solid angle when projected onto the cell's surface. See Eq. (A.29) in Appendix A on p. 188 and Refs. [75, 16] for a definition of the étendue. The solar photon flux per photon energy interval and per étendue on the Earth's surface is

$$n_{\gamma s}(E) = \frac{hcI_{AM1.5G}}{\pi \sin^2(0.266°)E^3} \tag{2.2}$$

Here, c is the vacuum velocity of light, h is the Planck constant, and k is the Boltzmann constant.

Light concentration has the practical advantage of collecting more solar power with less device area A_c than without concentration. Light concentrating with a lens is shown schematically in Figure 2.2. The cell "sees" the sun under the half-angle Φ_s and is therefore illuminated with the étendue $\mathcal{E}_s = A_c\,\pi\,\sin^2(\Phi_s)$. The half-angle of the sun's disk is 0.266° when observed from the Earth without the lens. The concentration factor $C = \sin^2(\Phi_s)/\sin^2(0.266°)$ has a maximum value $C_{max} = 4.6\times10^4$ that is reached for $\Phi_s = 90°$. At maximum concentration the cell only "sees" the sun. We assume there is an ideal reflector behind the cell. The cell emits luminescence light into the half-angle Φ_c. For $\Phi_c > \Phi_s$ luminescence light is emitted into the cold sky.

Figure 2.3 shows the optical absorption length $L_\alpha = \alpha_s^{-1}$ that extends from the nanometer range for ultraviolet light to the meter range for near-infrared light. The data shown in Figure 2.3 were derived from internal quantum efficiency spectra measured on high-efficiency Si solar cells [76, 22]. Using quantum efficiency measurements permits discrimination of weak active absorption by electron-hole pair generation from strong

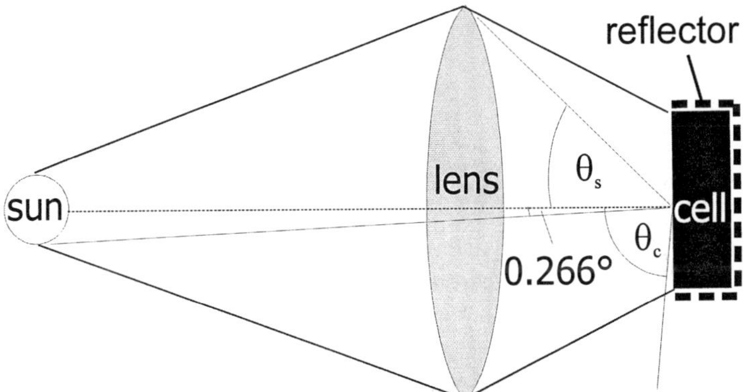

Figure 2.2. A lens concentrates the sunlight onto a solar cell. The cell "sees" the lens under a half-angle Θ_s while luminescence light leaves the sun under the half-angle Θ_c. The half-angle of the sun's disk is 0.266° without the lens.

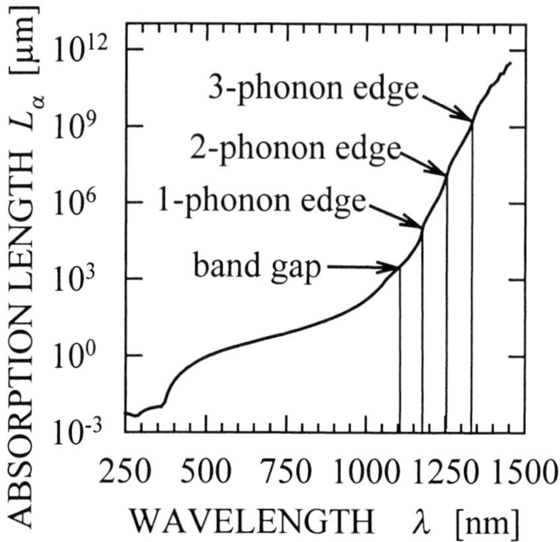

Figure 2.3. Absorption length $L_\alpha = \alpha_s^{-1}$ for Si with absorption coefficient α_s at 300 K. Sub-bandgap photogeneration is due to phonon-assisted absorption processes. Data from Ref. [22].

free carrier absorption [77]. The active absorption is also measured separately from free carrier absorption by an analysis of the photoluminescence spectrum of crystalline Si [78].

The bandgap of Si is $E_g = 1.12$ eV at a cell temperature $T_c = 300$ K [79]. The corresponding wavelength is $\lambda_g = 1108$ nm. The onset of absorption at the gap energy E_g is not sharp, since Si is an indirect semiconductor. The near-bandgap absorption is phonon-assisted. The steps in the sub-bandgap range in Figure 2.3 are due to the onset of the next order process with the participation of one more phonons. The smallest phonon-free direct band-to-band transition in crystalline Si is at about 3.4 eV, corresponding to a wavelength of 360 nm.

Photogenerated current density

The optical absorption $A(W_{eff},\lambda)$ of a cell of effective thickness W_{eff} determines the maximum short-circuit current density

$$j_{sc}^* = \frac{q}{hc} \int_0^{4035nm} \lambda \; A(W_{eff}, \lambda) \, I_{AM1.5G}(\lambda) \; d\lambda \tag{2.3}$$

We define the effective cell thickness $W_{eff} = V_c / A_c$ as the ratio of cell volume V_c to the macroscopic cell area A_c. This definition has the advantage that W_{eff} measures the Si consumption and that it is also applicable to cells with spatially inhomogeneous thickness. The asterisk in j_{sc}^* reminds us that the maximum current density is the photogeneration rate expressed in units of current density. Its value is in general larger than the short-circuit current density j_{sc}. Only a cell with no recombination losses has $j_{sc}^* = j_{sc}$.

High optical absorption $A(\lambda)$ in a wide spectral range is achieved by (i) a reduction of the cell's front surface reflectance R_f that is ideally zero, (ii) an enhancement of the back surface reflectance R_b that is ideally unity, and (iii) efficient light trapping that increases the light's path length in the cell.

Front surface reflectance

We only briefly discuss the reduction of the front surface reflectance, since this topic is not specific for thin-film cells.

Single or multi-layer antireflection coatings of ZnS, MgF_2, TiO_2, SiO_2, and Si_3N_4 with an optical thickness in the range of a quarter of the wavelength, are commonly used [80-83]. A reflectance as low as 2% from 440 to 960 nm was achieved with triple layer $ZnS/MgF_2/SiO_2$ systems on micro-grooved Si thin-film solar cells [82]. This low value includes the reflection from the silver-plated metal fingers.

Instead of optically thin layers, optically thick layers with a refractive index smaller than that of silicon are also used [84, 85]. The glass cover of solar modules is an example of an optically thick antireflection layer that is commonly applied.

Textured surfaces are frequently used for reflection control [86-90]. The reflectance is lower, the steeper the facets are [91]. In practice, a combination of all three measures is frequently used, e.g. an encapsulated textured Si solar cell with an antireflection coating.

Back surface reflectance

Using a back surface reflector as indicated in Figure 2.4a doubles the light path length. A high back reflectance is only required for long-wavelength light, since short-wavelength light is absorbed at its first pass through the cell. The internal reflectance of a bare Si/air interface is 0.35 at $\lambda = 1000$ nm, under normal incidence. Oblique incidence on the back reflector occurs if the front surface scatters the light. Then a Si/Air interface totally reflects all light that is incident at an angle larger than the critical angle of 16° for total internal reflection. For isotropic incidence onto the reflector a fraction $1 - 1/n_s^2 = 0.92$ is totally reflected, where n_s is the refractive index of Si. The small remainder falls into the loss cone and has a chance to leave the cell. A detached metal reflector with an optically thick gap between the open Si back surface and the reflector returns the escaping light back to the cell while maintaining the condition of total internal reflection for the majority of light rays. Reflectance values as high as $R_b = 0.998$ are feasible with this concept, as we show on p. 143.

Recently, we introduced a back surface reflector using a multi-layer of porous Si [29]. This concept, as well as more conventional reflectors using optically thin dielectric in-

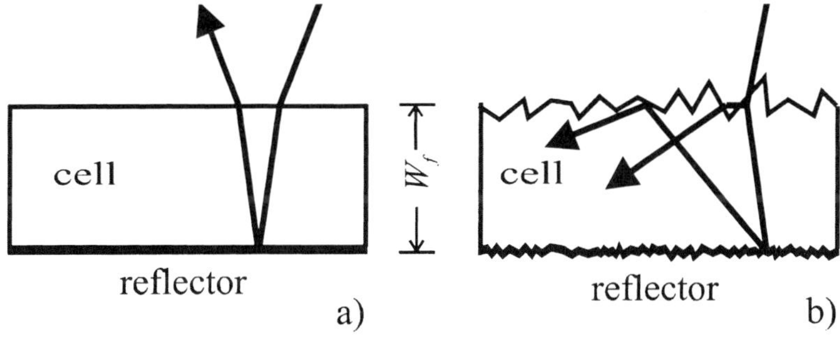

Figure 2.4. a) A back surface reflector doubles the path length of the light in a planar cell of thickness W_{eff}. b) A back surface reflector in combination with textured surfaces enables many double-passes through the cell due to total internal reflection.

terlayers between Si and a metal reflector, are discussed in Appendix A on p. 200.

Light trapping

The next section discusses various levels of light trapping, starting from planar cells without light trapping and ending with optimum geometrical light trapping.

2.1.2 Planar geometry

A planar thin film of thickness W_{eff} with zero front and zero back surface reflectance at all wavelengths λ exhibits a single pass of the light through the cell. We say such a cell has no light trapping (Model N). The optical absorption is

$$A(\lambda) = 1 - \exp(-\alpha_s(\lambda) W_{eff}) \tag{2.4}$$

The maximum short-circuit current density j_{sc}^* as calculated by Eq. (2.3) is shown in Figure 2.8 on p. 20 for Model N. A cell of thickness W_{eff} = 1 μm yields a maximum current density of j_{sc}^* = 12.1 mA cm^{-2}.

A sheet of Si with both sides polished and a surface reflectance of 0.35 also leads to an effective path length that is approximately a single pass.

2.1.3 Lambertian light trapping

Figure 2.4b shows a thin-film cell with rough surfaces. Oblique traversal increases the path length. More important, light rays that reach the rough front surface from inside the Si are likely to do so at angles more shallow than the critical angle for total internal reflection. These rays have at least one further pass through the cell.

By definition, Lambertian surfaces fully randomize the reflected and the transmitted light for all wavelengths. The photon flux density per solid angle is independent of direction and position. Cells with a Lambertian front surface are said to have Lambertian light trapping (Model L). Poruba et al. measured a close-to-Lambertian light scattering in microcrystalline Si cells which show a surface texture that originates from the dependence of the growth rate on the orientation of the crystallite [92]. Shimokawa et al. found a close-to-Lambertian reflectance of an Al_2O_3 ceramic substrate consisting of light scattering Al_2O_3 particles with a diameter of 0.5 μm [93].

Goetzberger discovered the benefit of Lambertian surfaces for thin-film solar cells [94]. The optical absorption of a cell with a Lambertian front surfaceis given by the simple analytic expression [342]

$$A = \frac{(1 - T_r)(1 + T_r) n_s^2}{n_s^2 - (n_s^2 - 1)\, T_r^2} \tag{2.5}$$

for zero front surface and unity back surface reflectance. In Appendix A on p. 182, we derive this result for the general case of a non-zero front reflectance R_f and non-unity back reflectance R_b. An expression for the carrier generation profile is also given in the appendix. The term

$$T_r = \exp(-\alpha_s W_{eff})(1-\alpha_s W_{eff}) + (\alpha_s W_{eff})^2 \int_{\alpha_s W_{eff}}^{\infty} t^{-1}\exp(-t)\,dt \tag{2.6}$$

describes the transmittance of fully randomized light through a planar film of thickness W_{eff} [111]. This planar film has zero front and back surface reflectance $R_f = R_b = 0$. Expanding the optical absorption expressed in Eq. (2.5) to the lowest order in $W_{eff}\,\alpha_s$ shows that the average path length is

$$\bar{l} = \lim_{\alpha \to 0}\left(\frac{A(\alpha)}{\alpha}\right) = 4\,n_s^2\,W_{eff} \tag{2.7}$$

a result first derived by Yablonovitch and Cody [95]. Here, n_s denotes the refractive index of the cell: that is 3.57 for Si at λ = 1000 nm. Hence we find $l = 51\ W_{eff}$ for crystalline Si: a 50 μm-thick cell with no back reflector and no light trapping absorbs light approximately as well as a 1 μm-thick cell with Lambertian light trapping.

Let us now discuss the physical origin of the path length enhancement factor $4\ n_s^2$ [95]. A factor of 2 is due to the oblique light traversal. For fully randomized light the average path length for one traversal is

$$2\pi \int_0^{\pi/2} \frac{W_{eff}}{\cos\vartheta}\cos\vartheta\,\sin\vartheta\;d\vartheta \Big/ 2\pi \int_0^{\pi/2} \cos\vartheta\,\sin\vartheta\;d\vartheta = 2\,W_{eff} \tag{2.8}$$

The back surface reflector makes the light travel upwards and downwards, which contributes another factor of 2 to the path length enhancement. The remaining factor n_s^2 is explained as follows: the density of photon states in the three-dimensional wave vector $\boldsymbol{k}$ space scales with n_s^3. The photon density is thus also proportional to n_s^3. The optical absorption is, however, proportional to the photon flux rather than the photon density. Since light propagates in Si at a speed of c/n_s, the optical absorption, and hence the average path length, are proportional to $c/n_s \times n_s^3 \propto n_s^2$. Thus, the path length enhancement $\bar{l} = 4\ n_s^2\ W_{eff}$ follows.

2.1.4 Geometrical light trapping

The surface of Si cells is often faceted by anisotropic etching [96-99] or by mechanical grinding [87, 90]. Etching of (100)-orientated Si wafers with KOH, for example, yields square-based pyramids with facets that are (111)-orientated and at an angle of 54.7° against the macroscopic cell surface. Figure 2.5 shows a wafer surface with photolithographically defined inverted pyramids. The base length of the pyramids is 13 μm and thus one order of magnitude larger than the wavelength. Geometrical optics is then adequate to model the absorption and reflection of the cell. An introduction to the ray-tracing analysis of geometrical light trapping schemes is given on p. 74. Geometrical light trapping in Si wafers has been studied extensively [89-90] and is reviewed in [102, 16, 103].

Campbell and Green demonstrated theoretically that certain surface textures may even perform better than Lambertian schemes [2]. This was calculated for the perpendicular groove geometry [104] that is sketched in Figure 2.6. Normal incidence, zero front, and unity back reflectance were assumed. Normal light incidence, however, is not represen-

Figure 2.5. SEM micrograph of a (100)-oriented Si wafer with regular inverted pyramids of period 13 μm. The pyramid facets have (111) orientation.

tative for the performance of a static module. Due to the movement of the Sun during the day and during the year, and due to light scattering in a cloudy sky, isotropic illumination is more relevant than normal incidence (see Figure A.7 on p. 193). We therefore focus on isotropic illumination here. It is not yet clear whether any geometrical light trapping scheme can outperform the Lambertian light trapping under isotropic illumination.

Due to present technological limitations the pyramids and grooves are more than 10 μm deep. The structure shown in Figure 2.6 has then to be thicker than 20 μm in order to avoid the formation of holes. A cell thickness of 20 μm is not required, however, for high optical absorption in thin-film cells [105, 94, 28]. The applicability of conventional surface texturing to thin-film cells is therefore limited.

Depositing the thin Si film onto a textured substrate avoids this problem. This approach was first demonstrated for amorphous Si cells [106]. We transferred this technique to crystalline Si using textured glass substrates [27, 107, 108, 28] or textured porous Si substrates [53]. Figure 2.7 shows a $W_f = 8$ μm-thick, crystalline Si film deposited onto a textured glass substrate. The film thickness W_f is measured perpendicularly to the facet surface. We use ion-assisted Si deposition (IAD) for the growth of crystalline Si at temperatures as low as 600°C [27]. The IAD technique is described in more detail in section 5.1.2 on p. 126. Such conformal films show efficient light trapping with groove-shaped [27, 108], pyramid-shaped [109, 107], and tetrahedral-shaped [107] substrates.

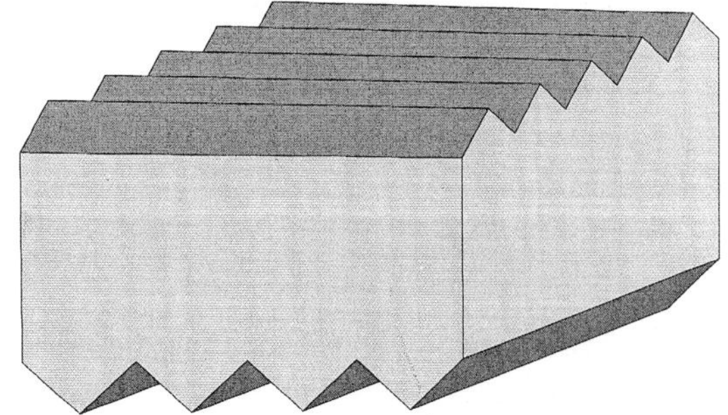

Figure 2.6. Perpendicular grooved cell for efficient light trapping.

In Appendix A on p. 194, we study basic film shapes with

two-, three-, four-, and six-sided structures. Steep facets turn out to perform best because they provide a larger Si volume for absorption (at constant film thickness W_f). In addition, steep facets reduce the front surface reflectance efficiently. A disadvantage of steep facets is an increased diode saturation current density due to an increased volume and an increased surface recombination. This disadvantage is, however, overcompensated for by an enhanced optical absorption (see section 5.4.4 on p. 153). The texture period is best chosen to have a value similar to the film thickness. For large periods as in Figure 2.7, the light only "sees" a planar film of Si and little internal randomization of light propagation occurs. Two-dimensional textures such as grooves perform worse than three-dimensional textures such as pyramids (see Figure A.12 on p. 196). The number of pyramid facets has little impact on the absorption properties.

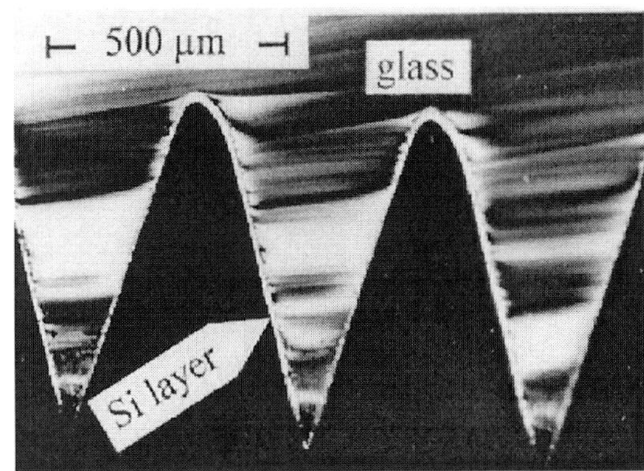

Figure 2.7. Optical microscope image of a 8 µm-thick microcrystalline Si layer on a V-structured glass substrate. The texture period is 500 µm.

The author introduced a novel technique, the porous Si (PSI) process, to fabricate monocrystalline conformal Si films with a thickness and texture period of a few microns [53]. A waffle-shaped film from this process is shown in Figure 5.8 on p. 128. We analyze the optical absorption of waffle-shaped Si films in section 5.3.2 on p. 141.

2.1.5 Optimum geometrical light trapping

The average path length $\bar{l} = 4\, n_s^2\, W_{eff}$ of the Lambertian light trapping scheme equals the theoretical maximum $\bar{l}_{max}$ of the average path length for any geometrical light trapping scheme under isotropic illumination [110, 16]. The proof was conducted in the framework of non-imaging geometrical optics for cells with non-reflecting interfaces. In Appendix A on p. 187, we show that $\bar{l}_{max} = 4\, n_s^2\, W_{eff}$ also holds for reflecting interfaces. The average path length $\bar{l}$ is, however, insufficient to characterize the light trapping performance. Obviously, an average path length $\bar{l}_{max}$ may be realized with various path length distributions $f(l)$. Higher-order moments of the path length distribution $f(l)$ have to be considered [16, 111]. Theoretically, the optimum optical absorption is reached for a texture with *all* of the isotropically incident rays having a path length l that equals the theoretical maximum $\bar{l}_{max} = 4\, n_s^2\, W_{eff}$. The absorption

$$A = 1 - \exp\left(-\alpha_s 4 n_s^2 W_{eff}\right) \tag{2.9}$$

is therefore unsurpassed by *any* geometrical light trapping scheme under isotropic illumination [112]. See Eq. (A.37) in Appendix A on p. 191 and its discussion for more details. It is presently not known how to construct a cell with the optimum light trapping and whether such a texture actually exists.

A cell with maximum absorption A at every wavelength λ given by Eq. (2.9) leads to a maximum photogenerated current density j_{sc}^* that cannot be exceeded by any geometrical texture. We call this limiting case optimum light trapping (Model O).

2.1.6 Short-circuit current limits

We consider the above-mentioned three levels of light trapping in our efficiency calculations: a planar cell, which means no light trapping (Model N), a cell with Lambertian light trapping (Model L), and a device with optimum light trapping (Model O). The three models yield the maximum short-circuit current densities j_{sc}^* shown in Figure 2.8. Selected values are tabulated in Table 2.1. The current density of a 1 μm-thick Si film is j_{sc}^* = 33.1 mA cm^{-2} for Lambertian light trapping (Model L), a value that agrees with previous calculations [19]. Without light trapping (Model N) the current is only one-third of that value. Optimum light trapping (Model O) yields a current density of 35.9 mA cm^{-2}. This value is only 8% higher than for Lambertian light trapping.

Figure 2.9 shows the optical absorption of thin-film cells with various levels of light trapping when plotted as a function of $\alpha_s W_{eff}$. For comparison we show the external

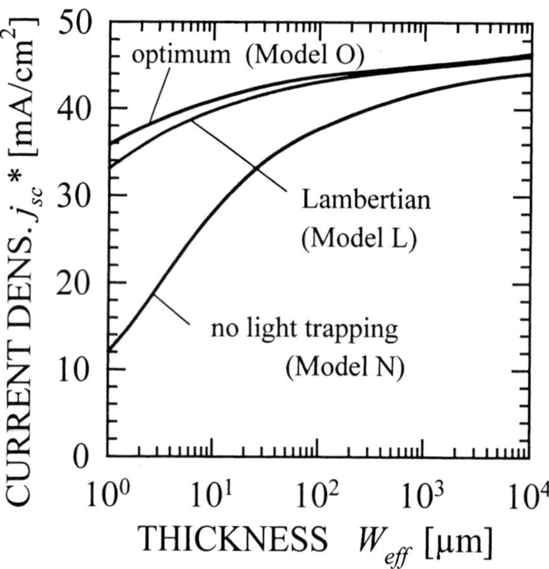

Figure 2.8. Maximum short-circuit current density j_{sc}^* for a silicon sheet of effective thickness W_{eff} without light trapping (Model N), with Lambertian light trapping (Model L), and with optimum light trapping (Model O).

Table 2.1. Photogenerated current densities for three levels of light trapping to an accuracy of ± 0.1 mA cm^{-2}.

Thickness W_{eff} [μm]	No light trapping (N) j_{sc}^* [mA cm^{-2}]	Lambertian (L) j_{sc}^* [mA cm^{-2}]	Optimum (O) j_{sc}^* [mA cm^{-2}]
1	12.1	33.1	35.9
10	28.3	39.7	41.0
100	37.8	43.1	43.7
300	40.1	43.9	44.4
500	40.9	44.3	44.6

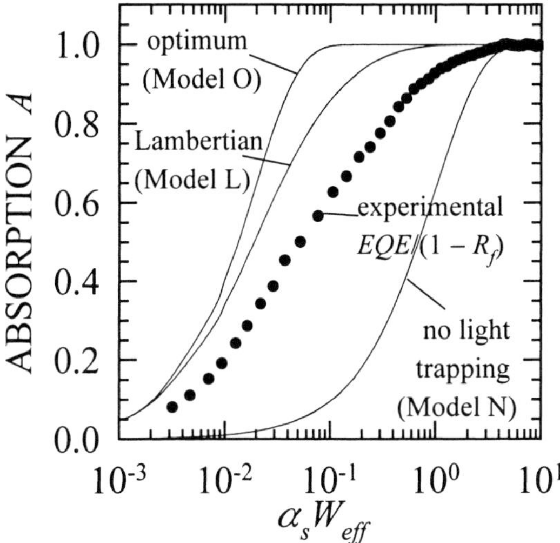

Figure 2.9. Theoretical optical absorption for Models N, L, and O plotted as a function of the product of film thickness W_{eff} and absorption coefficient α_s (lines). Experimental internal quantum efficiency $EQE/(1-R_f)$ (symbols) of a 20.6%-efficienct thin-film Si cell [113].

quantum efficiency $EQE_f(\alpha_s W_{eff})$ corrected with the front surface reflectance R_f of a 20.6%-efficient and 47 μm-thick thin-film cell, which Brendel et al. fabricated from thinned float-zone Si [113]. The cell has inverted pyramids on the front surface and a lapped and oxidized back surface covered with an Al film. The experimental data fall below the absorption expected for Lambertian light trapping. The deviation from the Lambertian behavior is not wholly due to optical losses, since the $EQE/(1-R_f)$ values are reduced by recombination losses and are therefore in general lower than the optical absorption A.

The comparison of ideal Lambertian light trapping with experimental data of record thin-film cells shows that Lambertian light trapping on the one hand is difficult to achieve practically and on the other hand is theoretically outperformed by specific geometrical schemes such as the perpendicular groove texture under normal illumination. Therefore, Lambertian light trapping serves well as a benchmark for good light trapping.

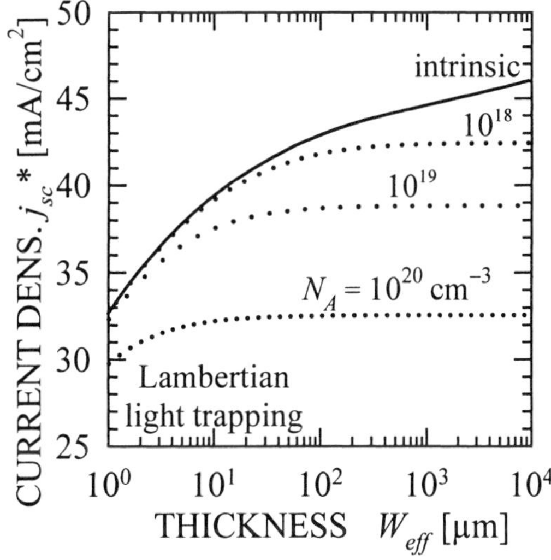

Figure 2.10. Maximum short-circuit current density j_{sc}^* for a silicon sheet of thickness W_{eff} with Lambertian light trapping. The reduction of the photogeneration j_{sc}^* due to free carrier absorption is shown for hole concentration N_A.

Thin-film solar cells require highly doped layers to provide a low resistance for current flow parallel to the film surface. Therefore we investigate the influence of free carrier absorption on j_{sc}^*. The free carrier absorption coefficient of holes is taken from Ref. [77]. According to Figure 2.10, the current loss is less than 0.8 mA cm^{-2} (2% relative) for thin ($W_{eff} \leq 50$ µm) films with a doping concentration $N_A \leq 10^{18}$ cm^{-3}. Hence, in terms of current loss the free carrier absorption is negligible in thin-film cells with Lambertian light trapping and is even less significant in cells without light trapping. However, free carrier absorption in the sub-bandgap range heats the cell and thus reduces the output power, an effect that is particularly important for cells used in outer space where cooling by convection is not possible.

2.1.7 Beyond geometrical optics

The existence of a theoretical limit to the average path length in geometrical light trapping schemes raises the question, of whether this limit is also valid in the framework of wave optics. We prove that the limit set by geometrical optics does not hold for wave optics. We conduct the proof by giving one specific example with an average path length $\bar{l} > \bar{l}_{max} = 4\, n_s^2\, W_{eff}$.

This scheme is sketched in Figure 2.11. The Si film is planar and of thickness $W_{eff} << \lambda$. The Si film has the optical constants n_s and α_s. We assume an ideal metal reflector below the Si film and an optically thick ($T >> \lambda$) transparent cover of refractive index $n_t > n_s$ above the Si film. The front surface of the cover has zero reflectance and scatters the light isotropically.

The light intensity impinging on the cover surface is I_o. For the weak absorption in the Si film the intensity of downwards propagating light in the cover is $n_t^2\, I_o$, as follows from the discussion of Lambertian light trapping schemes on p. 181 of Appendix A. Consider a ray of light with enhanced intensity $n_t^2\, I_o$ being reflected from inside the cover material at the thin Si film. We consider parallel polarization only. The reflectance of this ray is

$$R(\vartheta) = 1 - 4\frac{n_t^3}{n_s^3}\,\frac{\sin^2\vartheta}{\cos\vartheta}\,\alpha_s W_{eff} \tag{2.10}$$

due to the continuity of the normal component of the dielectric displacement current, and the constructive interference between incident and reflected p-polarized light [114]. The reflected intensity thus is $R\, n_t^2\, I_o$ and the optical absorption is

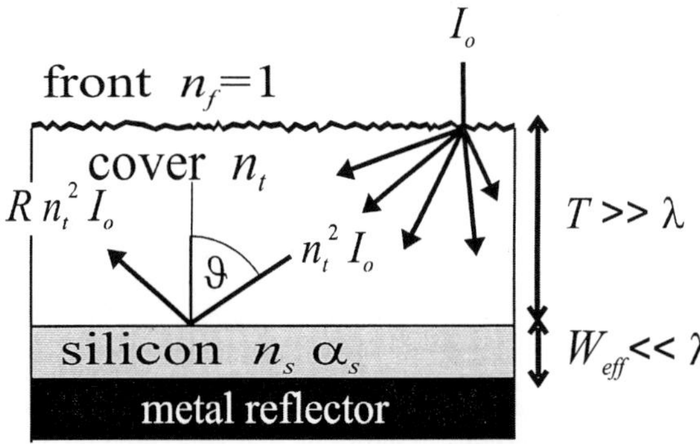

Figure 2.11. Non-geometrical light trapping scheme with a planar Si film of thickness W_{eff}, much thinner than the wavelength λ sandwiched between a cover of thickness $T >> \lambda$ that has an ideal metal reflector. The cover has a index of refraction $n_t > n_s$. The front surface of the cover scatters light isotropically.

$$A(\vartheta) = 4\frac{n_l^5}{n_s^3}\frac{\sin^2\vartheta}{\cos\vartheta}\alpha_s W_{eff} \tag{2.11}$$

Averaging this absorption over all directions we find

$$A = \frac{\int_0^{\pi/2} A(\vartheta)\cos\vartheta \sin\vartheta \, d\vartheta}{\int_0^{\pi/2} \cos\vartheta \sin\vartheta \, d\vartheta} = \frac{16}{3}\frac{n_l^5}{n_s^3}\alpha_s W_{eff} \tag{2.12}$$

To become more specific, let us now assume that the cover is made of intrinsic Ge that has a refractive index of n_l. The Ge cover is transparent for sub-bandgap light with $\lambda >$ 1.9 μm, which is weakly absorbed in a p-type Si film by free carrier absorption. For this case the absorption is

$$A = \frac{16}{3}\frac{4^5}{3.42^3}\alpha_s W_{eff} = 136\,\alpha_s W_{eff} \tag{2.13}$$

If we now consider that only half of the light is p-polarized, the resulting average path length expected is 68 W_{eff} and exceeds the geometrical limit of $4\times3.42^2\ W_{eff} = 46\ W_{eff}$.

Electron-hole pair generation in the Si film does not occur in this example. For this and other reasons our gedanken experiment has no practical relevance. The example does, however, prove that an increase of the optical absorption beyond the limits that are valid for geometrical optics is possible. The wave nature of light needs to be exploited in light trapping schemes that exhibit structures with sizes similar to or smaller than the optical wavelength.

2.2 Limitations imposed by radiative recombination

The photogeneration of electron-hole pairs is the fundamental process in solar cells. The inverse process, the recombination of an electron-hole pair and the emission of a luminescent photon, inevitably occurs also [115]. This band-to-band recombination process is sketched in Figure 2.12a. It obviously reduces the power output. A cell that has no other recombination processes but the inevitable radiative recombination has the maximum efficiency that is theoretically possible [14]. This section describes a model to calculate the efficiency of thin-film solar cells that are limited by radiative recombination. We include the other loss mechanisms that are shown in Figure 2.12 at a later stage of this study.

Infinite mobility assumption

In order to keep the mathematics simple, we assume infinite carrier mobility because this assumption implies flat quasi-Fermi levels, even for a finite carrier lifetime. The situation of flat quasi-Fermi levels is approximated in thin cells with a diffusion length that greatly exceeds the film thickness.

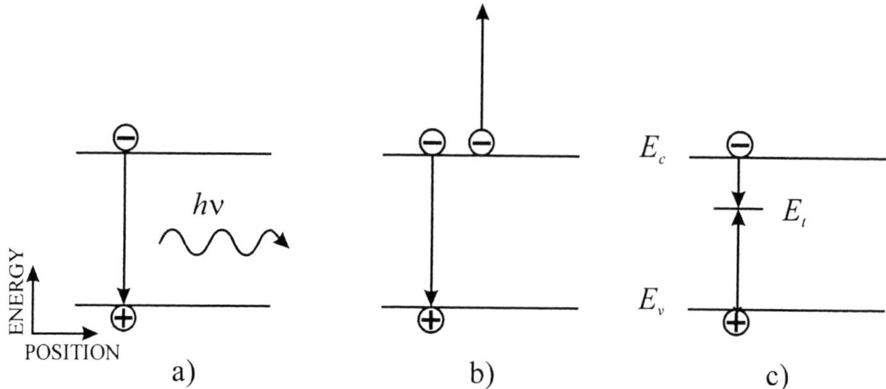

Figure 2.12. Recombination mechanisms in Si solar cells: a) radiative band-to-band recombination, b) Auger recombination, and c) Shockley-Read-Hall recombination via a bandgap state at energy E_t.

If the diffusion length is smaller than the device thickness W_{eff} required for sufficient optical absorption, it is still possible to have approximately flat quasi-Fermi levels by choosing special cell designs. The parallel multi-junction cell design [116] is illustrated in Figure 2.13a. The distance that the carriers have to diffuse is much smaller than the device thickness W_{eff}. Folding a thin layer in a V-shaped structure [27], as shown in Figure 2.13b, also de-couples the electrical collection thickness from the device thickness W_{eff}. We give a definition of the collection thickness on p. 145. The assumption of flat quasi-Fermi levels is approximately justified, provided the minority carrier diffusion length largely exceeds the carrier collection length and provided the surface recombination is small.

Current-voltage curve

The current

$$I(U) = R_{rad}(U) - G(U) \tag{2.14}$$

through a cell of area A_c depends on the voltage U and equals the difference between the radiative recombination current $R_{rad}(U)$ and the total generation current $G(U)$.

Generation current G

Our ansatz for the photogeneration rate permits us to model hypothetical solar cells that work under maximum light concentration and avoid thermalization losses of hot carriers. Such a general ansatz helps to quantify specific loss mechanisms by comparing calculated cell efficiency, e.g. without and with thermalization losses.

Photons γ_E of energy E with the flux $n_{\gamma s}(E)$ are absorbed with a probability equal to the optical absorption $A'(E, U)$ and generate $m(E)$ electron-hole pairs. A multiplicity factor $m(E) > 1$ accounts for the possibility that high-energy photons create more than one electron-hole pair in the reaction

$$\gamma_E \rightarrow m(E)[e+h] \tag{2.15}$$

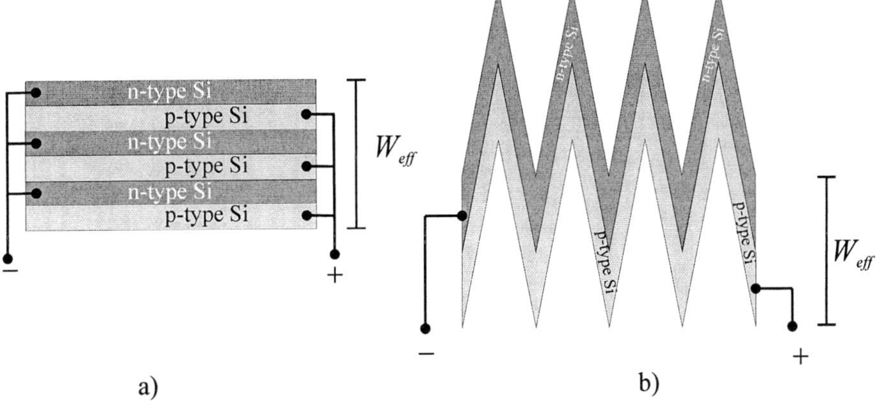

Figure 2.13. The parallel multi-junction solar cell (left) and the V-shaped structure with steep facets (right) both de-couple electrical and optical performance. While a large Si volume is available for optical absorption, the carriers only have a small distance to diffuse to the collecting junction. The effective film thickness W_{eff} is defined as the ratio of cell volume over macroscopic cell area.

The generation of a single electron-hole pair by two photons is modeled by $m < 1$. We do not specify the microscopic mechanisms for a quantum yield $m(E) \neq 1$ here. The generation current is

$$G(U) = q\,\mathcal{E}_s \int_0^\infty m(E)\, n_{\gamma s}(E)\, A'(E,U)\, dE \tag{2.16}$$

The optical absorption $A'(E, U)$ of the semiconductor changes with the quasi-Fermi level splitting U for various reasons, such as free carrier absorption, bandgap shrinkage, and a possible onset of laser action at very high U values with qU exceeding the smallest band-to-band excitation energy. Hence, we let the optical absorption A' depend on the cell voltage U.

Radiative recombination current R_{rad}

In a radiative band-to-band recombination process the energy of the electron-hole pair is transferred to a photon. In analogy to the ordinary radiative recombination process, the time reversal invariance of microscopic processes requires that the recombination reaction

$$m(E)[e + h] \rightarrow \gamma_E \tag{2.17}$$

which is the inverse process of reaction (2.15), also occurs. The emitted photon flux (photons per energy, time, and étendue)

$$n_{\gamma c}(E,U) = \frac{2}{c^2 h^3} E^2 \Big/ \left\{ \exp\left(\frac{E - \mu_\gamma(U)}{kT_c} \right) - 1 \right\} \tag{2.18}$$

of the cell at temperature T_c = 300 K is given by the generalized Planck-radiation law and depends on the chemical potential $\mu_\gamma(U)$ of the photons. The concept of photons having a non-zero chemical potential was introduced by Würfel [55]. The generalized Planck-radiation law expresses the chemical equilibrium between the electromagnetic photon field and the electron-hole gas in the solid. A chemical equilibrium establishes between the reactions (2.15) and (2.17). Therefore, the chemical potential μ_γ of the photon γ_E equals the sum of the chemical potential $\mu_{e/h}$ of all $m(E)$ generated electron-hole pairs [66, 18]. For infinite mobility, the chemical potential of a single electron-hole pair $\mu_{e/h}$ is related to the voltage U by $\mu_{e/h} = q\,U$ [54]. Then, the photon chemical potential

$$\mu_\gamma = m(E)qU \tag{2.19}$$

depends on the photon energy E, in contrast to cells without carrier multiplication ($m(E) = 1$), where $\mu_\gamma = q\,U$ holds. Every recombination process is accompanied by the emission of a photon, since non-radiative recombination processes are neglected. Each emitted photon γ_E contributes with the multiplicity $m(E)$ to the electronic recombination current

$$R_{rad}(U) = q\,\mathcal{E}_c \int_0^\infty m(E)n_{\gamma c}(E,V)\,A'(E,U)\,dE \tag{2.20}$$

The symbol $\mathcal{E}_c = A_c\,\pi\,\sin^2(\Phi_c)$ denotes the étendue of the luminescence light leaving the cell. The cell of area A_c emits photons into a half-angle Φ_c that may be larger than the angle Φ_s from which the sunlight is received (see Figure 2.2 on p. 13). Equation (2.20) includes the effect of the re-absorption of luminescence light: e.g., a black body with infinite absorption coefficient α_s and thus optical absorption $A' = 1$ will only emit radiation from its surface. This is exactly what is expressed by the generalized Planck law in Eq. (2.18).

The only voltage dependence of the optical absorption

$$A'(E,U) = \begin{cases} A(E) & \text{for } E \geq m(E)qU \\ \dfrac{A(E)}{A(E)-1} & \text{for } E < m(E)qU \end{cases} \tag{2.21}$$

that we account for is the onset of laser action at a chemical photon potential $\mu_\gamma = m(E)\,q\,U$ that is larger than the photon energy E. The absorption coefficient then becomes negative [69]. Let us assume that the absorption $A(E)$ is due to a path length l. The path length is identical for all rays of light in Model O and Model N. The absorption is $A(E,\ U < E/m(E)) = 1 - \exp(-\alpha(E)\,l)$. Changing $\alpha(E)$ to $-\alpha(E)$ yields the absorption $A'(E,\ U > E/m(E)) = 1 - \exp(\alpha(E)\,l) = A(E)/(A(E) - 1)$. Hence, we arrive at Eq.(2.21). For Lambertian light trapping (Model L) not all of the rays have the same path length. Equation (2.21) is thus an approximation.

Equation (2.20) represents the diode saturation current of a hypothetical, carrier-multiplying solar cell with the multiplicity $m(E)$, with only radiative recombination, and with the optical absorption $A(E)$.

Cell efficiencies

The cell's energy conversion efficiency

$$\eta(U) = -U\, I(U) / P_{inc} \tag{2.22}$$

is the ratio of the electrical output power $-U\, I(U)$ to the incident solar power

$$P_{inc} = \mathcal{E}_s \int_0^{\infty} E\, n_{\gamma s}(E)\, dE \tag{2.23}$$

In order to identify losses due to non-absorption, we also define the radiation power

$$P_{abs} = \mathcal{E}_s \int_0^{\infty} E\, n_{\gamma s}(E)\, A'(E,U)\, dE \tag{2.24}$$

that is absorbed in the cell. The cell efficiency, ignoring absorption losses, is then defined by

$$\eta_{abs}(U) = -U\, I(U) / P_{abs} \tag{2.25}$$

Unless stated otherwise, both efficiencies η and η_{abs} are given at the maximum-power point. Both efficiencies depend on the étendue ratio $\mathcal{E}_s / \mathcal{E}_c$. To conform with the usual notation, we assume that the cell emits luminescence radiation into a half-angle $\Phi_c = 90°$, which makes the efficiency a function of concentration C via the relation $\mathcal{E}_s / \mathcal{E}_c = C/C_{max}$.

We now apply the above model to calculate the efficiencies of a hypothetical and highly idealized thin-film photovoltaic device that has the optical absorption of crystalline Si. We add one loss mechanism after the other to identify the significance of each mechanism and to make the modeling more realistic. In section 2.3, we add non-radiative recombination mechanisms that, in practice, limit Si solar cell efficiencies.

2.2.1 Carnot efficiency

The photovoltaic device is a thermodynamic machine working between one heat reservoir, which is the sun at temperature T_s = 5780 K, and another heat reservoir, which is the cell's surroundings at temperature T_c = 300. The second law of thermodynamics limits the efficiency for the conversion from thermal to electrical power to the Carnot value

$$\eta_{Car} = \left(1 - \frac{T_c}{T_s}\right) = 94.8\% \tag{2.26}$$

The Carnot process working with an ideal gas can, in principle, realize this efficiency. The gas is expanded in thermal equilibrium with the high-temperature reservoir, then further expanded adiabatically, compressed in equilibrium with the low-temperature reservoir, and finally further compressed adiabatically to close the cycle.

Chemical equilibrium of the electron-hole gas with the photons emitted from the sun is reached if the cell at open-circuit only "sees" the sun. All rays that escape from the cell are directed back into the sun. We then have $\mathcal{E}_s = \mathcal{E}_c$, an equality that holds, e.g., for maximum concentration C_{max}. We assume maximum concentration to construct a hypothetical solar cell that will work with Carnot efficiency.

Entropy production by hot carriers generating phonons has to be suppressed to reach Carnot efficiency. We therefore de-couple the electron-hole gas from the lattice. Instead of generating phonons, the photons with an energy E generate

$$m(E) = E / E_g \tag{2.27}$$

electron-hole pairs as is required by energy conservation in our hypothetical device. For photons with energy $E > E_g$ the quantum yield exceeds unity. For sub-bandgap photons, a less-than-unity quantum yield means sub-bandgap absorption by multi-photon processes without phonon participation. Such processes are energetically necessary, if we assume the sub-bandgap absorption of Si while de-coupling the electron-hole gas and the lattice. We do not make any suggestions as to how such a process should work microscopically.

In a situation of chemical equilibrium between the electron-hole gas and the sun, a detailed balance between the solar photons irradiating the cell and the photons that are emitted from the cell prevails. This implies that no current flows in the external circuit of the cell and hence, the device works in open-circuit conditions.

We calculate the open-circuit voltage by numerically solving $I(U_{oc}) = 0$ with the current-voltage curve $I(U)$ from Eq. (2.14). The open-circuit voltage *is* $U_{oc} = (1.058 \pm 0.001)$ V for *all* three levels of light confinement (Models N, L, and O) and *all* cell thickness values ranging from $W_{eff} = 1$ µm to 10^4 µm. The energy corresponding to this voltage is a fraction

$$\frac{qU_{oc}}{E_g} = \frac{1.058\,\text{eV}}{1.12\,\text{eV}} = 94.5\% \tag{2.28}$$

of the energy gap E_g. The similarity of the numerical values for qU_{oc}/E_g and the Carnot efficiency suggests the relation

$$V_{oc} = \left(1 - \frac{T_c}{T_s}\right) E_g \tag{2.29}$$

It is indeed possible to derive this relation analytically, if the cell is illuminated with black body radiation instead of AM1.5G illumination. Hence, the deviation of 0.3% absolute of the Carnot efficiency (Eq. (2.26)) and the voltage to bandgap ratio (Eq. (2.28)) stems from the fact that the AM1.5G-spectrum deviates slightly from a black body spectrum. Please note that the cell does not convert *power* in open-circuit; it might, however, be considered as an *energy* converter. The open-circuit voltage U_{oc} is only the *potential* to perform electrical work. Hence, we constructed, in a gedanken experiment, thin-film photovoltaic devices that have the optical absorption coefficients of Si and convert radiation *energy* to electrical *energy* with Carnot efficiency. These devices work under optimum light concentration and redirect all non-absorbed photons, as well as all luminescence photons, back to the sun. In addition, the devices only exhibit radiative recombination, and ideal carrier multiplication avoids carrier thermalization. However, no or negligible solar *power* is converted. Strictly speaking, the commonly used term *energy conversion efficiency* is a misnomer. The sun irradiates the solar cell with solar radiation power and the cell generates electric power. What we are interested in is the *power conversion efficiency*.

The power which is re-directed to the sun is still available for solar power conversion. If we therefore subtract this power from the incident solar power P_{inc}, the devices can even covert a small amount of solar power to electrical power when operated at a voltage slightly below the open-circuit voltage. Such a cell converting power close to open-circuit conditions can be considered as a thermodynamic machine working very slowly to permit adiabatic processes.

2.2.2 Luminescence

In this section, we calculate the power efficiency η_{abs} of thin-film cells as defined in Eq. (2.25) on p. 27. The definition of η_{abs} ignores the non-absorbed photons. Non-absorption losses are thus excluded. As in the previous section, we consider optimum light concentration and optimum quantum yield $m(E)$ to exclude thermalization losses. Luminescence radiation is the only loss considered here.

The calculated power conversion efficiency η_{abs} is shown in Figure 2.14. Since non-absorption losses are ignored, optical absorption has a small impact on the conversion efficiency. All three light trapping models and all film thickness values have power conversion efficiencies η_{abs} ranging between 86 and 89%. Thinner cells with less efficient light trapping perform slightly better. The reason for this effect is that light trapping, as well as the use of thicker films, both enhance the emission of near-bandgap luminescence light. However, the corresponding increase of the absorption of visible

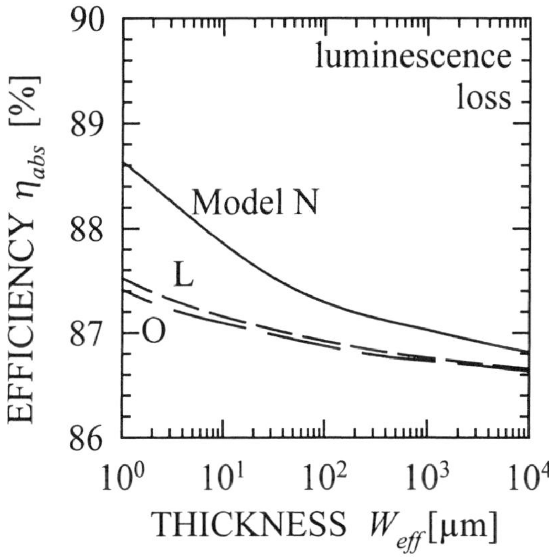

Figure 2.14. Power conversion efficiency limit η_{abs} for cells with the optical constants and the electronic bandgap of crystalline Si. Assumptions: radiative recombination only, no optical losses, maximum light concentration, and no thermalization losses. Model N: no light trapping; Model L: Lambertian light trapping; and Model O: optimum light trapping.

Table 2.2. Power conversion efficiency η_{abs} for cells of thickness W_{eff} with the optical constants and the electronic bandgap of crystalline Si. Assumptions: radiative recombination only, no optical losses, maximum light concentration, and no thermalization losses. Model N: no light trapping; Model L: Lambertian light trapping; and Model O: optimum light trapping.

	Efficiency η_{abs} [%] at W_{eff} = 1 μm	Efficiency η_{abs} [%] at W_{eff} = 10 μm	Efficiency η_{abs} [%] at optimum W_{eff}
Model N	88.6	87.9	> 88.6 at W_{eff} < 1 μm
Model L	87.5	87.2	> 87.5 at W_{eff} < 1 μm
Model O	87.4	87.1	> 87.4 at W_{eff} < 1 μm

light is not accounted for by the definition of η_{abs}. The efficiency η_{abs} thus decreases.

Efficiency values of cells 1 μm and 10 μm in thickness are listed in Table 2.2. A 1 μm-thick cell with Lambertian light trapping reaches an efficiency of 87.5% and loses 7% relative to the Carnot efficiency. Since power is now extracted, the thermodynamic machine now works at finite speed. Adiabatic processes are no longer possible, and thus the device fails to reach Carnot-efficiency.

2.2.3 Optical absorption

From a practical point of view, only the power conversion efficiency η defined relative to the full incident power P_{inc} is relevant. We thus have to consider the non-absorbed photons as a loss. The calculated η values are shown in Figure 2.15. For thick cells, the optical loss is primarily due to the non-absorption of sub-bandgap photons. For thin cells the non-absorption of photons with energies larger than the bandgap is also significant. Optical losses cause the efficiency η to increase with the film thickness, while η_{abs} discussed in the previous section decreases with the thickness W_{eff}.

A 1 μm-thick cell with Lambertian light trapping yields an efficiency of 52.2% under

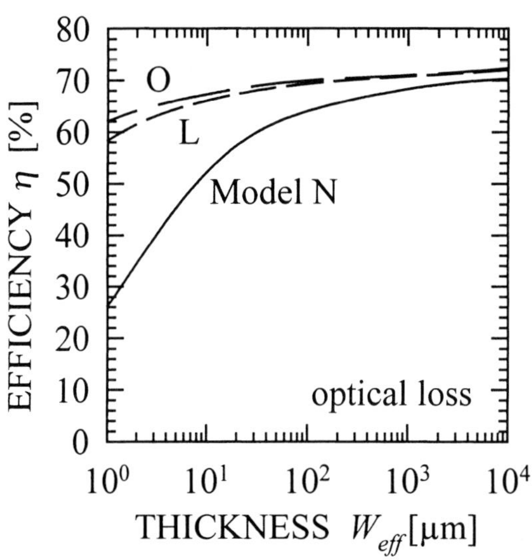

Figure 2.15. Power conversion efficiency limit η for cells with the optical constants and the electronic bandgap of crystalline Si. Assumptions: radiative recombination only, optical losses included, maximum light concentration, and no thermalization losses. Model N: no light trapping; Model L: Lambertian light trapping; and Model O: optimum light trapping.

maximum light concentration and without thermalization losses. For Lambertian light trapping, the optical losses are mainly sub-bandgap losses that reduce the efficiency by 40% relative to the case in which absorption losses are neglected. Non-absorption of solar photons is therefore one of the major losses in thin-film cells, even if efficient light trapping is used.

Table 2.3. Power conversion efficiency η for cells of thickness W_{eff} with the optical constants and the electronic bandgap of crystalline Si. Assumptions: radiative recombination only, optical losses included, maximum light concentration, and no thermalization losses. Model N: no light trapping; Model L: Lambertian light trapping; and Model O: optimum light trapping.

	Efficiency η [%] at W_{eff} = 1 μm	Efficiency η [%] at W_{eff} = 10 μm	Efficiency η [%] at optimum W_{eff}
Model N	26.2	58.3	62.0 at > 10^4 μm
Model L	52.2	66.1	67.3 at > 10^4 μm
Model O	70.3	72.0	72.3 at > 10^4 μm

2.2.4 Non-concentration

Under maximum concentration C_{max} = 4.6×10^4 the short-circuit current density is in the range of kA cm^{-2}, which leads to severe series resistance problems. In this section, we consider the same cell type of cell as in the previous section, but under only one sun of AM1.5G illumination (concentration C = 1). As before, our hypothetical photovoltaic device has the optical absorption coefficient of Si, is limited by radiative recombination, and has no thermalization losses, due to the assumption of an ideal quantum yield $m(E) = E/E_g$.

The calculated efficiencies are smaller than for maximum concentration because the cell now also sees the cold sky. A chemical equilibrium of the electron-hole gas with the sun is no longer feasible. The open-circuit voltage reduces when compared to the case of

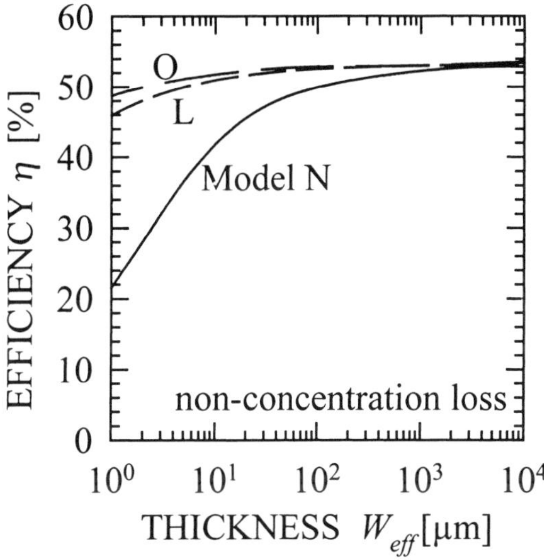

Figure 2.16. Power conversion efficiency limit η for cells with the optical constants and the electronic bandgap of crystalline Si. Assumptions: radiative recombination only, optical losses included, no light concentration, and no thermalization losses. Model N: no light trapping; Model L: Lambertian light trapping; and Model O: optimum light trapping.

full concentration, and thus the efficiency also decreases. A 1 μm-thick cell with Lambertian light trapping has a power conversion efficiency of 45.9% as listed in Table 2.4. This is a relative loss of 12% due to non-concentration when compared to maximum concentration. The efficiency of thick cells is 53% and is independent of the level of light trapping.

Table 2.4. Power conversion efficiency η for cells of thickness W_{eff} with the optical constants and the electronic bandgap of crystalline Si. Assumptions: radiative recombination only, optical losses included, no light concentration, and no thermalization losses. Model N: no light trapping; Model L: Lambertian light trapping; and Model O: optimum light trapping.

	Efficiency η [%] at W_{eff} = 1 μm	Efficiency η [%] at W_{eff} = 10 μm	Efficiency η [%] at optimum W_{eff}
Model N	21.5	41.8	52.9 at > 10^4 μm
Model L	45.9	50.9	53.4 at > 10^4 μm
Model O	48.7	51.7	53.5 at > 10^4 μm

2.2.5 Thermalization

All our previous efficiency calculations assumed an optimum quantum yield $m(E) = E/E_g$, which means the creation of as many electron-hole pairs as is energetically possible. Hence, a quantum efficiency exceeding unity is assumed for photons with energies larger than the bandgap. The experimental observation of pair generation with efficiencies above unity in p-n junctions was reported in the 1950s for germanium [117], for silicon [58], and indium antimonide [118]. The work by Christensen [119], Wilkinson et al. [120] and Kolodinski et al. [59] correlated this phenomenon to the band structure $E(\boldsymbol{k})$.

Figure 2.17 shows the experimental internal quantum efficiency *IQE* that we measure for a monocrystalline Si cell fabricated from float-zone Si wafer thinned to 50 μm. At

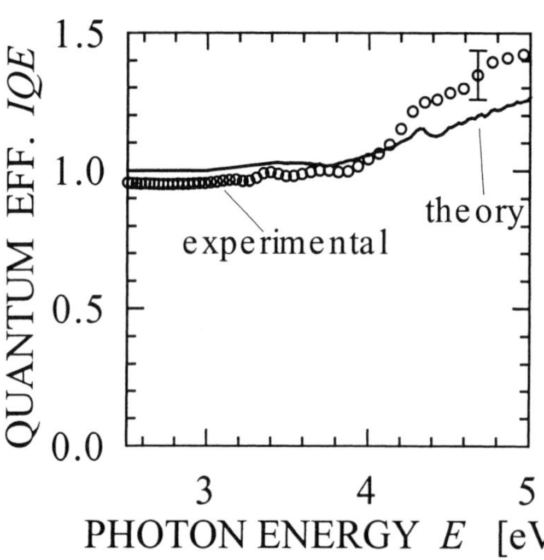

Figure 2.17. Internal quantum efficiency of a 50 μm-thick Si cell with inverted pyramid texture. The solid line is a simulation with a theory that accounts for the band structure of Si but neglects momentum conservation during impact ionization.

2.5 eV the value of *IQE* is 0.95, due to a minority carrier diffusion length $L > 200$ µm that exceeds the device thickness. For photon energies above a threshold value $E_t = 3.9$ eV, the experimental internal quantum efficiency exceeds unity and is approximately given by

$$IQE \approx 1 + \frac{E - E_t}{2\,E_g} \text{ for } 5\,\text{eV} > E > E_t = 3.9\,\text{eV} = 3.5\,E_g \tag{2.30}$$

The quantum efficiency exceeding unity is due to impact ionization [121, 59], which also plays an important role in device physics [122]. Theoretical investigations of impact ionization, especially in high electric fields, such as in reverse-biased p-n junctions, date back to the 1950s [123-126]. Several adjustable parameters were used at that time to describe the impact ionization in accordance with experimental data [124]. More exact analytical calculations of the impact ionization rate are only possible for simplified, e.g. parabolic, band structures [127]. In recent publications, the impact ionization probability was calculated including energy and momentum conservation for the realistic band structure and electron wave vectors in Si [128-132] as well as in other semiconductors, mainly GaAs [133-138]. The impact yield *N* in Si as calculated quantum mechanically by Sano and Yoshii compares well with the analytical theory of Alig, Bloom, and Struck (ABS) [139] that was used by Geist and Wang [121] to calculate the internal quantum efficiency of Si in the ultraviolet spectral range. Using the approach of Geist and Wang we calculated the theoretical quantum yield

$$m(E) = 1 + \int_{E=0}^{\infty} dE'\,P(E',E)\;N(E') \quad \text{for } E > E_g \tag{2.31}$$

We express the number of additional charge carrier pairs produced by impact ionization as the product of the probabilities $P(E', E)$ and the quantum yield $N(E)$ [140]. A photon of energy E has the probability $P(E', E)$ to create a primary electron or hole with kinetic energy E'. The impact yield $N(E')$ defines the number of secondary charge carriers that are impact-generated by a primary carrier with energy E'. Integrating over all energies and adding up the contributions of the valence and conduction bands, we obtain the quantum yield $m(E)$. When calculating $P(E', E)$, the transition matrix element M for direct photon absorption is assumed to be independent of the wave vector $\boldsymbol{k}$. Only direct electric dipole transitions with a change in angular momentum are allowed. We convert the integral of Eq. (2.31) into a sum of 4.6×10^4 wave vector k-points in the symmetry reduced Brillouin zone [141] and carry out the integration over the four highest valence and four lowest conduction bands that we calculate by the semi-empirical method of linear combination of atomic orbitals (LCAO) [142-144]. With this band structure, we calculate the quantum yield that is depicted as the solid line in Figure 2.17 [140]. The soft onset for $m(E) > 1$ is in agreement with the measured internal quantum efficiency *IQE*. Above $E = 4$ eV the experimental *IQE* values exceed the theoretical ones. One possible reason is an overestimation of $m(E)$ by the ABS-theory due to neglected momentum conservation. We draw the following conclusions from our experimental and the theoretical results:

- Internal quantum efficiencies exceeding unity by as much as 40% at E = 5 eV are feasible with Si cells. The size of this effect makes impact ionization – in principle – an important effect for power conversion with semiconductor cells.
- The threshold energy E_t = 3.9 eV for the soft onset of $IQE > 1$ is more than three times the bandgap energy. Although photons with an energy of $2E_g$ would energetically permit the generation of two electron-hole pairs, they are unlikely to do so in Si. The AM1.5G spectrum only contains photons with energies lower than 4.13 eV. Hence, for Si, there is negligible overlap of the solar spectrum with the region of quantum efficiencies exceeding unity. The current improvement in terrestrial Si cells is negligible and less than 0.1 mA cm^{-2}.
- Only half of the energy of photons with energy $E > E_t$ is converted into electron-hole pairs because $m \propto E/(2E_g)$ rather than $m \propto E/E_g$, as we assumed above for ideal carrier multiplication. A hot carrier is as likely to reduce its energy by phonon generation as by secondary pair generation. The power conversion of ultraviolet light with $E_\gamma > E_t$ is not particularly efficient in Si cells.

The efficiency improvement of thin-film Si cells due to carrier multiplication is thus negligible and thermalization remains a major unavoidable loss mechanism. We therefore now analyze thin-film Si cells without carrier multiplication having a quantum yield $m(E)$ = 1. Unity quantum yield implies full thermalization of hot carriers down to the band edges. We are considering a device with the optical absorption coefficient and the refractive index of Si that exhibits radiative recombination only.

The resulting thin-film efficiencies are shown in Figure 2.18. For all light trapping schemes the efficiencies have a maximum at an optimum cell thickness, as was observed before [19]. This maximum is a consequence of the optical absorption loss that decreases with the thickness and the radiative recombination loss that increases with the film thickness. The maximum would be more pronounced if photon recycling was neglected.

A 1 µm-thick cell with Lambertian light trapping has an optimum conversion efficiency of 27.1%. This is a relative reduction of 40% when compared to the case without any thermalization loss. At an optimized thickness of W_{eff} = 133 µm the efficiency is

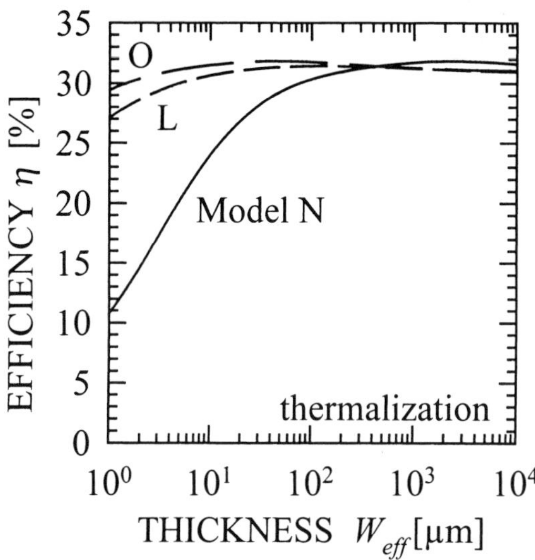

Figure 2.18. Power conversion efficiency limit η for cells with the optical constants and the electronic bandgap of crystalline Si. Assumptions: radiative recombination only, optical losses included, no light concentration, thermalization losses included. Model N: no light trapping; Model L: Lambertian light trapping; and Model O: optimum light trapping.

31.4%. That value is slightly larger than the optimum efficiency of 30% calculated previously [14] because we use an AM1.5G spectrum that contains less ultraviolet light than the black body spectrum. Further efficiency values are listed in Table 2.5.

Table 2.5. Power conversion efficiency η for cells of thickness W_{eff} with the optical constants and the electronic bandgap of crystalline Si. Assumptions: radiative recombination only, optical losses included, no light concentration, thermalization losses included. Model N: no light trapping; Model L: Lambertian light trapping; and Model O: optimum light trapping.

	Efficiency η [%] at W_{eff} = 1 μm	Efficiency η [%] at W_{eff} = 10 μm	Efficiency η [%] at optimum W_{eff}
Model N	10.7	24.0	31.8 at 2371 μm
Model L	27.1	30.6	31.4 at 133 μm
Model O	29.4	31.5	31.8 at 50 μm

2.3 Limitations imposed by non-radiative recombination

We are now halfway through on our tour from an ideal photovoltaic device working at Carnot efficiency to a realistic Si cell. In this section, we add non-radiative recombination mechanisms that are more important in Si cells than radiative recombination [19, 56]. For Auger, surface, and grain boundary recombination we deduce the recombination rates from experiments. These recombination processes are then included in our model to calculate more realistic cell efficiencies.

2.3.1 Auger recombination

Auger recombination [145] and radiative recombination are *intrinsic recombination mechanisms* that cannot be avoided and thus limit the power conversion efficiency. As depicted in Figure 2.12b on p. 24, the energy of a recombining electron-hole pair is transferred to a third particle (electron or hole) that is excited from the band edge high into the band. We first describe how to measure Auger recombination in Si and then apply a parameterization of the measured injection level-dependent recombination rate to our efficiency calculations. A pre-condition for the measurement of Auger recombination is the suppression of other recombination mechanisms, such as surface recombination.

Suppression of surface recombination

The fraction of the total recombination that is due to surface recombination is reduced by choosing thick samples. A different approach is to reduce surface recombination to a negligible level.

The application of corona charges in the field of semiconductor characterization is well known [146-150]. However, recently we introduced the temporal suppression of surface recombination by deliberately depositing charged air molecules onto an oxidized Si wafer with a corona discharge [45, 151].

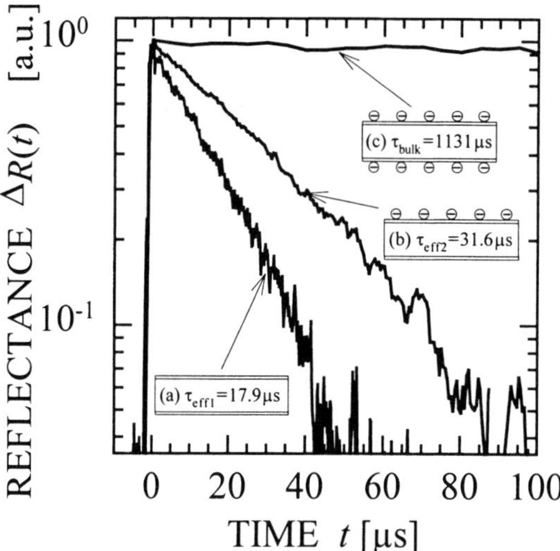

Figure 2.19. Transient microwave reflection of an oxidized 1 Ω cm float-zone Si wafer of 280 μm thickness. The effective minority carrier lifetime increases from 17 μs to 1113 μs when charging both oxide surfaces with corona charges of density 5×10^{-12} cm^2. After Ref. [45]

The transient microwave reflection technique is commonly used to measure minority carrier lifetime in Si [152-154, 151, 155]. Figure 2.19 shows a transient microwave reflectance measurement on an oxidized 1 Ω cm Si wafer. The effective lifetime increases from a value of 17.9 μs for the as-oxidized sample, through 31.6 μs for the charged on one side wafer, to 1.1 ms for charged on both sides wafer. Charging reduces the surface recombination velocity down to values as small as 1 cm s^{-1} [45]. With such well passivated surfaces, the measured effective lifetime equals the bulk lifetime of the Si wafer. Under these conditions, the determination of Auger recombination lifetimes becomes feasible, provided the bulk material quality is sufficiently high to have negligible bulk recombination via defects. This technique of temporary surface passivation with corona charges is now being used frequently [156, 157].

Auger lifetime measured on bulk-dominated samples

Altermatt et al. collected lifetime data [21] that were measured either on thick samples or on electrostatically passivated samples. The electrostatic passivation was done with corona charges or with fixed charges in the passivating silicon nitride layer [160]. The data shown in Figure 2.20 are corrected for residual Shockley-Read-Hall recombination and radiative recombination. The net Auger recombination current

$$R_{Aug} = \left[C_p\left(p^2 n - p_o^2 n_o\right) + C_n\left(n^2 p - n_o^2 p_o\right)\right] W_{eff} A_C \,. \tag{2.32}$$

is the total recombination current minus the Auger generation current at thermal equilibrium. The Auger recombination contains a first contribution from the interaction of two holes with one electron and a second contribution from two electrons interacting with one hole. This ansatz assumes independent free particles. The multiplication by the cell volume $W_{eff}\times A_c$ converts the recombination rate to a recombination current. The coefficients C_p and C_n were determined from photoluminescence decay measurements [158, 161, 162], transient photocurrents [163], and lateral diffusion length measurements in transistors [164]. The values of C_n and C_p that were determined by various authors agree to better than a factor of 2. We tabulate literature values for the Auger recombination coefficients in Appendix B on p. 214.

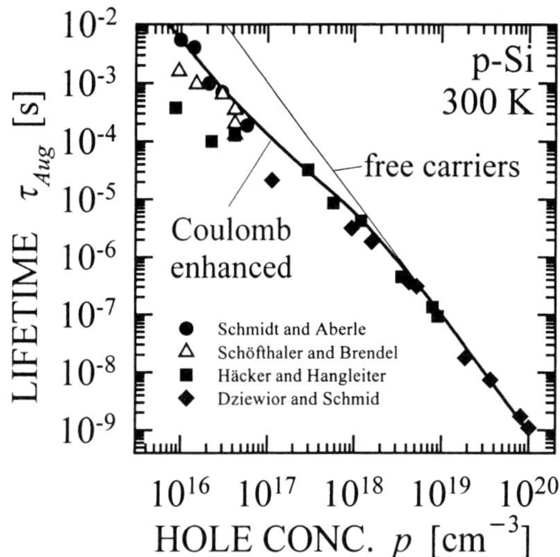

Figure 2.20. Measured Auger lifetime for p-type Si (symbols) [158, 159, 45, 160] under low injection conditions. The thin solid line applies for free carriers. The thick solid line is an empirical upper limit to the experimental data [21].

For p-type material under low-injection conditions the Auger-limited lifetime is

$$\tau_{Aug} = A_C W_{eff} (n - n_o) / R_{Aug} = 1/(C_p p^2) \tag{2.33}$$

The process involving two electrons is negligible as the electrons are rare. The lifetime is then inversely proportional to the square of the hole concentration p. For high hole concentrations $p > 5\times10^{18}$ cm^{-3}, the data in Figure 2.20 are well described by the free particle ansatz (2.33). This square dependence was used by Dziewior and Schmid to determine the most widely used Auger recombination coefficients C_p and C_n [158]. Hangleiter and Häcker explained the departure of experimental lifetime data from this square-relation for small doping concentrations. Coulomb attraction, which is shielded at high hole concentrations, becomes significant at small hole concentrations. The Coulomb interaction enhances the probability for an electron to be close to a hole and vice versa [165, 159]. In contrast, carriers of identical polarity repel each other. The three carriers that interact in an Auger recombination process consist of two attracting pairs with opposite polarity and one repelling pair of the same polarity. Thus the attractive interaction dominates, enhances recombination, and causes the lifetimes in Figure 2.20 to fall below the free carrier line.

In high-injection conditions, the neutrality relation $n = p$ results in an Auger lifetime

$$\tau_{Aug} = \frac{1}{(C_n + C_p)p^2} \tag{2.34}$$

Holes and electrons are now present in equal quantities and the Coulomb enhancement is therefore expected to be different from the low injection case. Sinton and Swanson [166] derived the ambipolar Auger coefficient from transient and steady-state open-circuit voltage decay measurements on solar cells that were designed to make Auger recombination the dominating process. Yablonovitch and Gmitter [167] performed photoconduction decay measurements on intrinsic Si wafers with surfaces passivated by an

HF-treatment. For our efficiency estimates we use a parameterization of the Auger recombination rate [71] that accounts for the Coulomb enhancement and for the injection level dependence. This parameterization is given by Eq. (B.11) on p. 215 of Appendix B.

Auger-limited efficiency

Adding the Auger recombination current R_{Aug} from Eq. (2.32) to the current-voltage curve yields

$$I(U) = R_{rad}(U) + R_{Aug}(U, N_A) - G(U) \tag{2.35}$$

Assuming infinite mobility and spatially homogeneous doping, the carrier concentrations n and p are constant throughout the volume. The values of n and p depend on the splitting $qU = E_{Fn} - E_{Fp}$ of the quasi-Fermi levels and the doping concentration N_A via Eqs. (B.3) and (B.4) on p. 210. When calculating the cell efficiency η, we use an optimized base doping that is p-type and around $N_A = 10^{17}$ cm^{-3} for 1 µm-thick films, and around $N_A = 10^{15}$ cm^{-3} for 1000 µm-thick films. The resulting cell efficiencies are shown in Figure 2.21. The injection level at the maximum-power point ranges from 0.2 to 0.4 in units of the doping concentration N_A. However, the impact of the base doping on the limiting efficiency is small. Keeping the base doping fixed to $N_A = 10^{16}$ cm^{-3} reduces the efficiencies plotted in Figure 2.21 by less than 0.2% absolute for all thickness values W_{eff} that we investigated.

Selected efficiency values are listed in Table 2.6. An efficiency of 24.5% is feasible for a 1 µm-thick Si cell with Lambertian light trapping and no defect recombination. This clearly demonstrates that the thickness of conventional solar cells with a value of typically $W_{eff} = 300$ µm is more than two orders of magnitudes larger than is physically necessary. Please note the enormous impact of light trapping on the efficiency of thin films. Without light trapping (Model N), the efficiency of a thin-film cell is 8.7% at best. Light trapping almost triples the cell efficiency. The efficiency reduction due to Auger recombination is 10% relative to the case with no Auger recombination.

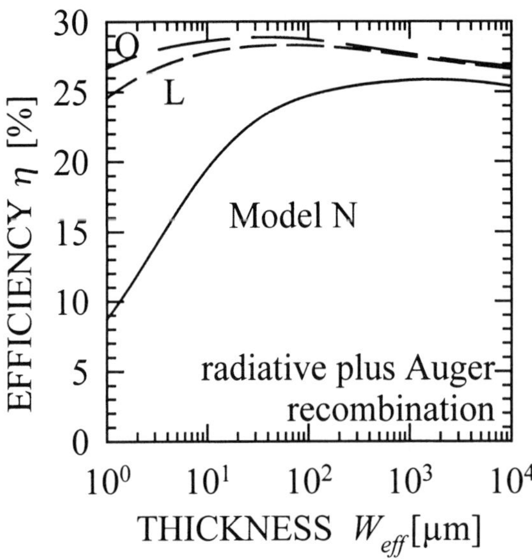

Figure 2.21. Limiting power conversion efficiency η for crystalline Si cells. Assumptions: radiative recombination only, optical losses included, no light concentration, thermalization losses included, Auger recombination included. Model N: no light trapping; Model L: Lambertian light trapping; and Model O: optimum light trapping.

Table 2.6. Limiting power conversion efficiency η for crystalline Si cells of thickness W_{eff}. Assumptions: radiative recombination only, optical losses included, no light concentration, thermalization losses included, Auger recombination included. Model N: no light trapping; Model L: Lambertian light trapping; and Model O: optimum light trapping.

	Efficiency η [%] at W_{eff} = 1 μm	Efficiency η [%] at W_{eff} = 10 μm	Efficiency η [%] at optimum W_{eff}
Model N	8.7	19.5	25.9 at 1780 μm
Model L	24.5	27.8	28.3 at 64 μm
Model O	26.7	28.6	28.9 at 36 μm

Figure 2.22 shows the generation and recombination current densities for planar solar cells (Model N) and Lambertian cells (Model L). Light trapping enhances the interaction of the radiation field with the cell. This increases radiative recombination, whereas

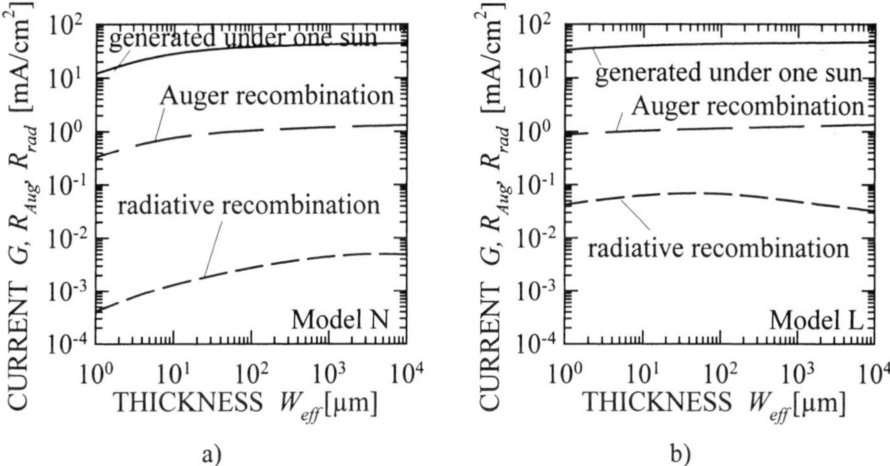

Figure 2.22. Generation and recombination current densities at the maximum-power point for Si thin-film cells: a) without light trapping (Model N); b) cells with Lambertian light trapping (Model L). Light trapping enhances the relative importance of radiative recombination.

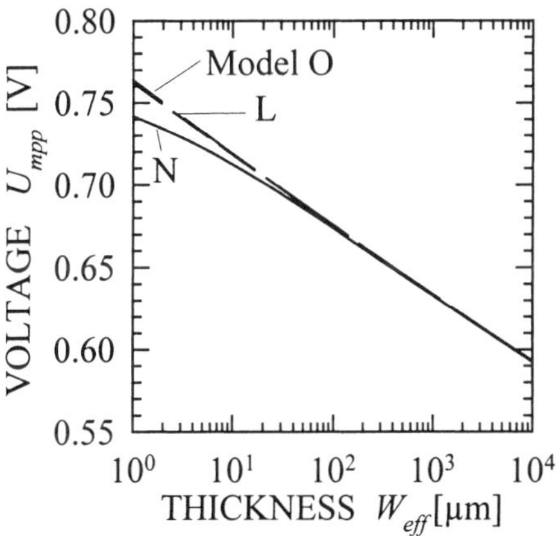

Figure 2.23. Voltage at the maximum-power point of solar cells limited by Auger recombination and radiative recombination. Model N: no light trapping; Model L: Lambertian light trapping; Model O: optimum light trapping.

Auger recombination is approximately independent of light trapping.

Figure 2.23 shows the cell voltages at the maximum-power point. The voltage decreases with the layer thickness W_{eff}, because the volume-integrated Auger recombination increases with the thickness due to our assumption of flat quasi-Fermi levels. An improved light trapping increases the carrier generation rate per cell volume and thus increases the cell voltage [168]. The voltages are smaller than 0.76 V for all thickness values above 1 μm. The smallest phonon-assisted band-to-band excitation we account for has an energy of 0.85 eV (corresponding to 1450 nm). Thus, laser action does not occur in our model of thin-film Si cells.

The efficiency limits for thin-film cells displayed in Figure 2.21 clearly exceed 20% and are thus encouraging for the development of thin-film cells. On the other hand, comparing these numbers with the experimental efficiencies that we discuss in Chapter 3 indicates that loss mechanisms that are important in practice are still missing in our model.

2.3.2 Surface recombination

Extrinsic recombination of electron-hole pairs via defect levels in the gap is, in principle, avoidable by reducing the defect density to close to zero. However, in experimental cells, this so-called Shockley-Read-Hall (SRH) recombination [169, 170] is the dominating loss. The SRH recombination is sketched Figure 2.12c on p. 24 and occurs in the volume of the Si material (bulk recombination at point defects and dislocations), in the surface of the solar cell (surface recombination at dangling Si bonds), and in the boundaries of Si grains (grain boundary recombination). This section is concerned with the measurement and the modeling of SRH recombination at surfaces.

Definition

The surface recombination velocity (SRV)

$$S = \frac{U_{sur}}{\Delta n_{sur}} \tag{2.36}$$

quantifies the surface's recombination activity and depends on the surface recombination rate U_{sur} divided by the excess carrier concentration Δn_{sur}. Both values are taken at the surface, which is located at $Z = 0$ in Figure 2.24. Surface charges in general bend the bands, and thus create a space charge region that extends from the surface at $Z = 0$ to $Z = W_{scr}$. The effective surface recombination velocity

$$S_{eff} = \frac{U_{sur}}{\Delta n_{W}} \tag{2.37}$$

is defined with the excess carrier concentration Δn_W taken at the edge of the space charge region at $Z = W_{scr}$. Here, the surface recombination rate U_{sur} contains all recombination losses that occur in the volume $Z < W_{scr}$. The effective SRV S_{eff} is conveniently used in the boundary conditions of the minority carrier diffusion equation that holds in the quasi-neutral Si region.

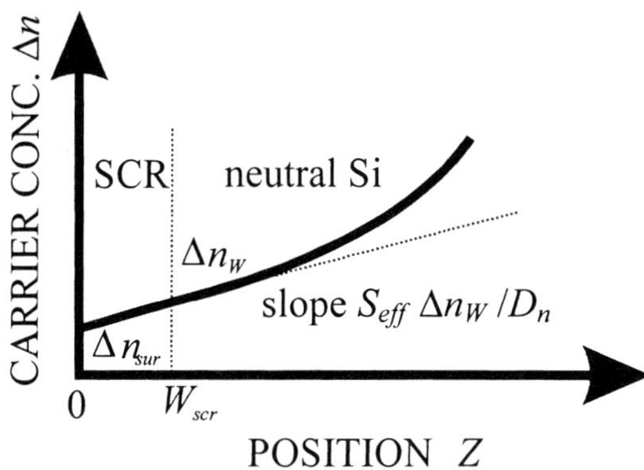

Figure 2.24. Excess minority carrier concentration as a function of position. The slope at the edge of the space charge region defines the effective surface recombination velocity S_{eff}.

Stable surface passivation

Growing an oxide on a Si surface forms Si–O bonds with bonding and anti-bonding states in the valence and the conduction band, respectively. Similarly, dangling Si bonds are passivated by hydrogenation. Again, the Si–H bonds move the electronic states out of the energy gap.

A second approach to achieve a small surface recombination velocity S is the reduction of the concentration of either the electrons or the holes. Since electrons *and* holes are required for recombination, the absence of one partner reduces the surface recombination velocity. In contrast to the corona technique discussed above, the reduction of the carrier concentration has to have long-term stability.

A permanent reduction of the electron concentration at the surface is, for example, achieved by introducing a highly p^+-type doped surface layer. In p-type Si the p^+-type surface layer is a so-called back surface field [171-173]. For n-type Si, a p^+-type layer also reduces the surface recombination. If not contacted, this layer is called a floating junction [174, 175]. If contacted, it still passivates, while also functioning as a collecting junction.

Measuring surface recombination

The SRV in solar cells has been derived from quantum efficiency measurements [176, 177], from measurements of the transient photoconductance decay (PCD) [176, 178], from modulated free carrier absorption data [179], and from admittance spectroscopy [180]. All these techniques are modulation techniques that apply a small modulated excitation superimposed on a bias illumination of bias voltage to define the working point of the devices. We show on p. 86 that it is necessary to distinguish carefully between small signal measurements that yield a differential SRV $S_{eff,dif}$ and large signal measurements that permit the extraction of the actual SRV S_{eff} [30].

Figure 2.25 shows the effective SRV S_{eff} that we extracted from the quantum efficiency data shown in Figure 3.16 on p. 87. The surface analyzed is a lapped and chemically etched back surface of a high-efficiency thin-film Si cell. The surface has a 89 nm-thick thermal SiO_2 layer that is covered by Al. Holes in the oxide permit a contact from the Al to the Si. The measured SRV S_{eff} decreases by two orders of magnitude with the injection level increasing from below $\Delta n_W = 10^9\ cm^{-3}$ to above $10^{14}\ cm^{-3}$.

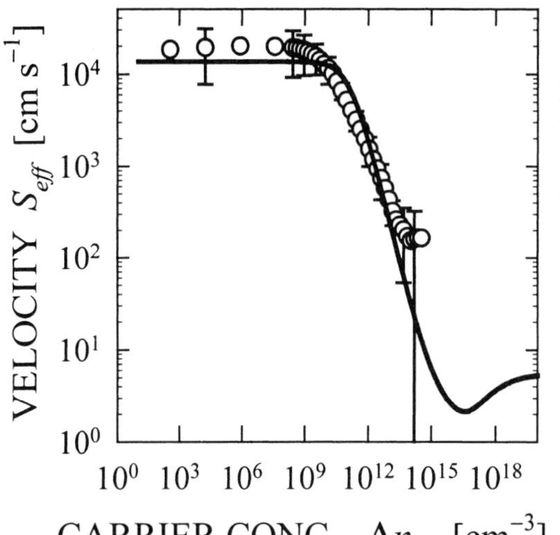

Figure 2.25. Injection level dependence of the recombination velocity at the point-contacted back surface of the Si cell. Symbols are measured values, while the solid line is a simulation result. The injection level at the maximum-power point is $n_W = 5\times10^{13}$ cm^{-3}.

Modeling the injection dependence

The impact of surface band bending on the interface recombination velocity was first pointed out for grain boundary recombination [181]. Band bending enhances the minority carrier concentration at the interface and thereby increases the recombination rate. The situation is similar at semiconductor surfaces or at interfaces with dielectric layers [182]. In addition to asymmetric capture cross-sections [183], band bending [184, 176, 185] causes a dependence of the recombination velocity (SRV) on the injection level. The extended Shockley-Read-Hall recombination model explains the injection level dependence of the SRV S [184]. This model was successfully applied to monocrystalline Si surfaces passivated with silicon dioxide [176, 179, 178] and silicon nitride [186, 187].

The solid line in Figure 2.25 shows the extended SRH model when applied to our thin-film high-efficiency Si cell. As analyzed in detail in Appendix B on p. 237, the decrease in SRV is caused by the change from an electron-limited recombination at low injection to a hole-limited recombination process at high injection [188]. The SRV extracted from the IQE measurement is, in general, less accurate than an SRV measured by the photoconductance decay (PCD) technique. Particularly or small SRV (S_{eff} < 1000 cm s^{-1}) the PCD technique is more sensitive, since it operates under open-circuit conditions thus leading to greater recombination rates than for quantum efficiency measurements that are performed under short-circuit conditions.

Silicon oxide vs. silicon nitride passivation

Surface passivation with an oxide layer requires the application of high temperatures around 1000°C. Metal contacts would contaminate the Si at these temperatures and must therefore be formed after oxidation. The oxide has to be opened either by costly processes, such as photolithography, or by mechanical grinding to permit electric contacting.

Here, Si nitride offers an attractive alternative. Silicon nitride is deposited in a plasma at low temperatures (< 400°C). Deposition on top of a partially metallized Si surface is possible. Surface passivation with Si nitride was extensively investigated by the ISFH research team [189-191, 187]. These authors found that Si wafers with a resistivity of 0.7

to 1.5 Ω cm are more efficiently passivated by Si nitride than by Si non-alnealed oxide layers [190, 192].

The origin of the excellent passivation obtained is the reduction of the interface state density in combination with a large concentration of positive surface charges that are positioned within 20 nm of the interface within the silicon nitride [193]. The charges originate from the so-called K^+ centers, which consist of Si dangling bonds with nitrogen back bonds. The amount of positive charge stored in the K^+ centers is greater in the dark, and negligible under illumination by one sun [187]. The positive interface charge under illumination originates from the P_{ox}-states (see Figure B.9 on p. 223 for the various types of surface states). These charges repel holes from the surface and thereby reduce the surface recombination rate. For thin-film crystalline Si, passivation by nitride is *the* choice for surface passivation since optical index matching and bulk hydrogen passivation are also feasible.

Figure 2.26 shows measured effective surface recombination velocities for nitride-passivated Si wafers of various resistivites. The experimental data and the interface parameters, such as interface state density and capture cross-sections, are taken from Ref. [187]. The experimental data are reasonably described by our modeling with the extended SRH model.

Interestingly, the SRH theory predicts a small and constant $S_{eff} = 12$ cm s^{-1} for p-type Si wafers with a low resistivity of 0.2 Ω cm (dashed line in Figure 2.26). In low-resistivity material the surface potential is small (here $\Psi s = 0.02$ V), even at low injection. The small surface potential Ψs can hardly be further reduced by a transition from low to high injection. In consequence, S_{eff} is small and constant. If such a small S_{eff} value on a low-resistivity Si could be verified experimentally, this finding is important for thin-film crystalline Si cells because highly doped Si layers are required to reduce the lateral resistance.

Our analysis of surface recombination rates at the Si/nitride interface shown in Figure B.13 on p. 228 demonstrates that the silicon nitride interface investigated here does not have a transition from electron to hole-limited recombination. The injection level de-

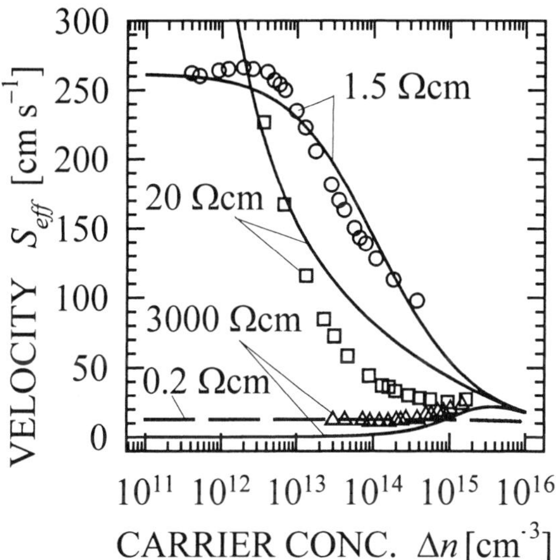

Figure 2.26. Measured effective surface recombination velocities of silicon nitride films deposited by low frequency plasma deposition (full symbols) and by remote plasma deposition (open circles). The resistivity of the substrate is marked. Experimental data and interface parameters for the simulation from Ref. [187].

pendence of S originates solely from the band bending induced by the surface charges.

Surface recombination at the Si/SiO_2 and Si/Si_3N_4 interfaces has been reviewed recently [194].

Efficiency limits for surface recombination velocity of 100 cm s^{-1}

The effective SRV is 200 cm s^{-1} at the maximum-power point of our 20.6%-efficient thin-film cell [113]. Even smaller SRVs, below 50 cm s^{-1}, have been reported for silicon nitride surfaces [190]. Hence, we assume a surface recombination velocity of 100 cm s^{-1}, which is a realistic choice for a well passivated thin-film cell surface. This value is a factor of 10^5 better than the thermal velocity of the electrons that is an upper bound for S_{eff}. We neglect injection level dependence in our efficiency calculations and assume a Si resistivity of 1 Ω cm. Including surface recombination, the current-voltage curve then becomes

$$I(U) = R_{rad}(U) + R_{Aug}(U) + R_{sur}(U) - G(U) \tag{2.38}$$

with

$$R_{sur} = 2S_{eff}(n - n_o)A_C \tag{2.39}$$

The factor 2 accounts for front and back surface. Here, n is the electron concentration and n_o the equilibrium electron concentration in the dark with no voltage applied. We assume a p-type semiconductor.

The efficiencies displayed in Figure 2.27 are *not* fundamental efficiency limits, since we assume an SRV S_{eff} = 100 cm s^{-1} that is typical for the present status of surface passivation technology. With this assumption, surface recombination is higher than radiative recombination and Auger recombination. The curves no longer exhibit a maximum. Please remember that the infinite mobility assumption implies constant excess carrier concentrations in the cell. In thicker cells, this constant carrier concentration is lower due to a lower averaged carrier generation rate. This smaller excess minority carrier concentration $n - n_o$ reduces the surface recombination current R_{sur}.

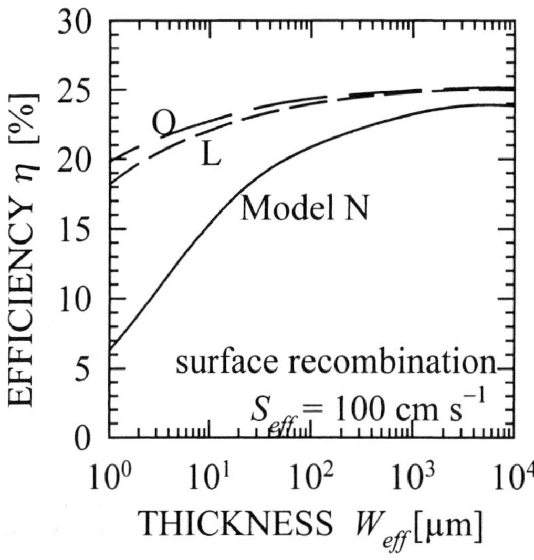

Figure 2.27. Power conversion efficiency η for crystalline Si cells. Assumptions: radiative recombination only, optical losses included, no light concentration, thermalization losses included, Auger recombination included, SRV S_{eff} = 100 cm s^{-1}, and acceptor concentration $N_A = 10^{16}$ cm^{-3}. Model N: no light trapping; Model L: Lambertian light trapping; and Model O: optimum light trapping.

Table 2.7. Power conversion efficiency η for crystalline Si cells of thickness W_{eff}. Assumptions: Radiative recombination only, optical losses included, no light concentration, thermalization losses included, Auger recombination included, SRV S_{eff} = 100 cm s^{-1} and acceptor concentration N_A = 10^{16} cm^{-3}. Model N: no light trapping; Model L: Lambertian light trapping; and Model O: optimum light trapping.

	Efficiency η [%] at W_{eff} = 1 µm	Efficiency η [%] at W_{eff} = 10 µm	Efficiency η [%] at optimum W_{eff}
Model N	6.3	15.3	23.9 at 5600 µm
Model L	18.2	22.0	25.0 at 6900 µm
Model O	19.8	22.7	25.2 at 7500 µm

Efficiency values are listed in Table 2.7. With S_{eff} = 100 cm s^{-1}, the efficiency is 6% for a 1 µm-thick cell without light trapping. With Lambertian light trapping an efficiency of 18.2% is feasible. Again, light trapping triples the cell efficiency. The relative loss due to surface recombination is 30% when compared to a cell with zero SRV. For all thickness values the cell is in low injection conditions with injection levels in the region of 10^{-3}. Our model yields the same efficiency values for all cells of identical S_{eff}/N_A ratio, because increasing the doping decreases the excess minority carrier concentration at constant quasi-Fermi level splitting (constant U_{mpp}). Hence, increasing the doping is a measure to reduce surface recombination losses only provided the S value increases sub-linearly with the acceptor concentration N_A. Although the measurement of the doping dependence of surface recombination velocities is a difficult matter experimentally, there are indications that the enhancement of S with doping is sub-linear in doping concentration [195]. Hence, a high surface doping is advantageous.

The efficiency of thick Si cells with Lambertian light trapping (Model L) is 24.6% for W_{eff} = 420 µm according to Figure 2.27. That value is close to today's record efficiency of 24.7% [196].

2.3.3 Grain boundary recombination

Most of the thin-film approaches use polycrystalline Si. Defects at grain boundaries provide additional locations for recombination. The grain structure of the Si film depends critically on the substrate and the growth technique. Grain sizes ranging from 10 nm to a few 1 cm are used in crystalline Si photovoltaics. A columnar grain structure is usually preferred because the carriers do not have to cross a grain boundary on their way to the collecting junction. Figure 2.28 shows a model of a single columnar grain with four grain boundaries that perpendicularly intersect the junction. Similar to surfaces, the recombination activity of a grain boundary is quantified by the effective grain boundary recombination velocity

$$S_{grb} = \frac{U_{grb}}{2\Delta n_W} \tag{2.40}$$

where U_{grb} is the interface recombination rate at the grain boundary and Δn_W is the excess minority carrier concentration at the edge of the space charge region. The factor 2 accounts for the fact that the grain boundary collects carriers from both sides. The effec-

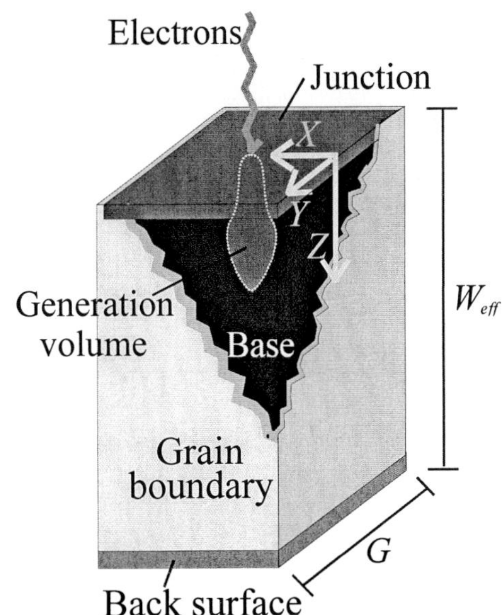

Figure 2.28. Carriers generated by an electron beam are collected by the junction. Generation in a volume near grain boundaries reduces the photocurrent.

tive grain boundary recombination velocity S_{grb} depends on the defect concentration, the capture cross-sections, the doping, and the injection level.

Measurement of grain boundary recombination

The grain boundary recombination velocity S_{grb} is typically determined from experiments that generate electron-hole pairs by excitation with light (LBIC) or electrons (EBIC) at a variable distance from the grain boundary. Figure 2.29 shows an EBIC mapping in the *XY*-plane measured for a thin-film cell fabricated by high-temperature chemical vapor deposition on a Si-seeded glass substrate [197]. The seed was formed by solid-phase crystallization of highly P-doped amorphous Si.

We measure the current profile along the *X*-direction marked by the white line in Figure 2.29. Fitting theoretical profiles with multi-dimensional transport models to the experimental data yields the minority carrier diffusion length L_b in the grain and the grain boundary recombination velocity S_{grb}. The standard evaluation technique assumes an infinitely thick cell [198]. In thin-film cells, however, recombination at the back surface is also important.

Three-dimensional analytical model for carrier transport to the grain boundaries

A carrier may recombine in the grain, at the back of the cell, or at either of the four grain boundaries depicted in Figure 2.28. In order to cope with these more general cases, we develop an analytical model that calculates the local carrier concentration at any point of the neutral base using a Fourier decomposition of the minority carrier diffusion equation [34].

The simulation of an LBIC or EBIC measurement is greatly simplified by the reciprocity theorem for charge collection [37], which relates the excess minority carrier concentration calculated under a voltage bias to the local carrier collection efficiency under short-circuit conditions. As we show on p. 57, the reciprocity theorem not only

Figure 2.29. The electron beam-induced current mapping in the *XY*-plane at an electron energy of 5 keV for a thin-film cell directly deposited by chemical vapor deposition on a seeded glass substrate. A particular line scan used for quantitative evaluation is marked in white. (Photo: J. Krinke, University of Erlangen-Nuremberg). Reprinted from Ref. [197].

holds for the case of Boltzmann statistics, but may also be used for the case of Fermi statistics.

Figure 2.30 shows the local carrier collection efficiency at $Y = 0$ calculated with our model and the reciprocity theorem. The grain size G equals the film thickness W_{eff} and the minority carrier diffusion length L. All three values are 5 μm, corresponding to one of the larger grains in Figure 2.29. The collection efficiency η_c is unity at the junction because there is no chance for recombination prior to reaching the junction.

The value of η_c does not change in the X-direction in Figure 2.30a, due to the vanishing grain boundary recombination $S_{grb} = 0$. The carrier transport is one-dimensional. The

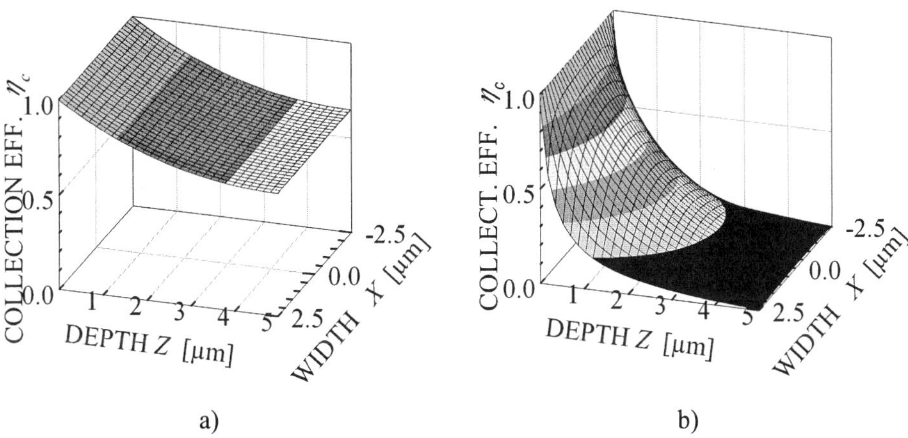

Figure 2.30. Carrier collection efficiency η_c in a square silicon grain of size $G = 5$ μm with diffusion length $L = 5$ μm and thickness $W = 5$ μm. The collection efficiency is shown in the plane $Y = 0$ (see coordinate system in Figure 2.28). a) Grain boundary recombination velocity $S_{grb} = 0$ and back surface recombination velocity $S_b = 0$; b) $S_{grb} = 10^6$ cm s^{-1} and $S_b = 10^6$ cm s^{-1}.

back surface recombination velocity S_b is also zero.

Figure 2.30b shows the carrier collection efficiency for the same grain as in a) but now with a large grain boundary recombination velocity $S_{grb} = 10^6$ cm s^{-1} and a large back surface recombination velocity $S_b = 10^6$ cm s^{-1}. The collection efficiency at the junction is still unity. The average collection efficiency in the grain decreases compared to case a). The local collection efficiency η_c decreases in the X-direction towards the grain boundaries. It is this change in local collection efficiency with distance from the junction that permits the extraction of grain boundary recombination velocities from EBIC measurements.

The derivation of the analytic equations that were used to calculate the local carrier collection efficiency are given in section B.4.2 of Appendix B on p. 232.

Applications of the transport model

The change in extracted current density under electron-beam radiation is calculated by multiplying η_c from Figure 2.30 by the carrier generation rate g and then integrating over the generation volume. For EBIC measurements, the generation volume has the shape of a peach, as is shown schematically in Figure 2.28. The values of S_{grb} and L are then deduced from fitting the above theory to the current profiles along the white line in Figure 2.29. Typical S_{grb} values we find for the above thin-film cell and many other Si cells range from $S_{grb} = 10^4$ cm s^{-1} to 5×10^5 cm s^{-1} [197, 199].

Representative grain size

Polycrystalline cells consist of grains of different sizes. Figure 2.31 shows the grain size distribution measured for a phosphorus-doped Si layer on glass that was crystallized in the solid phase at 600°C for 12 h. The EBIC image from Figure 2.29 is measured on a cell that we fabricate by Si epitaxy on this type of seed layers [197]. Bergmann and Krinke [200] found that the log-normal probability distribution [201]

$$P(G)=\frac{1}{(2\pi)^{1/2}\sigma G}\exp\left(-\frac{1}{2}\left(\frac{\ln(G/\overline{G})}{\sigma}+\frac{\sigma}{2}\right)^2\right) \tag{2.41}$$

describes the experimental grain size distribution. The measured distribution is best fitted with an average grain size of $\overline{G} = 3.1 \pm 0.2$ µm and a standard deviation $\sigma = 0.60 \pm 0.04$. Let us assume that all grains have the same intra-grain diffusion length L and the same grain boundary recombination velocity S_{grb}. What is then the best choice for a representative effective grain size G_{eff} to model the polycrystalline device with only a single grain size G_{eff}? In Appendix C on p. 256 we show that this grain size distribution is best modeled by an effective representative grain size that is

$$G_{eff} = \overline{G}^{3/8}\,\overline{G}_{aw}^{\;5/8} \tag{2.42}$$

where $\overline{G}_{aw} = \overline{G}\exp(2\sigma^2)$ denotes the area-weighted average grain size. The value of G_{eff} is thus always larger than the average grain size and smaller than the average area-weighted grain size.

For the distribution of Figure 2.31 we calculate an effective grain size of G_{eff} = 4.9 µm. The value is, in this case, of the same order of magnitude as the average grain

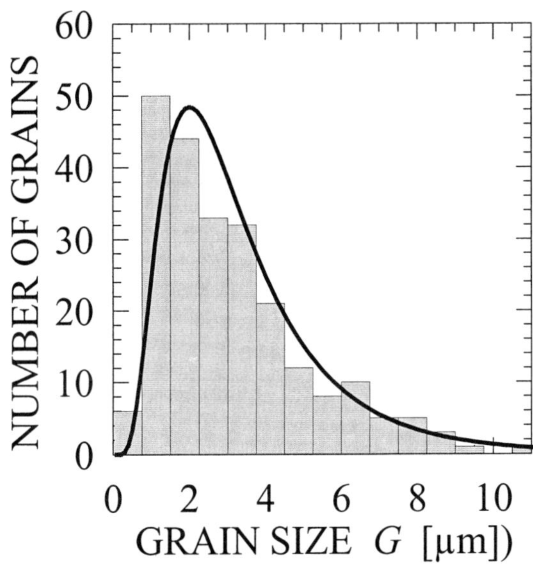

Figure 2.31. Bars: measured grain size distribution of an amorphous Si film that was crystallized at 600°C for 12 h. The phosphorus concentration is 1.4×10^{20} cm^{-3}. Solid line: fit of a log-normal distribution from Eq. (C.38) to measured data. Data from Ref. [200].

size. Depending on the parameter σ, larger differences between average grain size and effective grain size are possible.

Modeling grain boundary recombination

Our modeling of grain boundary recombination follows the same line of argument as the modeling of surface recombination. We assume grain boundaries in Si with energy-independent interface state density D_{it} and identical electron and hole capture cross-sections $\sigma_n = \sigma_p = 10^{-14}$ cm^2. All states below the midgap are assumed to be donors while those above midgap are acceptors. The neutrality level Φ_o is then at the midgap, as indicated by the broken line in Figure 2.32. The space charge region of the p-type semiconductor is positively charged. Band bending adjusts to a value at which the charges at the interface and in the two space charge regions balance each other. The band bending enhances the minority carrier concentration at the grain boundary relative to the minority carrier concentration at the edge of the space charge region. The recombination velocity $S_{grb} = S_o \exp(q\,\Psi_{sur}/kT)$ increases exponentially with the surface band bending potential Ψ_{sur}. Increased doping shifts the Fermi level away from the midgap towards the majority band and charges the grain boundary. The larger the interface state density, the smaller the shift of the Fermi level. Therefore, the quasi-Fermi level is fully pinned to the neu-

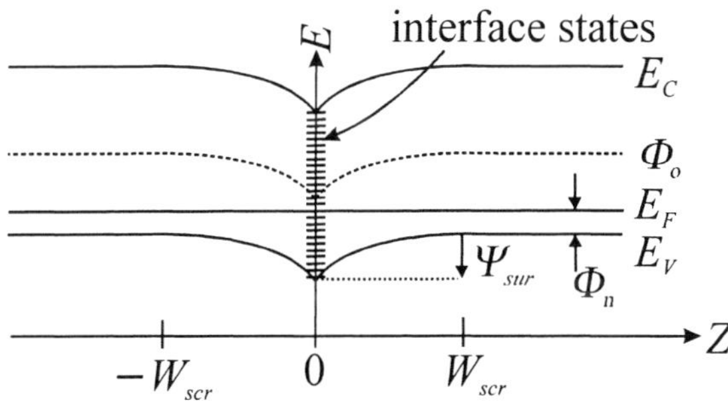

Figure 2.32. Energy diagram of a grain boundary in p-type Si. The positive space charge is compensated by negative interface charges. The neutrality level Φ_o is at the midgap.

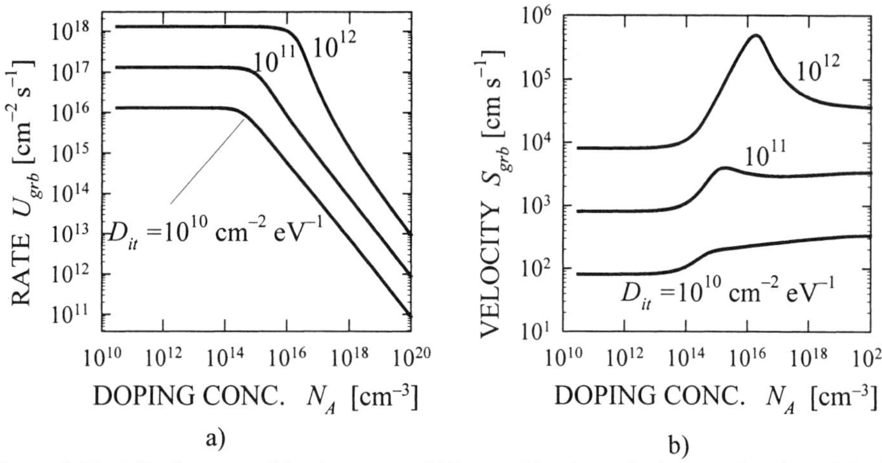

Figure 2.33. a) Surface recombination rate and b) recombination velocity as a function of doping concentration and interface state density D_{it}. The splitting of the quasi-Fermi levels is 0.5 V.

trality level in the limit of infinitely high interface state densities.

Using the extended Shockley-Read-Hall model [184] (see section B.3 in Appendix B on p. 220) we calculate the grain boundary recombination velocity as a function of base doping concentration. The simulation results are shown in Figure 2.33a for a fixed quasi-Fermi level splitting of 0.5 V. The recombination rate varies linearly with interface state density D_{it}. The recombination rate U_{grb} is largest for small doping concentrations because the semiconductor is then in high injection conditions. Electrons and holes have equal concentrations and are captured at equal rates, which maximizes the recombination rate. Band bending is negligible at low doping. With increasing doping, the minority carrier concentration, and hence the capture rate, decrease and thus limit the recombination rate. Following this argument, high doping should be preferred. In reality, the situation is more complex because high doping implies high electric fields that for sufficiently large values enhance defect recombination by tunneling [202, 35].

The corresponding recombination velocities are shown in Figure 2.33b and range from S_{eff} = 100 cm s^{-1} to the thermal limit of 10^7 cm s^{-1}. The recombination velocity is smallest for very low doping N_A with negligible attraction of the minority carriers by band bending. The peak in S_{grb} is caused by the doping dependence of the band bending that also exhibits a maximum. Please note that the smallest S_{grb} values occur for low doping. That is where the recombination rate is largest! Hence grain boundary recombination velocity (or, similarly, surface recombination velocity) is *not* the parameter that needs to be optimized. The quantity that has to be minimized is the grain boundary recombination current density U_{grb}. See p. 229 in Appendix B for details. For a typical interface state density D_{it} = 10^{12} cm^{-2}eV^{-1} [203] the value for S_{grb} ranges from 10^4 cm s^{-1} to 10^6 cm s^{-1}, in agreement with our measurements [197, 199].

A major difference between grain boundary and surface recombination is that the band bending of grain boundaries is solely due to charges in the interface states. No fixed charges in a dielectric layer exist. Therefore, the band bending decreases at much lower injection levels, leading to an enhanced recombination rate when compared to dielectrically passivated surfaces.

There are two aspects that limit the applicability of the extended SRH model to grain boundary recombination: first, at grain boundaries, the defect density is much higher (D_{it}

$= 10^{13}$ cm^{-2} eV^{-1}) than at passivated surfaces ($D_{it} = 10^{10}$ cm^{-2} eV^{-1}). Therefore, the average distance d between two defects is of the order of $d = (D_{it}\,E_g)^{-0.5} = 3$ nm. Hence, interaction of individual defects becomes possible, a fact that is not accounted for by the extended SRH model. Secondly, high defect densities imply high currents into the grain boundary. Hence, our assumption of flat quasi-Fermi levels becomes questionable. The simulation results are therefore only qualitatively correct for high surface recombination rates.

Efficiency estimate for small-grained cells with $G/W_{eff} = 0.1$

For an efficiency estimate including grain boundary recombination we consider columnar, grained material with a grain size G that is a factor of 10 smaller than the thickness W_{eff}. The grain boundary recombination velocity is $S_{grb} = 10^4$ cm s^{-1}, a value that is typical for $N_A = 10^{16}$ cm^{-3} doped p-type Si with grain boundaries passivated by hydrogenation [203, 197, 199]. The current-voltage curve including grain boundary recombination is

$$I(U) = R_{rad}(U) + R_{Aug}(U) + R_{sur}(U) + R_{grb}(U) - G(U) \tag{2.43}$$

with

$$R_{grb} = \frac{4\,W_{eff}\,A_C}{G}\,S_g\,(n - n_o) \tag{2.44}$$

We consider a p-type semiconductor with a spatially homogeneous acceptor concentration of $N_A = 10^{16}$ cm^{-3}. The symbols n and n_o denote the minority carrier concentrations in steady-state and at equilibrium, respectively, while A_c is the cell's surface area. The factor 4 accounts for the four sides of the grain. The factor W_{eff}/G describes the enhancement of the grain boundary area by reducing the grain size.

The resulting efficiencies, again calculated in the framework of infinite mobility, are displayed in Figure 2.34. Please note that our assumption of constant carrier concentration throughout the cell, in general, overestimates the impact of grain boundary recombination. What we calculate is a worst-case scenario for the impact of grain boundary recombination on cell performance that, nevertheless, indicates the size of the efficiency losses.

At a thickness $W_{eff} = 1$ µm the efficiency triples from $\eta = 4\%$ for a planar cell to 12%, if Lambertian light trapping is introduced. Grain boundary recombination reduces the efficiency by 34% relative to the case without grain boundary recombination. Table 2.8 lists some of the efficiency values displayed in Figure 2.34.

In the conclusions on p. 166, we analyze the relative significance of the various loss mechanisms considered in this chapter.

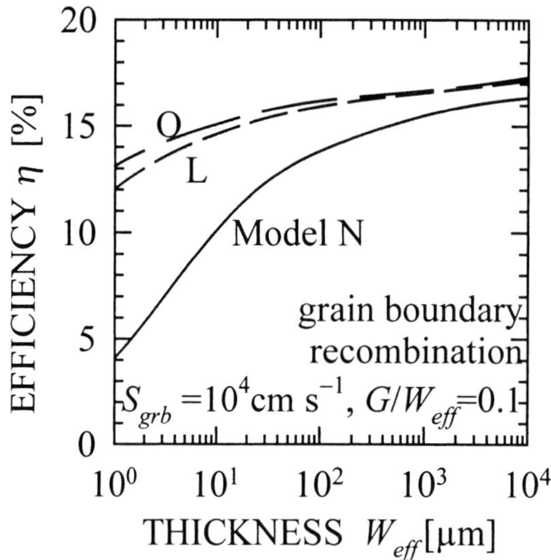

Figure 2.34. Power conversion efficiency η for crystalline Si cells. Assumptions: radiative recombination only, optical losses included, no light concentration, thermalization losses included, Auger recombination included, SRV S_{eff} = 100 cm s^{-1} and acceptor concentration $N_A = 10^{16}$ cm^{-3}, grain size G ten times smaller than film thickness W_{eff}, grain boundary recombination velocity S_{grb} = 10^4 cm s^{-1}. Model N: no light trapping; Model L: Lambertian light trapping; and Model O: optimum light trapping.

Table 2.8. Power conversion efficiency η for crystalline Si cells of thickness W_{eff}. Assumptions: radiative recombination only, optical losses included, no light concentration, thermalization losses included, Auger recombination included, SRV S_{eff} = 100 cm s^{-1} and acceptor concentration N_A = 10^{16} cm^{-3}, grain size G ten times smaller than film thickness W_{eff}, grain boundary recombination velocity S_{grb} = 10^4 cm s^{-1}. Model N: no light trapping; Model L: Lambertian light trapping; and Model O: optimum light trapping.

	Efficiency η [%] at W_{eff} = 1 µm	Efficiency η [%] at W_{eff} = 10 µm	Efficiency η [%] at optimum W_{eff}
Model N	4.1	10.1	>16.3 at >10^4 µm
Model L	12.0	14.6	>17.2 at >10^4 µm
Model O	13.1	15.1	>17.3 at >10^4 µm

3 Advanced quantum efficiency analysis

The current-voltage curve $j(\Phi, U)$ measured under an illuminating photon flux Φ yields the power conversion efficiency η, the open-circuit voltage U_{oc} and the short-circuit current density j_{sc}. If all recombination currents rates are linear in the excess carrier concentrations, the cell current

$$j(\Phi,U) = j(\Phi,0) + j(0,U) \tag{3.1}$$

obeys the superposition principle, where U denotes the forward injection voltage. The total current is the sum of a current with only light and a current with only voltage excitation. At the maximum-power point voltage U_{mpp}, the cell is forward biased, and thus carriers are injected everywhere into the cell by the current contribution $j(0,U)$. Carriers recombine simultaneously in all regions of the device. The current-voltage curve is thus of little help when searching for the location of dominant carrier recombination.

In contrast, quantum efficiency is measured under short-circuit conditions. Device areas that are not illuminated do not contribute to the measurement signal. Thus, the localization of areas of high recombination becomes possible.

External quantum efficiency

The external quantum efficiency EQE is the probability of an incident photon contributing one electron to the short-circuit current of the cell. Figure 3.1 illustrates the change in short-circuit current Δj due to a change $\Delta\Phi_\lambda$ of the illuminating photon flux. Hence, the definition of the external quantum efficiency is

$$EQE(\lambda) = \frac{\Delta j}{q\Delta\Phi_\lambda} \tag{3.2}$$

wherein q denotes the elementary charge. Using a monochromatic photon flux $\Delta\Phi_\lambda$ provides a depth resolution in the Z-direction since the optical penetration depth L_α increases monotonically with the wavelength (see Figure 2.3 on p. 14). Scanning the (X,Y)-position of the illumination spot across the sample provides a lateral resolution in addition, which is of particular interest for polycrystalline thin-film cells.

The ratio $\Delta j/(q\,\Delta\Phi_\lambda)$ often depends on the flux $\Delta\Phi_\lambda$. The EQE is then not well defined unless the flux is sufficiently reduced to make EQE independent of $\Delta\Phi_\lambda$. To keep in mind that the change in photon flux $\Delta\Phi_\lambda$ needs to be small and thus induces a small change in current Δj, we denote both quantities with a Δ. As described in Appendix C on p. 241, the EQE measurement is commonly performed with the lock-in technique. A stationary bias light or a stationary bias voltage defines the working point on the current-voltage curve. Despite the bias, the cell is short-circuited for the modulated current contribution.

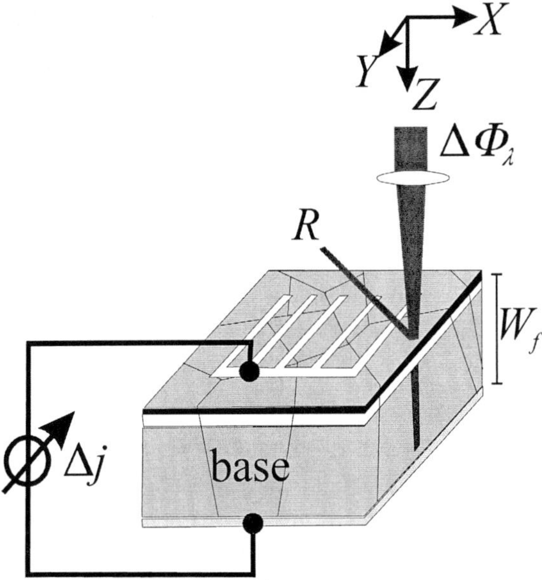

Figure 3.1. Short-circuited polycrystalline cell under monochromatic illumination. The cell reflectance is R at wavelength λ.

Internal quantum efficiency

The external quantum efficiency EQE depends on the optical properties: light that does not enter the cell, or light that leaves the cell again after entering, does not contribute to the current. The extraction of electronic recombination parameters from an external quantum efficiency spectrum is thus difficult. The internal quantum efficiency is defined to have a quantity that depends less strongly on the optical design than the external quantum efficiency. The internal quantum efficiency is

$$IQE(\lambda) = \frac{EQE(\lambda)}{1 - R(\lambda) - T(\lambda)} \tag{3.3}$$

wherein R is the hemispherical reflectance and T the hemispherical transmittance of the cell. The internal quantum efficiency is the probability of a photon of wavelength λ which is neither reflected from the cell nor transmitted through the cell contributing one electron to the short-circuit current. In most cases, the transmittance T is zero due to an opaque optical back reflector.

For short-wavelength light with an absorption length $\alpha_s^{-1} \ll W_f$ smaller than the cell thickness W_f, the IQE spectrum is, in fact, independent of the reflectance R, provided no light is absorbed in the antireflection coating.

For long-wavelength light with an absorption length $\alpha_s^{-1} \gg W_f$ larger than the film thickness, the photogeneration rate becomes spatially homogeneous. If the parasitic optical absorption, e.g. absorption by free carriers or the back reflector, is absent, the quantity $(1 - R - T)$ equals the active absorption caused by electron-hole pair generation. In that case, the internal quantum efficiency as defined in Eq. (3.3) equals the average carrier collection probability η_c, which is a purely electronic property. The IQE data will be constant and independent of wavelength.

In all practical cases parasitic absorption exists and the value of $(1 - R - T)$ is larger than the active absorption. Parasitic absorption reintroduces a dependence of the measured *IQE* on optical properties.

Light trapping is another source of the dependence of the quantum efficiency *IQE* on the optical design. Two cells with different light trapping may cause optical absorption close to or far from the collecting junction. Hence, these two cells have a different *IQE*, although the optical absorption and the minority carrier diffusion length may be equal for both cells.

The above examples show that, in general, the internal quantum efficiency depends on the optical design of the cell. This requires the simultaneous analysis of the experimental reflectance data *R*, transmission data *T*, and *IQE* data. A reliable analysis of the quantum efficiency is impossible for thin-film cells without simultaneously analyzing the optical cell design.

3.1 Definition of effective diffusion lengths

The standard internal quantum efficiency analysis [36] derives a diffusion length L_Q from the slope of the inverse quantum efficiency IQE^{-1} plotted versus the optical absorption length α_s^{-1}. For a thick monocrystalline Si cell ($\alpha_s^{-1} \ll W_f$), L_Q equals the minority carrier diffusion length L in the base. For a thin polycrystalline cell, such as depicted in Figure 3.1, the value of L_Q is, in general, not equal to the bulk diffusion length L in the grains, since recombination at the grain boundaries and at the back of the cell also occurs. The quantity L_Q is thus an effective diffusion length. We distinguish three types of effective diffusion lengths.

3.1.1 Quantum efficiency diffusion length L_Q

For absorption lengths α_s^{-1} larger than the emitter thickness W_e, we find two linear regimes when plotting IQE^{-1} versus the optical absorption length α_s^{-1}[24]. This is shown schematically in Figure 3.2. The first linear regime at $\alpha_s^{-1} \ll W_f$ defines the above-mentioned effective diffusion length L_Q by the relation

$$\left.\frac{dIQE^{-1}}{d\alpha_s^{-1}}\right|_{\alpha_s^{-1}=0} = \frac{1}{L_Q} \tag{3.4}$$

The derivative is taken at small absorption lengths (compared to film thickness) as indicated by the index ($\alpha_s^{-1} = 0$). We name L_Q the effective *quantum efficiency* diffusion length since it is generally derived from quantum efficiency data. The index Q stands for quantum efficiency. Please note that L_Q is often also denoted by L_{eff}.

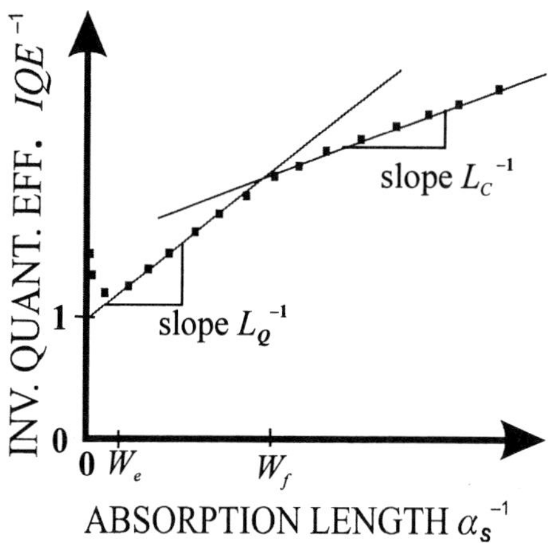

Figure 3.2. The two linear regimes in the $IQE^{-1}(\alpha_s^{-1})$ plot define the effective diffusion lengths L_Q and L_C. The symbols W_e and W_f denote the emitter and the film thickness, respectively.

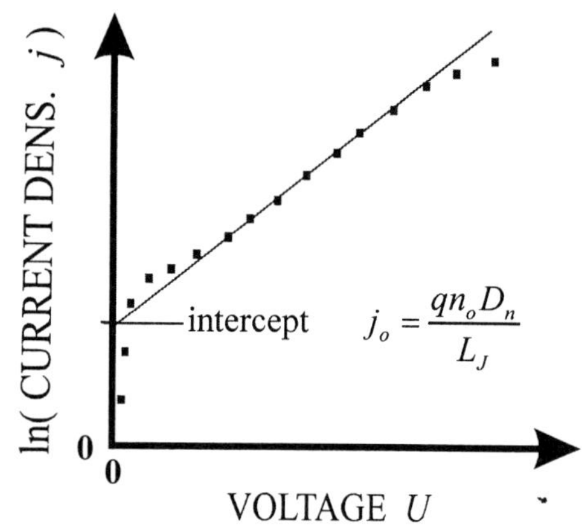

Figure 3.3. The linear regime in the semi-logarithmic current-voltage plot defines the diode saturation current, which is related to the effective diffusion length L_J.

3.1.2 Collection diffusion length L_C

The photogeneration rate is spatially homogeneous for absorption lengths α_s^{-1} that are much larger than the cell thickness W_f. Then, the cell current and thus the internal quantum efficiency *IQE* are linear in the absorption coefficient α_s . This weak absorption limit defines a second effective length L_C by the relation

$$\left.\frac{dIQE^{-1}}{d\alpha_s^{-1}}\right|_{\alpha_s^{-1}=\infty} = \frac{1}{L_C} \tag{3.5}$$

The derivative is taken at large absorption lengths (compared to film thickness) as indicated by the index ($\alpha_s^{-1} = \infty$). We name this length effective *collection* diffusion length.

3.1.3 Current-voltage diffusion length L_J

The semi-logarithmic current-voltage curve of a Si diode often shows a linear regime at intermediate voltages where recombination in the low-injected base dominates. This is shown schematically in Figure 3.3. The intercept of a linear fit with the current axis yields the diode saturation current

$$j_o = \frac{qD_n n_o}{L_J} \tag{3.6}$$

if recombination in the emitter is negligible. Here, q is the elementary charge, n_o the equilibrium electron concentration, and D_n the diffusion coefficient in p-type Si. If n_o and D_n are known, the measured j_o value *defines* an effective diffusion length L_J. By definition, L_J yields the correct saturation current when inserted into the standard Eq.

(3.6). We call L_J an effective *current-voltage* diffusion length, since it is derived from a current-voltage measurement.

3.1.4 Interrelation of L_Q and L_J

In practice, an accurate determination of L_J from the current-voltage characteristic as indicated in Figure 3.3 is only possible to an accuracy of about one order of magnitude. In addition, the minority carrier diffusion coefficient D_n needs to be known to determine L_J from j_o. The measurement of the quantum efficiency diffusion length L_Q, in contrast, is more accurate. Quantum efficiency measurements also give access to locally resolved information.

The question arises, whether there is a general relation between the quantity L_Q and the quantity L_J. For a planar monocrystalline thin-film cell with the junction on the front surface and without any spatial inhomogeneities $L_{Q,mono} = L_{J,mono}$ indeed holds [24]. We can thus accurately measure L_Q and infer L_J. The formulas for $L_{Q,mono}$ and $L_{J,mono}$ are given in the Appendix C on p. 247. For the general case of a spatially inhomogeneous and surface textured semiconductor the relation of L_Q and L_J has been only recently clarified by Brendel and Rau [204]. In the subsequent sections, we conduct the proof that $L_Q = L_J$ also holds in the general case of an inhomogeneous and non-planar semiconductor.

Basore conducted the proof of $L_{Q,mono} = L_{J,mono}$ by explicitly solving the transport equations in the quasi-neutral base region [24]. This approach is not possible in the general case since the cell geometry is not specified. Similarly, the position of the grain boundaries is arbitrary; unknown bandgap fluctuations, as well as band bending at interfaces, may occur. We thus have to derive the relation of L_Q and L_J from more general principles.

We proceed in two steps. In a first step, we derive the reciprocity theorem of charge carrier collection [37] from the principle of detailed balance and generalize it to Fermi statistics. The reciprocity theorem relates collection conditions to injection conditions. This is helpful for our problem since L_Q is measured under carrier injection while L_C is measured under carrier collection conditions. In a second step, the generalized reciprocity theorem is applied to derive the relation of L_Q and L_J [204].

3.2 Reciprocity theorem for charge carrier collection

The reciprocity theorem, as originally derived by Donolato, states that the local carrier collection efficiency

$$\eta_c(\boldsymbol{r}) = \frac{u_D(\boldsymbol{r})}{u_D(\boldsymbol{r}_{jct})} \tag{3.7}$$

equals the excess minority carrier concentration $u_D(\boldsymbol{r})$ of the voltage-biased device in the dark, when normalized to the excess minority carrier concentration $u_D(\boldsymbol{r}_{jct})$ at the junction [37]. Here, the excess minority carrier concentration

$$u(\boldsymbol{r}) = \frac{\delta n(\boldsymbol{r})}{n_o(\boldsymbol{r})} = \exp\left(\frac{E_{Fn} - E_{Fp}}{kT}\right) - 1 \tag{3.8}$$

is given in units of the local equilibrium carrier (here electrons) concentration $n_o(\boldsymbol{r})$. The excess concentration u is determined solely by the splitting of the quasi-Fermi level for electrons in the conduction band E_{Fn} and holes in the valence band E_{Fp}.

The original version of the reciprocity theorem as derived by Donolato [37] holds for devices of arbitrary shape and for minority carrier lifetimes that vary spatially. However, the minority carrier concentration has to be linear with respect to a variation of the carrier generation profile and the carriers have to obey Boltzmann statistics. These requirements exclude all devices with injection level-dependent recombination, with high doping concentrations or states near the midgap.

3.2.1 Derivation from detailed balance

Following the work of Rau and Brendel [42], we derive the generalized reciprocity theorem in two steps. In a first step, we consider the solar cell to be discretized into separate sites, which consist of either local sites or discrete energy states. In a second step, we consider the continuous carrier dynamics.

Discrete case

We describe the electronic transport in a solar cell by means of transition of carriers between discrete states. This situation can be understood as a discretization of a solid-state cell, as well as a description of a solar cell based only on discrete quantum states, as is the case for electrochemical solar cells [205]. We choose this example because, on the one hand, it is instructive, simple and general and, on the other hand, it demonstrates all the mathematical ingredients necessary to prove the reciprocity theorem.

Consider the abstract scheme of a solar cell as illustrated in Figure 3.4, in which charge transport occurs along discrete sites. Notice that within this section we need not refer to charge carriers. Therefore, we simply talk about particles which can be generated and annihilated at the given sites. Let the transition probability of these particles from a site i to a site j be T_{ij}. Then it holds from the principle of detailed balance [44] for the equilibrium transition rates R_{ij}^0, R_{ji}^0 from site i to site j and vice versa

$$R_{ij}^0 = T_{ij} n_i^0 = T_{ji} n_j^0 = R_{ji}^0 \tag{3.9}$$

wherein n_i^0 denotes the equilibrium particle concentration at site i. Note that, when talking about discrete quantum states, n_i^0 is identical to the equilibrium occupation probability of state i, since the density of states is, by definition, unity for discrete states. Equation (3.9) assumes a transition rate from the initial state i to the final state j that is independent of the occupation n_i^0 of the final state. This simplification holds for electrons as a good approximation, if the final states are almost empty. For semiconductors, this is exactly the case when Boltzmann statistics applies. For the generation and recombination of particles the relationship

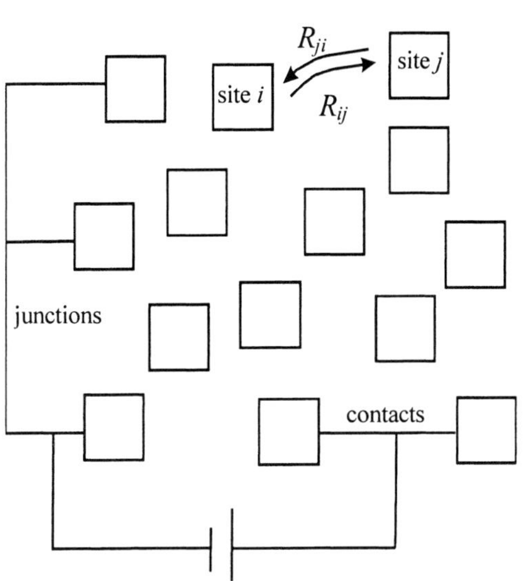

Figure 3.4. Schematic representation of a solar cell consisting of discrete sites, some of them working as junction sites and some others working as back contact sites. The quantities R_{ij} and R_{ji} denote the mutual transition rates from site i to j and vice versa. Figure reproduced from Ref. [42] with permission.

$$g_i^0 = \frac{n_i^0}{\tau_i} \tag{3.10}$$

follows from the principle of detailed balance, in which g_i^0 is the thermal equilibrium generation rate and τ_i the particle lifetime at site i.

Let us now turn to the non-equilibrium situation and assume that the transition probabilities T_{ij} are independent of deviations from equilibrium. With Eqs. (3.9)and (3.10), it follows for the steady-state excess particle concentration δn_i that

$$\delta g_i = \frac{\delta n_i}{\tau_i} + \delta n_i \sum_j T_{ij} - \sum_j \left(T_{ji} \delta n_j\right) \tag{3.11}$$

where δg_i is the excess generation rate at site i. Now, we rewrite the above equation with $u_i := \delta n_i / n_i^0$ and use Eq. (3.10) to find

$$\delta g_i = g_i^0 u_i + u_i \sum_j T_{ij} n_i^0 - \sum_j \left(T_{ji} n_j^0 u_j\right) \tag{3.12}$$

Let us now rewrite Eq. (3.12) in operator form

$$\delta g_i = \sum_j \Theta_{ij} u_j \tag{3.13}$$

with

$$\Theta_{ij} = \delta_{ij}\left(g_i^0 + \sum_k n_i^0 T_{ik}\right) - T_{ji} n_j^0 \tag{3.14}$$

where δ_{ij} is the Kronecker symbol. We notice that by Eq. (3.9) Θ_{ij} is a symmetric operator acting on the vector $\boldsymbol{u}$.

Now consider two specific vectors $\boldsymbol{u}^{(k)}$ and $\boldsymbol{u}^{(l)}$ that are the solutions of Eq. (3.13) for unity excess carrier generation at arbitrary but fixed sites k and l, respectively. It holds that

$$\delta_{ik} = \sum_j \Theta_{ij} u_j^{(k)} \tag{3.15}$$

and

$$\delta_{il} = \sum_j \Theta_{ij} u_j^{(l)} \tag{3.16}$$

Consider further the mutual projection of $\boldsymbol{u}^{(k)}$ and $\boldsymbol{u}^{(l)}$ on the sites l and k, i.e. the excess concentration $u_l^{(k)}$ ($u_k^{(l)}$) at site l (k) resulting from generation of particles at site k (l). It holds that

$$u_l^{(k)} = \sum_i u_i^{(k)} \delta_{il} = \sum_i u_i^{(k)} \sum_j \Theta_{ij} u_j^{(l)} = \sum_{j,i} \Theta_{ji} u_i^{(k)} u_j^{(l)} = \sum_j \delta_{jk} u_j^{(l)} = u_k^{(l)} \tag{3.17}$$

i.e. particle generation at a site k will lead to the same change of the normalized excess particle concentration at site l as does generation at site l for the concentration at site k. The symmetry relationship (3.17) is a direct consequence of Onsager's principle of micro-reversibility and will be the central tool to derive the reciprocity theorem for carrier collection in solar cells in the following subsection.

Let us now come back to the case of a solar cell and let the particles be minority carriers. From the symmetry of Eq. (3.17) an important conclusion can be drawn: if we describe the excess minority carrier concentration δn_j at site j by a shift ΔE_{fj} of the minority carrier quasi-Fermi level according to

$$\delta n_j = n_j^0 \left\{\exp\left(\Delta E_{Fj} / kT\right) - 1\right\} \tag{3.18}$$

with kT as the thermal energy, we find

$$\Delta E_{Fj}^{(k)} = kT \ln\left(u_j^{(k)} + 1\right) \tag{3.19}$$

Finally, we may rewrite Eq. (3.17) as

$$\Delta E_{Fl}^{(k)} = \Delta E_{Fk}^{(l)} \tag{3.20}$$

i.e. the generation δ_{ik} of carriers at site k will lead to the same shift $\Delta E_{Fl}^{(k)}$ of the Fermi level at site l as vice versa. We will therefore denote reciprocity relationships of the form (3.20) as *the reciprocity theorem for Fermi levels*. Notice that it was assumed in Eq. (3.18) that the occupation probability of all sites is described by Boltzmann statistics.

The assumption of Boltzmann statistics also enters into our initial description of the transition rates and finally leads to the fact that the reciprocity theorem for Fermi levels holds for the normalization $u_i = \delta n_i / n_i^0$. All previous authors describing reciprocity relationships for solar cells have used this normalization [37-41]. It will turn out that, for *small* deviations from thermal equilibrium, the form (3.20) also holds for Fermi-Dirac statistics, whereas the form (3.17) has to be modified by another normalization.

Let us now describe the reciprocity principle including currents within solar cells. This type of reciprocity principle relies sensitively on the proper description of junctions and junction currents. First, we will investigate the dark carrier concentration $\boldsymbol{u}^D$ of the solar cell, where the non-equilibrium generation rate at all non-junction sites is zero and the same voltage U_{jct} is applied to all junction sites, such that the vector components are

$$u_j^D = u_{jct} = \exp(qU_{jct} / kT) - 1 \quad \text{if } j \in JCT \tag{3.21}$$

for all indices j belonging to the set JCT of junction sites. For this specific solution we write

$$\sum_j \Theta_{ij} u_j^D = \begin{cases} \delta g_i^D \text{ for } i \in JCT \\ 0 \quad \text{for } i \notin JCT \end{cases} \tag{3.22}$$

where δg_i^D stands for the excess junction generation rate (i.e. the junction current) at $i \in JCT$ necessary to maintain the junction non-equilibrium concentration u_j^D. The generation rate components δg_i^D vanish at all non-junction sites $i \notin JCT$. The index D denotes the dark situation. By the linearity of Eq. (3.13) we may write u_D by the superposition of solutions $\boldsymbol{u}^{(l)}$ arising from the respective application of a voltage to only one of the junctions site

$$u_j^D = \sum_{l \in JCT} \delta g_l^D u_j^{(l)} \tag{3.23}$$

From Eq. (3.21) it holds for all junction sites that

$$u_j^D = \sum_{l \in JCT} \delta g_l^D u_j^{(l)} = u_{jct} \quad \text{for } j \in JCT \tag{3.24}$$

Now we consider the illuminated short-circuit case. Volume photogeneration occurs exclusively at a single $k \notin JCT$ with a rate $\delta g_k^{L(k)}$. The index $L(k)$ stands for light-induced generation at site k. The elements of $\boldsymbol{\delta g}^{L(k)}$ are zero for all other non-junction sites. However, to keep the cell under short-circuit conditions, non-zero elements $\boldsymbol{\delta g}^{L(k)}$ are necessary for the junction sites $l \in JCT$. These quantities can be looked at as the negative short-circuit currents at the respective junction sites l. There, it holds, because of the short-circuit condition that

$$u_l^{L(k)} = \delta g_k^{L(k)} u_l^{(k)} + \sum_{l' \in JCT} \delta g_{l'}^{L(k)} u_l^{(l')} = 0 \quad \text{for } l \in JCT, k \notin JCT \tag{3.25}$$

Resolving Eq. (3.25) for $u_l^{(k)}$ yields

$$u_l^{(k)} = -\sum_{l' \in JCT} \frac{\delta g_{l'}^{L(k)}}{\delta g_k^{L(k)}} u_l^{(l')} \text{ for } l \in JCT \tag{3.26}$$

Now we turn to the dark excess carrier generation at site *k*. From Eq. (3.23) and by the use of the reciprocity theorem for Fermi levels (Eq. (3.17)), we find

$$u_k^D = \sum_{l \in JCT} \delta g_l^D u_l^{(k)} \tag{3.27}$$

By substituting Eq. (3.26) into (3.27), we derive

$$u_k^D = -\sum_{l,l' \in JCT} \delta g_l^D \frac{\delta g_{l'}^{L(k)}}{\delta g_k^{L(k)}} u_l^{(l')} \tag{3.28}$$

Again using the reciprocity theorem for Fermi levels together with Eq. (3.24) yields

$$u_k^D = -\sum_{l' \in JCT} \frac{\delta g_{l'}^{L(k)}}{\delta g_k^{L(k)}} \sum_{l \in jct} \delta g_l^D u_{l'}^{(l)} = -\sum_{l' \in JCT} \frac{\delta g_{l'}^{L(k)}}{\delta g_k^{L(k)}} u_{jct} \tag{3.29}$$

With the definition $f_{l'}^{(k)}$ of the collection efficiency, i.e. the probability of a carrier photogenerated at a site *k* to be collected by the junction *l'*

$$f_{l'}^{(k)} := -\frac{\delta g_{l'}^{L(k)}}{\delta g_k^{L(k)}} \tag{3.30}$$

we finally derive the reciprocity theorem for charge carrier collection

$$\frac{u_k^D}{u_{jct}} = \sum_{l' \in JCT} f_{l'}^{(k)} = f_C^{(k)} \tag{3.31}$$

in which $f_C^{(k)}$ denotes the total collection probability for a photogenerated charge carrier at site *k*. For small deviations from the thermal equilibrium, we derive from Eq. (3.18) that

$$u_k = \frac{\Delta E_{Fk}}{kT} \tag{3.32}$$

Hence, the reciprocity theorem for small excitation conditions is

$$f_C^{(k)} = \frac{\Delta E_{Fk}}{qU_{jct}} \tag{3.33}$$

Notice that for the *third* term in Eq. (3.29) and, consequently, for Eq. (3.31) we have assumed that the same voltage is applied to all junctions. However, the dark carrier concentration at site k for a situation where different voltages are applied to the junctions can be derived from the *second* term of Eq. (3.29) as

$$u_k^D = \sum_{l' \in JCT} f_{l'}^{(k)} u_{l'}^D \tag{3.34}$$

with

$$u_{l'}^D = \exp(qU_{l'}/kT) - 1 \qquad l' \in JCT \tag{3.35}$$

and $U_{l'}$ being the voltage applied to junction l'.

Equation (3.31) is the discrete form of Donolato's reciprocity theorem for the collection of photogenerated charge carriers and Eq. (3.34) its generalization for the case of different voltages applied to different junctions. Here, we have derived the theorem directly from the principle of detailed balance. In the next section, we will show that the essential ingredient to derive the reciprocity theorem in the continuous case, i.e. in which charge carrier transport is described by a differential operator, is the self-adjointness of this differential operator, and that therefore the reciprocity theorem is also equivalent to the principle of detailed balance.

Continuous case

The most general equation to describe minority carrier transport with a linear equation was derived by Del Alamo and Swanson [206]

$$-\nabla\{D(\boldsymbol{r})n_0(\boldsymbol{r})u(\boldsymbol{r})\} + \frac{n_0(\boldsymbol{r})}{\tau(\boldsymbol{r})}u(\boldsymbol{r}) = g(\boldsymbol{r}) \tag{3.36}$$

It holds for the excess minority n when normalized to the equilibrium value n_o; see Eq. (3.8). The diffusion coefficient D, the lifetime τ, and the carrier generation rate g depend on the position $\boldsymbol{r}$. The probability of a charge carrier photogenerated at an arbitrary site $\boldsymbol{r}$ in the cell to be collected at the junction located at site $\boldsymbol{r}_{jct}$ is $\eta(\boldsymbol{r})$. In the previous section, we restricted ourselves to the case of discrete sites (or states) within a solar cell. Electronic transport in semiconductors is, however, described by differential equations, like Eq. (3.36), acting on a continuous space. Let $\boldsymbol{A}$ be an operator defined on a function space F on domain D. Let $\boldsymbol{A}$ further be symmetric, i.e. it holds for $u_1, u_2 \in F$

$$\int_D (Au_1)u_2 d\boldsymbol{r} = \int_D u_1(Au_2)d\boldsymbol{r} \tag{3.37}$$

Notice that the self-adjointness of the operator $\boldsymbol{L}$ defined by

$$Lu(\boldsymbol{r}) = -\nabla(D(\boldsymbol{r})n_0(\boldsymbol{r})\nabla u(\boldsymbol{r})) + \frac{n_0(\boldsymbol{r})}{\tau(\boldsymbol{r})}u(\boldsymbol{r}) \tag{3.38}$$

making up Eq. (3.36) was implicitly or explicitly used by all authors of Refs. [37-41] to prove the reciprocity theorem. In the next step, we show that every symmetric (i.e. self-adjoint and real) operator defined on a function space fulfills Eq. (3.17). Consider now the analog to Eq. (3.13) $Au = g$, and two special solutions (i.e. Green functions), $Au^{(k)} = \delta(\boldsymbol{r} - \boldsymbol{r}_k)$ and $Au^{(l)} = \delta(\boldsymbol{r} - \boldsymbol{r}_l)$. Then

$$\begin{aligned} u^{(k)}(\boldsymbol{r}_l) &= \int_D \delta(\boldsymbol{r} - \boldsymbol{r}_l) u^{(k)}(\boldsymbol{r}) d\boldsymbol{r} \\ &= \int_D A u^{(l)}(\boldsymbol{r}) u^{(k)}(\boldsymbol{r}) d\boldsymbol{r} \\ &= \int_D u^{(l)}(\boldsymbol{r}) A u^{(k)}(\boldsymbol{r}) d\boldsymbol{r} \\ &= \int_D u^{(l)}(\boldsymbol{r}) \delta(\boldsymbol{r} - \boldsymbol{r}_k) d\boldsymbol{r} = u^{(l)}(\boldsymbol{r}_k) \end{aligned} \tag{3.39}$$

follows from Eq. (3.37), i.e. photogeneration at site $\boldsymbol{r} = \boldsymbol{r}_k$ leads to the same relative $\delta n/n_o$ in particle concentration at site $\boldsymbol{r} = \boldsymbol{r}_l$ as vice versa. On the one hand, we may now follow the line of argument given above to see that Eq. (3.16) is equivalent to the principle of detailed balance as formulated in Eq. (3.9). On the other hand, we have already shown that the reciprocity theorem follows from Eq. (3.17) for charge collection in the case of discrete states. Hence, it follows in the same way from Eq. (3.37) for the continuous case. Thus, the above result provides an alternative proof for the reciprocity theorem which is not even restricted to the specific case of the generalized diffusion equation Eq. (3.36), but holds for all differential equations made up by self-adjoint operators.

Moreover, reciprocity does not only hold for either a continuous or a discrete situation, but also for combinations of both cases, i.e. for cross products of discrete and continuous spaces. As applications one may think about solar cells including interface states or about the impurity photovoltaic effect where the short-circuit current is enhanced by additional absorption by states within the forbidden gap of the semiconductor material [207, 208]. In the latter case, the collection probability of charge carriers generated into these states is also proportional to the relative change of dark occupation caused by application of a voltage to the junction as expressed by Eq. (3.31).

3.2.2 Generalization to Fermi statistics

In situations in which states in the forbidden gap play a dominant role and for highly doped cell regions with the quasi-Fermi levels close to the band edges it is necessary to consider Fermi-Dirac statistics. Since our derivation of the reciprocity relation is not founded on particular transport equations but on the principle of detailed balance, we follow the same line of argument to show [42] that the reciprocity theorem (3.7) also holds for Fermi-Dirac statistics. The occupation probability according to Fermi-Dirac statistics is

$$n_i = \frac{1}{\exp\left(\left(E_i - E_{Fi}\right)/kT + 1\right)} \tag{3.40}$$

wherein E_i is the energy of the state i (here E_i is *not* the intrinsic Fermi energy) and E_{Fi} is the quasi-Fermi energy. Then, the reciprocity theorem reads

$$f_C^{(k)} = \frac{\nu_k^D}{\nu_{jct}} \tag{3.41}$$

with the normalization of Eq. (3.8) being replaced by

$$\nu_k = \delta n_k \left\{ \frac{1}{n_k^0} - \frac{1}{1 - n_k^0} \right\} \tag{3.42}$$

For small deviations from the thermal equilibrium we find, by expanding Eq. (3.40), the relation

$$\nu_k = \frac{\Delta E_{Fk}}{kT} \tag{3.43}$$

Since Eq. (3.43) derived for Fermi statistics is identical to Eq. (3.32) for Boltzmann statistics, the new form

$$f_C^{(k)} = \frac{\Delta E_{Fk}^{(jct)}}{qU_{jct}} \tag{3.44}$$

of the reciprocity theorem is independent of the underlying statistics. The collection probability of a charge carrier generated at a any site k of the solar cell equals the relative shift $\Delta E_{Fk}^{(jct)}$ of the minority carrier Fermi level at this site caused by a voltage U_{jct} at the junction.

3.3 Applications of the generalized reciprocity theorem

Our first application of the generalized reciprocity theorem is to clarify the relation of the quantum efficiency diffusion length L_Q and the current-voltage diffusion length L_Q. The theorem is also useful to vastly simplify analytical and numerical transport simulations. We discuss this second application on p. 69.

3.3.1 General equality of L_Q and L_J

Consider the two-dimensional band diagram as depicted in Figure 3.5. We assume a p-n junction which is sufficiently flat fo the cell to be described in Cartesian coordinates (X, Y, Z) where Z is the direction of the incident light and the junction is located at $Z = 0$. The valence and conduction bands, and consequently the equilibrium minority carrier concentration n_o, are allowed to fluctuate at the junction and in the cell. The emitter region is assumed to be thin enough to neglect the absorption of photons there. With the assumption that light absorption occurs only by generation of one electron-hole pair per photon, the photogeneration rate is given by $\alpha_s \exp(-\alpha_s Z)$. We define the local quantum efficiency by

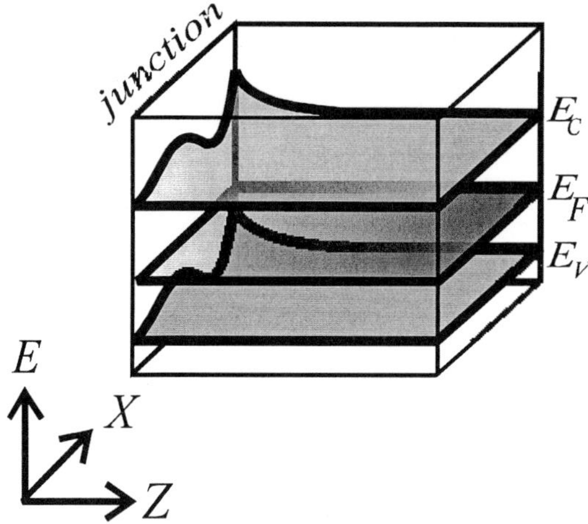

Figure 3.5. Energy band diagram for the solar cell region close to the junction at Z = 0. The conduction band energy E_C and the valence band energy E_V fluctuate relative to the equilibrium Fermi E_F. The third direction in space, the Y-axis, is not shown. Figure after Ref. [204].

$$IQE_l(X,Y,\alpha_s) = \int_{Z=0}^{\infty} dZ\ \eta_c(X,Y,Z)\,\alpha_s\,\exp(-\alpha_s Z) \tag{3.45}$$

$\eta_c(X,Y,Z)$ being the local collection probability for minority carriers generated at $\boldsymbol{r}$ = (X,Y,Z). Equation (3.45) is also valid for non-zero back surface reflectance, as we assume an absorption length α_s^{-1} that is sufficiently smaller than the film thickness W_f. The index l indicates that $IQE_l(X,Y,\alpha_s)$ is a local property at the position (X,Y) of the junction. The assumed strong optical absorption guarantees a locally one-dimensional transport in the direction of the Z-axis.

Note that with Eq. (3.45) the local quantum efficiency

$$IQE_l(X,Y,\alpha_s) = \alpha_s\ \mathrm{L}\{\eta_c(\boldsymbol{r})\} \tag{3.46}$$

is the Laplace transform L of $\eta_c(X,Y,Z)$ [209]. The possibility of extracting the local carrier collection efficiency η_c from the inverse Laplace transform of a measured quantum efficiency is discussed in Appendix C on p. 254.

In agreement with the definition of the (global) quantum efficiency diffusion length L_Q in Eq. (3.4), we define the *local* diffusion length L_{Ql} by

$$\frac{1}{L_{Ql}} = \left.\frac{\partial IQE_l^{-1}(X,Y,\alpha_s)}{\partial \alpha_s^{-1}}\right|_{\alpha_s^{-1}=0} \tag{3.47}$$

We first consider local properties and then turn to the global relationship of L_Q and L_J.

Equality of local effective diffusion lengths L_{Ql} and L_{Jl}

We use properties of Laplace transforms to prove that the local effective diffusion length always fulfills the relation

$$\frac{1}{L_{Ql}} = -\left.\frac{\partial\eta_c(X,Y,Z)}{\partial z}\right|_{Z=0} \tag{3.48}$$

With the definition of the local quantum efficiency IQE_l from Eq. (3.45) and the final value theorem for Laplace transforms [210] we find

$$\lim_{\alpha_s\to\infty} IQE_l(X,Y,\alpha_s) = \lim_{\alpha_s\to\infty}\alpha_s \mathrm{L}\{\eta_c(\boldsymbol{r})\} = \lim_{Z\to+0}\eta_c(\boldsymbol{r}) \tag{3.49}$$

The local quantum efficiency for infinitely strong absorption equals the collection probability at the junction ($Z = 0$). From the differentiation theorem for Laplace transforms [210] we obtain

$$\mathrm{L}\{\frac{\partial\eta_c(\boldsymbol{r})}{\partial Z}\} = \alpha_s \mathrm{L}\{\eta_c(X,Y,Z)\} - \eta_c(X,Y,0) = IQE_l(X,Y,\alpha_s) - \eta_c(X,Y,0) \tag{3.50}$$

The final value theorem for Laplace transforms and the use of Eq. (3.50) give us

$$\begin{aligned}\left.\frac{\partial\eta_c}{\partial Z}\right|_{Z=0} &= \lim_{Z\to+0}\frac{\partial\eta_c}{\partial Z} = \lim_{\alpha_s\to\infty}\left[\alpha_s L\left\{\frac{\partial\eta_c}{\partial Z}\right\}\right] \\ &= \lim_{\alpha_s\to\infty}\left[\alpha_s\left(IQE_l(X,Y,\alpha_s) - \eta_c(X,Y,0)\right)\right] \\ &= \lim_{\alpha_s^{-1}\to 0}\left[\frac{IQE_l(X,Y,\alpha_s) - \eta_c(X,Y,0)}{\alpha_s^{-1}}\right]\end{aligned} \tag{3.51}$$

at the junction ($Z = 0$). We may write, with the help of Eq. (3.48),

$$\left.\frac{\partial\eta_c(\boldsymbol{r})}{\partial Z}\right|_{Z=0} = \lim_{\alpha_s^{-1}\to 0}\left[\frac{IQE_l(X,Y,\alpha_s) - IQE_l(X,Y,0)}{\alpha_s^{-1}}\right] = \left.\frac{\partial IQE_l(X,Y,\alpha_s)}{\partial\alpha_s^{-1}}\right|_{\alpha_s^{-1}=0} = \frac{1}{L_{Ql}} \tag{3.52}$$

Thus, Eq. (3.48) is proven. The inverse global effective diffusion length

$$\frac{1}{L_Q} = \left.\frac{\partial IQE}{\partial\alpha_s^{-1}}\right|_{\alpha_s^{-1}=0} = \frac{1}{A_c}\int_{A_c}\left.\frac{\partial IQE_l(X,Y)}{\partial\alpha_s^{-1}}\right|_{\alpha_s^{-1}=0} dA = \frac{1}{A_c}\int_{A_c}\frac{1}{L_{Ql}}dA \tag{3.53}$$

is the spatial average of the inverse local effective lengths L_{Ql}. Now, we calculate the local diffusion current

$$j_D(x,y) = -q\,n_0(\boldsymbol{r})\,D(\boldsymbol{r})\left.\frac{\partial u^D(\boldsymbol{r})}{\partial Z}\right|_{z=0} \tag{3.54}$$

according to the generalized diffusion equation

$$-\nabla\{D(\boldsymbol{r})n_0(\boldsymbol{r})\nabla u(\boldsymbol{r})\}+\frac{n_0(\boldsymbol{r})}{\tau(\boldsymbol{r})}u(\boldsymbol{r})=g(\boldsymbol{r}) \tag{3.55}$$

derived by Alamo and Swanson [206]. With the reciprocity theorem (3.7) and Eq. (3.52) we obtain

$$\begin{aligned} j_D(X,Y) &= -q\,n_0(X,Y,0)\,D_n(X,Y,0)\,u_{jct}^{D}\left.\frac{\partial\eta_c(\boldsymbol{r})}{\partial Z}\right|_{Z=0} \\ &= \frac{q\,n_0(X,Y,0)\,D_n(X,Y,0)}{L_{Ql}(X,Y)}\left\{\exp\left(\frac{qU_{jct}}{kT}\right)-1\right\} \end{aligned} \tag{3.56}$$

Hence, the local dark current density at each location in the junction can be described by the same local diffusion length as is measured by quantum efficiency (LBIC). In analogy to Eq. (C.11), we define a local effective diffusion length L_{Jl} that determines the local forward saturation current density

$$j_o(X,Y)=\frac{q\,n_o(X,Y,0)\,D_n(X,Y,0)}{L_{Jl}(X,Y)} \tag{3.57}$$

A comparison of Eq. (3.56) and Eq. (3.57) yields the equality $L_{Ql} = L_{Jl}$ even for cases for which all quantities entering in Eq. (3.56) may fluctuate in the junction area $Z = 0$.

Equality of global effective diffusion lengths

The number of equilibrium minority carriers n_o may change by orders of magnitude in the presence of potential fluctuations which result, e.g., from doping fluctuations and from grain boundaries penetrating the junction. Let us now assume that the spatial fluctuations of the quantities entering in Eq. (3.57) are statistically independent. Then, we write for the total saturation current over junction area A_c

$$A_c\, j_o = q\int n_o dP(n_o)\ \int D_n\, dP(D_n)\ \int_{A_c} dX\, dY\, L_{Ql}^{-1}(X,Y) = \frac{q\overline{n_o}\,\overline{D_n}}{L_Q} \tag{3.58}$$

$dP(E_C)$ and $dP(D)$ being the probability distributions for the n_o and the diffusion coefficient D_n. The bar denotes the spatial average taken over the whole junction area. Let us now write the diode current density in the usual form

$$j_o = \frac{q\overline{D}\,\overline{n_o}}{L_J} \tag{3.59}$$

to find

$$L_Q = L_J. \tag{3.60}$$

Hence, the global diffusion length L_Q from a quantum efficiency measurement equals the diffusion length L_J that determines the diode saturation current.

Significance of the definition of effective parameters in inhomogeneous media

Note that this equality depends on the definition of the "effective parameters" by Eq. (3.59). If we had used the definition

$$j_o = \frac{q\overline{D}\,N_C \exp\left(E_F - \overline{E_C}\right)}{L_J} \tag{3.61}$$

the equality $L_Q = L_J$ does *not* hold and is replaced by $L_Q > L_J$ in the case of fluctuating conduction band energy [33]. Here, N_C, E_F, and E_C are the effective density of states in the conduction band, the Fermi energy, and the conduction band energy, respectively.

3.3.2 Quantum efficiency spectra

Simplified calculation of local carrier collection efficiency $\eta_c(\mathbf{r})$

The reciprocity theorem simplifies the *analytical* calculation of the local carrier collection efficiency $\eta_c(\mathbf{r}) = \Delta E_f / (qU_{jct})$ that can now be determined from solving the transport equations in the dark. This is much simpler than the conventional approach that first has to determine a Green $G(\mathbf{r}, \mathbf{r}_o)$ function for the inhomogeneous differential transport equation. Using the reciprocity theorem, it is sufficient to find a solution of the homogeneous differential equation.

Similarly, $\eta_c(\mathbf{r})$ can be determined *numerically* with the aid of the reciprocity theorem from a single finite-element solution of the transport equation in the dark. This is particularly helpful for the analysis of spatially heterogeneous solar cells, as was demonstrated by an application of the reciprocity theorem to the analysis of electronic transport in polycrystalline solar cells [211]

Simplified calculation of quantum efficiency IQE

Knowing $\eta_c(\mathbf{r})$, the internal quantum efficiency

$$IQE(\lambda) = \int_{V_c} \eta_c(\mathbf{r})\, g(\mathbf{r},\lambda)\, dV_c \tag{3.62}$$

is easily calculated as the generation-weighted average of the local carrier collection efficiency $\eta_c(\mathbf{r})$ over the cell volume V_c. Here, we assume a unity absorbed photon flux $\Phi = \int_{V_c} g(\mathbf{r})\, dV_c = 1$. The determination of a quantum efficiency spectrum is thus reduced to the purely optical problem of determining the generation rate $g(\mathbf{r},\lambda)$ at various wavelengths λ. The conventional approach is to solve the transport equations for each generation profile $g(\mathbf{r},\lambda)$ separately. Hence, fitting of experimental data with theoretical simulations is greatly accelerated using the reciprocity theorem.

3.4 Limiting recombination parameters derived from L_Q

The experimental determination of the effective quantum efficiency diffusion length L_Q is simple when compared to the extortion of recombination parameters (such as the minority carrier diffusion length L) from a measured quantum efficiency spectrum. However, L_Q is an effective diffusion length and does not describe only recombination the bulk, but also recombination at surfaces and grain boundaries. We therefore raise the question of what information on the recombination parameters we can draw from a single measured value of L_Q in monocrystalline and in polycrystalline cells.

3.4.1 Monocrystalline Si

We consider a planar, non-textured, and monocrystalline cell of thickness W_f that has the bulk diffusion length L and the back surface recombination velocity S_b. A small length L_Q may be caused by a small bulk diffusion length L or by a large surface recombination velocity S_b. A unique determination of both these parameters is therefore not possible. Instead, minimum and/or maximum values of L and S_b can be deduced from the (measured) value of L_Q. These limiting values are depicted in Figure 3.6 [212]. The coordinates are given in reduced units $l = L/W_f$, $l_Q = L_Q/W_f$ and $s_b = S_b\, W_f/D_n$ using W_f as the length scale and W_f^2/D_n as the time scale.

Electronically thick films ($W_f > L_Q$)

If L_Q is smaller than W_f (that is, $l_Q < 1$), the injected carriers ($L_Q = L_J$) do not reach the back of the cell and we cannot expect to learn anything on the back surface recombination s_b from l_Q. We do, however, get maximum and minimum possible values for the bulk diffusion length $l = L/W_f$. The maximum value $l_{max} = L_{max}/W_f$ occurs if the back surface recombination is infinitely large. The value l_{max} is thus the solution of

$$l_Q(l_{\max}) = \lim_{s_b \to \infty} \{ \, l_Q(l_{\max}, s_b) \, \} \tag{3.63}$$

The minimum value l_{min} is realized for zero back surface recombination and is thus the solution of

$$l_Q(l_{\min}) = \lim_{s_b \to o} \{ \, l_Q(l_{\min}, s_b) \, \} \tag{3.64}$$

Case of electronically thin films ($W_f < L_Q$)

For the opposite case $l_Q > 1$, the injected carriers reach the back of the cell. Hence, the maximum possible value of the back surface recombination $s_{b,max}$ occurs only if the cell has the minimum possible bulk recombination, that is, $l = \infty$. The value of $s_{b,max}$ is therefore the solution of

$$l_Q(s_{b,\max}) = \lim_{l \to \infty} \{ l_Q(l, s_{b,\max}) \} \tag{3.65}$$

The effective diffusion length l_Q could also originate from pure bulk recombination, thus giving a lower limit l_{min} that is the solution of Eq. (3.63).

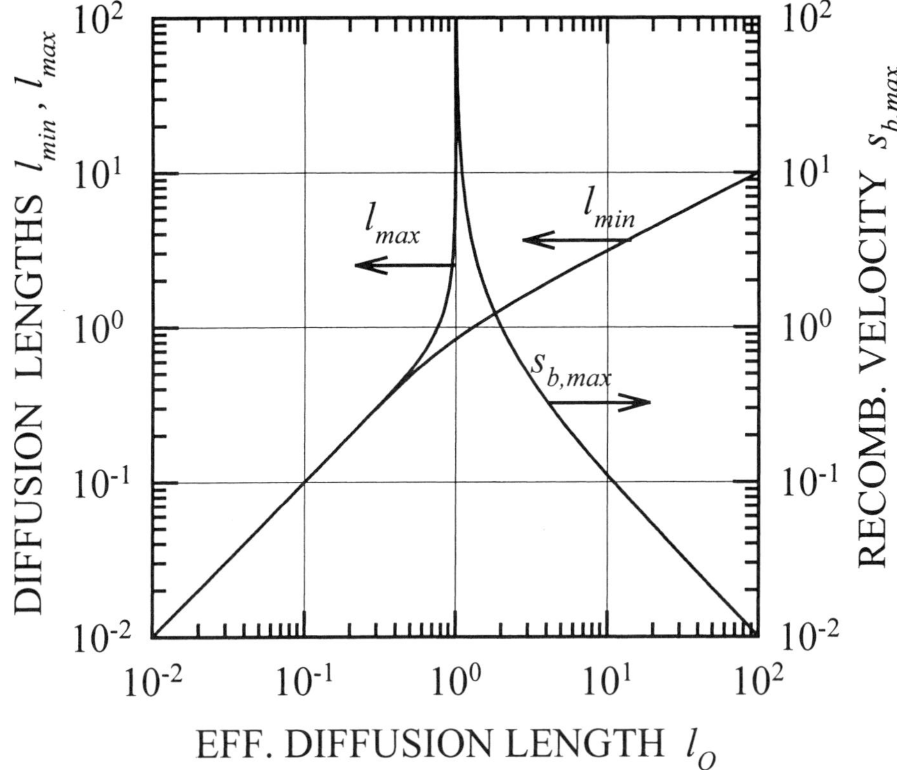

Figure 3.6. For a measured reduced quantum efficiency diffusion length $l_Q = L_Q / W_f < 1$ the graph allows us to read off a lower limit $l_{min} = L_{min} / W_f$ and an upper limit $l_{max} = L_{max} / W_f$ of the bulk diffusion $l = L / W_f$. For $l_Q = L_Q / W_f > 1$, a lower limit l_{min} of the diffusion length and an upper limit of the back surface recombination $s_b = S_b W_f / D_n$ is given.

Formulas

Analytical equations for the quantum efficiency diffusion length L_Q and also for the collection diffusion length L_C of monocrystalline films with homogeneous doping are given in Appendix C on p. 247. Approximations to the functions shown in Figure 3.6 are given in Ref. [212].

3.4.2 Polycrystalline Si

A polycrystalline film with columnar and square grains is shown schematically in Figure 2.28 on p. 46. Grain boundary recombination has to be considered here, and thus three recombination parameters have an impact on L_Q: the intra-grain diffusion length L, the grain boundary recombination velocity S_{grb} and the back surface recombination velocity S_b.

To simplify the analysis, we consider a thick polycrystalline cell $W_f >> L$. The back surface of the device must play neither an optical nor an electronic role for the results of this analysis to be applicable. This assumption eliminates the two parameters film thickness W_f and SRV S_b from the analysis. We remain with only one length scale, that is, the grain size G, and two recombination parameters L and S_{grb}. We conduct the analysis in terms of the reduced quantities $l_Q = L_Q/G$, $l = L/G$, and the reduced grain boundary recombination velocity $s_{grb} = S_{grb}\,G/D_n$.

It is possible to extract the limiting values l and s_{grb} from a measured value of l_Q in exactly the same manner as for monocrystalline material. The determination of these limiting values is done in the following steps:

(i) Analytically solving the minority carrier diffusion equation for the forward biased grain in the dark. This solution is derived in Appendix B on p. 231. Using the reciprocity theorem, this solution gives the local carrier collection efficiency.

(ii) Analytically calculating the internal quantum efficiency *IQE* from the local collection efficiency $\eta_c(Z)$, using Eq. (3.62) on p. 69.

(iii) Analytically calculating the quantum efficiency diffusion length L_Q from the defining Eq. (3.4) on p. 55. We thus arrive at an analytical expression $l_Q = l_Q(l, s_{grb})$ that is derived in Appendix C on p. 248.

(iv) Following the same line of argument as for monocrystalline Si but formally replacing the back surface recombination velocity s_b by the grain boundary recombination velocity s_{grb} yields the limiting recombination parameters l_{min}, l_{max}, and $s_{grb,max}$ in thick polycrystalline Si cells.

The solid lines in Figure 3.7a show $l_{max}(l_Q)$, $l_{min}(l_Q)$, and $s_{max}(l_Q)$ as functions of the experimentally accessible quantity l_Q. The dotted lines apply for a two-dimensional grain geometry, e.g. for a grain which is much larger in the X-direction than in the Y-direction. Thin-film cells with seed layers from zone-melting-recrystallization have this two-dimensional grain geometry.

We calculate $l_{Q,\infty} = \lim_{l\to\infty, s_{grb}\to\infty}\left\{l_Q\left(l, s_{grb}\right)\right\} = 0.072$ for large grain boundary recombination and negligible intra-grain recombination. Thus in the worst case grain boundaries reduce the effective diffusion length L_Q to about 10% of the grain diameter. For the 3D and the 2D geometry the plot has two regions: one for electronically large and one for electronically small grains. These regions permit different information to be deduced from $l_Q = L_Q/G$.

Electronically large grains ($G > L_Q$)

For large grain size G (that is, $l_q = L_Q/G < 0.072$ for square grains), a large fraction of injected carriers do not reach the grain boundary. Thus, we cannot deduce the grain boundary recombination velocity s_{grb} in this region. Instead, we derive an upper bound l_{max} and lower bound l_{min} for the intra-grain diffusion length l that dominates the recombination in this region of the plot.

Electronically small grains ($G < L_Q$)

In the opposite case of electronically small grains, grain boundary recombination becomes dominant. Using Figure 3.7, we derive a maximum value for the (reduced) recombination velocity $s_{grb} = S_{grb}\,G/D_n$ from a measured value of the (reduced) quantum efficiency diffusion length $l_Q = L_Q/G$.

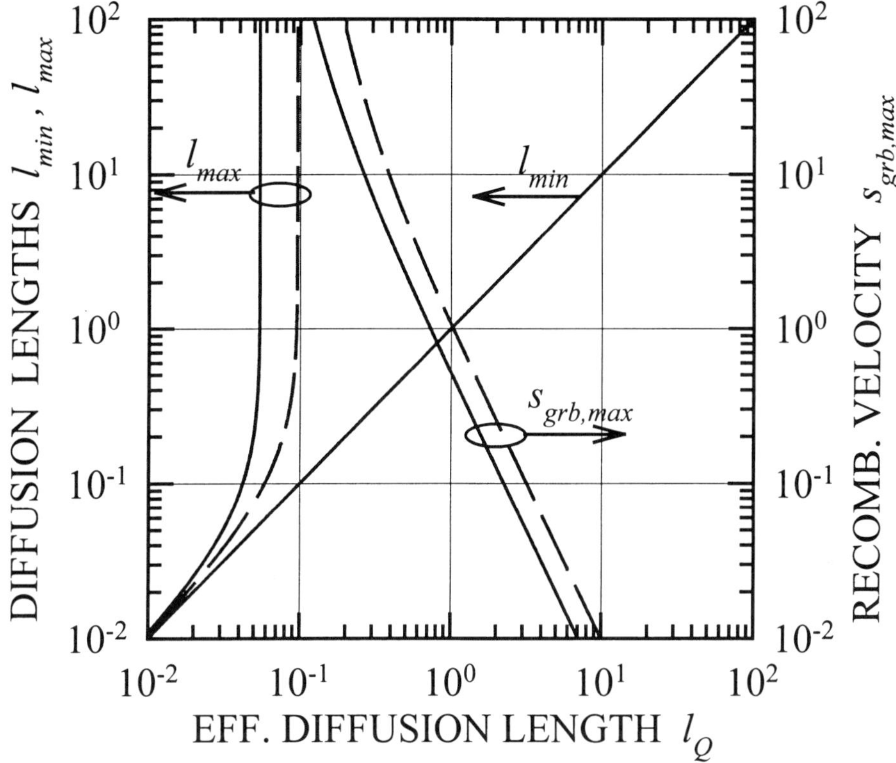

Figure 3.7. Limiting values l_{min}, l_{max} and $s_{grb,max}$ that are compatible with a measured value of the effective quantum efficiency diffusion length l_Q. Solid lines apply for a three-dimensional geometry with square grains while the dashed lines are valid for two-dimensional grains.

3.5 Analytical quantum efficiency model for thin films

In this section, we present a model to simulate the quantum efficiency of monocrystalline thin-film cells. We shall use this model to fit experimental quantum efficiency data. Choosing an analytical model guarantees a fast simulation and thus the fitting of experimental data in a short time.

The model we develop is applicable to cells similar to that depicted in Figure 3.9. The cell has an emitter of thickness W_e that follows the surface texture with a typical feature size p. This feature size p is much smaller than the film thickness W_f. In contrast to the previously published analytical models, we include a highly doped Si substrate region of thickness W_{sub}, the so-called back surface field, into the model. In addition, we include light trapping.

We will apply this model in section 3.5.3 on p. 83 to a W_f = 47 μm-thick, 20.6%-efficiency crystalline Si cell from thinned float-zone Si with inverted pyramids of texture period p = 13 μm. The assumption of a small texture size permits us to simulate the electronic transport one-dimensionally in the Z-direction perpendicular to the macroscopic cell surface. The first step towards modeling the quantum efficiency is the simulation of the photogeneration.

3.5.1 Modeling the photogeneration rate

The performance of a textured cell depends on many parameters, such as the shape and the size of the texture, the illuminating spectrum, the direction of light incidence, the state of polarization of the light, the dielectric coatings applied to the cells, the optical constants of the cell and the encapsulant. Therefore, it is almost impossible to describe the light trapping performance of a specific texture by a priori analytical modeling: computer simulations are necessary. We first calculate the photogeneration rate by ray-tracing. We then develop an analytical model that permits faster fitting of optical reflectance data. We compare the carrier generation profiles determined with this simpler analytical model with those determined by comprehensive ray-tracing.

3.5.1.1 Numerical ray-tracing model

Ray-tracing software for solar cell applications

The authors of the commercially available ray-tracing program TEXTURE [213] were the first to realize the necessity for an optical simulation program that is adapted to the requirements of solar cell simulations. Neither TEXTURE nor any of the studies on light trapping in solar cells accounted for the combined dependence of the propagation of light on the angle of incidence, the wavelength, the optical properties of dielectric multi-layers, and the polarization. To meet the need for a program that accounts for all these dependencies, we developed the solar cell ray-tracing program SUNRAYS [26], which is available for the photovoltaic community [214] and is in use in many of the laboratories developing Si solar cells. Shortly afterwards, a software package RAYN that uses the same Monte Carlo approach as SUNRAYS, and incorporates the Phong model for light scattering at lapped Si surfaces, was developed [215]. The ray-tracing program SONNE [216] divides all possible passes into classes. For each class the reflection and transmission coefficients are calculated only once, which accelerates the simulation.

Tracing algorithm of SUNRAYS

A comprehensive introduction to ray-tracing was given by Glassner [217]. Whatever tracing algorithm is used, the concept is always to trace individual light rays through the solar cell according to the laws of geometrical optics until they eventually leave the cell or are absorbed. Within the volume of a thick piece of material (e.g. the cover-glass or the Si film) we ignore the wave nature, which means the phase, of the ray of light. SUNRAYS is therefore an incoherent simulator [218]. At every interface between different materials (e.g. the interface between the cover-glass and the Si film) the light ray is either reflected, refracted, or absorbed. The probability of each event is calculated from the angle of incidence, the state of polarization, and the optical properties of the dielectric multi-layer system at the interface [218]. Interference effects are fully accounted for at these interfaces.

To simulate polarization effects, SUNRAYS calculates the probability that a photon is found in either of the two polarization eigenstates of a forthcoming reflection. With these probabilities, we take the ray as actually being s- or p-polarized. Tracing many rays yields, on the average, the result for the mixed polarization state. The intensity of a ray decreases exponentially when passing through the volume of an absorbing material. More details on the algorithm are given in Ref. [214].

We use the optical constants of Si as compiled in Ref. [22]. The optical constants of SiO_2 and Al are compiled in Ref. [219]. Periodic textures are particularly simple to analyze because the analysis is restricted to a small unit cell. A periodic surface texture that consists of only a few facet orientations has a geometry that is well defined by only a few parameters. Random textures are much more demanding. SUNRAYS utilizes a simple approach for random textures: the ray position (not the direction) is randomized every time the ray passes a plane in the middle of the thin film [214]. Rodriguez found this approach to be sufficiently accurate and his results with a model that explicitly accounts for a random surface agree with simulations performed with the program SUNRAYS [220].

Experimental test of ray-tracing simulations

The literature on light trapping in Si solar cells offers a large number of theoretical studies e.g. Refs. [2, 221, 86, 222]. In contrast, only a few direct comparisons of ray-tracing simulations and experimental data have been published. The first of these tests was conducted using our ray-tracing program SUNRAYS, to simulate Si films with periodic inverted pyramids [26]: Two silicon samples, one 45 μm thick and the other 122 μm thick, are fabricated by thinning (100)-oriented, p-type Si wafers of 0.35 Ω cm specific resistance. The thinned region has a final size of 25×25 mm^2. A texture period of p = 13 μm is defined by photolithography. The width of the flat ridge tops between the pyramids is 1.6 μm. The unit cell of this texture is shown in Figure 3.8a. A 109 nm-thick

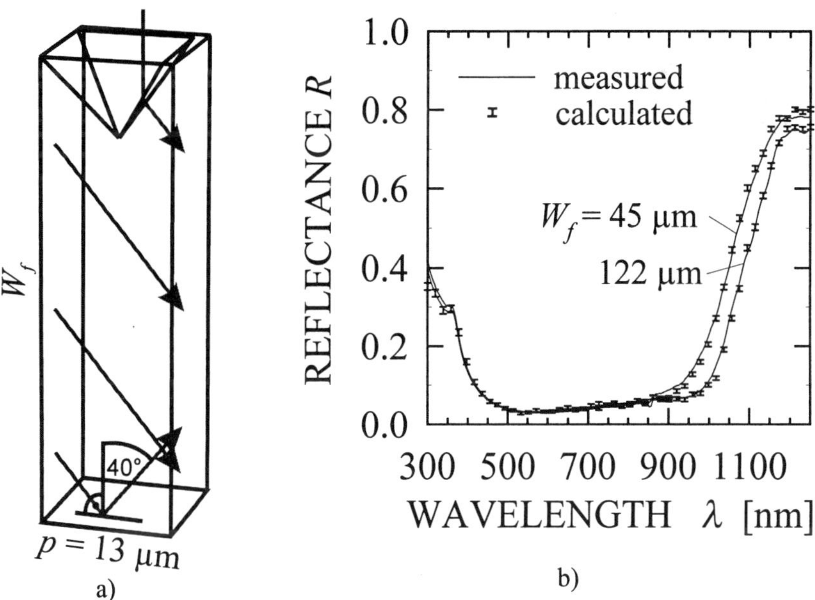

Figure 3.8. a) Unit cell of regular inverted pyramid texture. W_f is the film thickness, p is the texture period. The flat ridge tops originate from the oxide mask used for photolithography. b) Measured and calculated hemispherical reflectance agrees well for silicon samples with inverted pyramids of period p = 13 μm of thickness W_f = 45 μm and W_f = 122 μm textured with inverted pyramids.

SiO_2 film grows on the (111)-oriented facets in an oxidation furnace. At the same time, a 86 nm-thick SiO_2-film grows on the (100)-surfaces. The oxide on the flat back surface is coated with Al. Light that impinges normally onto the macroscopic cell surface reaches the back surface at an angle that depends on the index n_f of the medium in front of the cell and the index n_s of the Si substrate. For $n_f = 1$ and $n_s = 3.65$ we find

$$\arccos\left(1/\sqrt{3}\right) - \arcsin\left(n_f \sin\left(\arccos\left(1/\sqrt{3}\right)/n_s\right)\right) \approx 39.6° \tag{3.66}$$

This angle is larger than the critical angle of total internal reflection. Therefore, this light is redirected into the cell by total internal reflection. A mirror coating on the back of the cell is not required for the ray's first reflection at the back surface.

Figure 3.8b shows that measured and calculated spectra agree well for both samples without adjustment of fitting parameters. After subtraction of the free carrier absorption the program SUNRAYS determines a maximum short-circuit current density of 38 mA cm^{-2} for the 45 μm-thick cell, thus demonstrating the feasibility of efficient light trapping in thin silicon films. Because of the good agreement between simulation and experiment we conclude that in terms of integrated hemispherical reflection and in terms of absorption spectra the diffraction phenomena are negligible for a pyramid base length of 13 μm. Nevertheless, weak interference effects are observable by eye inspection in sunlight.

Ray-tracing is applicable to a wide class of textures that are coarse compared to the wavelength. In this work we will also apply ray-tracing to a few micron-thick conformal films as discussed in Appendix A on p. 194. The applicability of incoherent ray-tracing to film thickness values of only a few microns is discussed in Appendix A on p. 191. Ray-tracing also permits us to calculate multi-dimensional carrier generation profiles. For example in Ref. [223] a two-dimensional carrier generation rate was calculated for a V-shaped texture using the ray-tracing program SUNRAYS.

3.5.1.2 Analytical model

Ray-tracing is a computing-intensive task and fitting experimental reflectance spectra by ray-tracing simulations is impractical for routine IQE analysis. On the other hand, the IQE is not sensitive to the smaller details of the generation profile and we may hope to calculate IQE spectra with sufficient accuracy with an *analytical* approximation $g(Z)$ to the carrier generation profile. We derive such an analytical generation profile for cells with a thickness W_f that is much larger than the texture size p.

We use the optical model illustrated in Figure 3.9 [24]. Monochromatic light with absorption coefficient α_s is partially reflected at the front with a reflectance R_f. Light passes at an angle φ through the emitter and the space charge region of total thickness $W_e + W_{scr}$. The angle of light propagation is ϑ_1 in the base and the substrate that together have the thickness $W_{bas} + W_{sub}$. For the first pass through the cell, the light transmitted is attenuated by a factor

$$T_1 = \exp\left[-\alpha_s (W_e + W_{scr})/\cos(\varphi) - \alpha_s (W_{bas} + W_{sub})/\cos(\vartheta_1)\right] \tag{3.67}$$

The back surface of the substrate reflects light with a reflectance R_{b1} into an effective mean angle ϑ_2. Similarly to the first pass, the second pass also has different directions of light propagation in the emitter and in the base. However, due to the attenuation of the

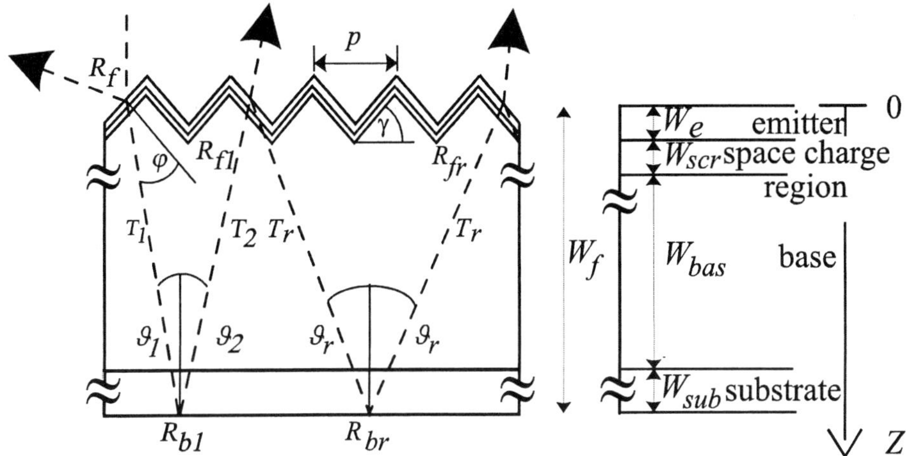

Figure 3.9. Schematic representation of a textured solar cell with a base of thickness W_{bas} and a highly doped substrate of thickness W_{sub}. The thickness W_e of the emitter and W_{scr} of the space charge region scr are much smaller than the scale p of the texture. The average facet angle is γ. A one-dimensional projection is shown on the right, with emitter and space charge region out of scale.

light intensity in the base and the substrate, the resulting electronic current from the emitter and the space charge region is much smaller for the second pass than for the first. Hence, we use the approximation of equal propagation direction in the emitter and the base region for the second and all consequent passes. Then, the attenuation is

$$T_2 = \exp\left(-\alpha_s W_f / \cos(\vartheta_2)\right) \tag{3.68}$$

for the second pass, wherein $W_f = W_e + W_{scr} + W_{bas} + W_{sub}$ is the total device thickness. The first internal reflectance R_{fl} randomizes the direction of light propagation. We assume that the light passes through the cell at an average angle ϑ_r for the third and all following turns. For each turn, the electron-hole pair generation attenuates the light intensity by the factor

$$T_r = \exp\left(-\alpha_s W_f / \cos(\vartheta_r)\right) \tag{3.69}$$

while optical losses, due to non-unity back surface reflectance R_{br} and non-unity internal front surface reflectance R_{fr}, decrease the light intensity without electron-hole pair generation. The subscript r stands for randomized light. We obtain the generation profile

$$\begin{aligned} g(Z) = (1 - R_f)\Bigg[& \Theta(W_e + W_{scr} - Z)\frac{\alpha_s}{\cos(\varphi)} \exp\left(-\frac{\alpha_s Z}{\cos(\varphi)}\right) \\ & + \Theta(Z - W_e - W_{scr})\frac{\alpha_s}{\cos(\vartheta_1)} \exp\left(-\frac{\alpha_s (W_e + W_{scr})}{\cos(\varphi)}\right) \exp\left(-\frac{\alpha_s (Z - W_e - W_{scr})}{\cos(\vartheta_1)}\right) \end{aligned} \tag{3.70}$$

$$+R_{b1}T_1\frac{\alpha_s}{\cos(\vartheta_2)}\exp\left(-\frac{\alpha_s(W_f-Z)}{\cos(\vartheta_2)}\right)$$

$$\left.+T_1R_{b1}T_2R_{f1}\frac{\alpha_s}{\cos(\vartheta_r)}\frac{\exp\left(-\frac{\alpha_s Z}{\cos(\vartheta_r)}\right)+T_rR_{br}\exp\left(-\frac{\alpha_s(W_f-Z)}{\cos(\vartheta_r)}\right)}{1-R_{br}R_{fr}T_r^2}\right]$$

as the final result from the sum of a geometrical series for all internal reflection processes. Here, $\Theta(Z)$ is the Heaviside function, which is unity for positive Z and zero elsewhere.

Reflection, absorption, and average path length

The theoretical function for the reflectance

$$R=1-(1-R_f)\left\{1-T_1R_{b1}T_2(1-R_{f1})-\frac{T_1R_{b1}T_2R_{f1}R_{br}(1-R_{fr})T_r^2}{1-R_{br}R_{fr}T_r^2}\right\} \quad (3.71)$$

and the active absorption

$$A=(1-R_f)\left\{(1-T_1)+R_{b1}T_1(1-T_2)+\frac{T_1R_{b1}T_2R_{f1}(1-T_r)(1+R_{br}T_r)}{1-R_{br}R_{fr}T_r^2}\right\} \quad (3.72)$$

can then be taken from Ref. [24]. The average path length of weakly absorbed light is

$$\bar{l}=\lim_{\alpha_S\to 0}\left\{\frac{A}{\alpha_S}\right\}=(1-R_f)\left\{\frac{1}{\cos(\vartheta_1)}+\frac{R_{b1}}{\cos(\vartheta_2)}+\frac{R_{b1}R_{f1}(1+R_{br})}{\cos(\vartheta_r)(1-R_{br}R_{fr})}\right\}W_f \quad (3.73)$$

Equations (3.70) to (3.72) describe the optical properties of the cell, as required to calculate the quantum efficiency. The values of the generation profile $g(Z)$, the total reflectance R, and the active absorption A depend on nine parameters: the external front surface reflectance R_f, the angles of light propagation φ for the first pass through the emitter, the angle ϑ_1 for the first pass through the base, the angle ϑ_2 for the second pass through the base, the angle ϑ_r for the third and subsequent passes through the base when the light is fully randomized, the first back reflectance R_{b1}, the second and subsequent back reflectance values R_{br}, the first internal reflectance R_{f1} and finally, the internal front surface reflectance R_{fr} of the second and subsequent passes. Not all of these nine parameters can be determined from a fit of Eq. (3.71) to the measured hemispherical reflectance spectrum.

The following paragraphs explain how to adjust most of these parameters to approximate and physically sensible values. We exemplify this simplification with a thin-film high-efficiency cell with inverted pyramids on the front surface and a lapped back surface.

Front surface reflectance R_f

We take the front surface reflectance R_f equal to the *measured* reflectance in the wavelength region of strong absorption (absorption length $L_\alpha = \alpha_s^{-1} << W_f$). In the wavelength range of weak absorption ($L_\alpha > W_f$), the front surface reflectance R_f is extrapolated from the reflectance measured in the wavelength range of strong absorption ($L_\alpha << W_f$).

Angles φ and ϑ_1 of light propagation at the first pass

A cell with inverted pyramids has the facet angle $\gamma = 54.7°$. Monochromatic light illuminates the cell at normal incidence. The angles φ and ϑ_1 of light propagation are determined by Snell's law of refraction and from the facet angle γ. The dispersion of the refractive index results in a wavelength dependence of the angles φ and ϑ_1.

Back surface reflectance values R_{b1} and R_{br}

The back surface reflectance R_{b1} after the first pass and the reflectance R_{br} after subsequent passes through the cell differ in general. In a first approximation, however, we use $R_{b1} = R_{br} = R_b$ in order to reduce the number of parameters. The back surface reflectance is also different for cells with a polished, textured, or rough back surface. The back surface of most experimental cells will be neither ideally flat nor completely rough. Let us assume that a fraction Λ of the back surface area is an ideal Lambertian reflector while a fraction $(1 - \Lambda)$ is ideally flat. Then, the back surface reflectance $R_b = (1 - \Lambda)R_{bs} + \Lambda R_{bd}$ is a weighted average of the specular back reflectance R_{bs} of flat regions and the diffuse reflectance R_{bd} of the rough back surface. Here we assume, however, that $R_{bs} = R_{bd} = R_b$ to further reduce the number of free parameters. As a consequence of this approximation, the Lambertian character Λ has no influence on the back surface reflectance R_b; however, Λ will be shown to have a strong impact on the average angle ϑ_2 and the first internal reflectance R_{f1}.

Angle ϑ_r of light propagation for the third and subsequent passes

Textures have a strong tendency to randomize the directions of light propagation within the cell [95, 2, 111]. Following Basore [24], we assume that the light is fully randomized after the second internal reflection at the front surface, and stays randomized for all further passes. The transmittance T_r of randomized light has to be averaged over all directions and is given by (A.12) on p. 183. Such an isotropic distribution of the directions of light propagation produces a non-exponentially decaying generation profile. In order to apply the generation profile $g(Z)$ from Eq. (3.70), we define the average angle ϑ_r as the solution of Eq. (3.69), with T_r from Eq. (A.12). The average angle ϑ_r depends on the absorption coefficient α_s. This α_s dependence represents an extension of Basore's model to arbitrary wavelengths.

Angle ϑ_2 of light propagation for the second pass

For the first pass we assume only one direction of light propagation, which is described by the angle ϑ_1. For the third and all following passes, the model assumes complete randomization. It therefore seems natural to express the parameters of the second pass as weighted averages of these two extreme cases. The weighting factors are i) the energy flux of the light that is randomized after the first reflection at the back surface and ii) the energy flux of the light that is specularly reflected at the back of the cell. Consequently, the transmittance for the second pass is

$$T_2 = \frac{\Lambda R_{bd} T_r + (1-\Lambda) R_{bs} T_{2s}}{\Lambda R_{bd} + (1-\Lambda) R_{bs}} \tag{3.74}$$

with $T_{2s} = \exp(-\alpha_s W_f / \cos(\vartheta_1))$ being the transmittance for the specularly reflected light which obeys the law of reflection and hence propagates at an angle equal to ϑ_1. The mean effective angle ϑ_2 follows from the solution of Eq. (3.68), with T_2 from Eq. (3.74).

Second and subsequent internal front surface reflectance R_{fr}

We determine the internal reflectance R_{fr} with our Monte Carlo ray-tracing program SUNRAYS [26, 214]. SUNRAYS finds $R_{fr} = 0.928 \pm 0.001$ for rays at a wavelength of $\lambda = 1000$ nm which are isotropically incident onto the inner surface of regular inverted pyramids with 109 nm of SiO_2 as antireflection coating (ARC). The reflectance values are similar for wavelengths other than $\lambda = 1000$ nm because the index of refraction for Si is almost constant in the near-infrared spectral range. For very thin cells, light rays with much shorter wavelengths than $\lambda = 1000$ nm can perform two and more passes through the cell. Then, the wavelength dependence of the internal reflectance for randomized light should be considered. In Appendix A on p. 182 we show that R_{fr} always has a value close to $1 - 1/n_S^2 = 0.921$, with $n_s = 3.57$ denoting the index of refraction in Si at $\lambda = 1000$ nm.

First internal front surface reflectance R_{f1}

The first internal front reflectance R_{f1} has a contribution R_{fs} from the light reflected specularly at the back and a contribution R_{fr} from the light randomized by the first back reflectance. The weighted reflectance

$$R_{f1} = \frac{\Lambda R_{bd} T_r R_{fr} + (1-\Lambda) R_{bs} T_{2s} R_{fs}}{\Lambda R_{bd} T_r + (1-\Lambda) R_{bs} T_{2s}} \tag{3.75}$$

contains only known parameters, except for the specular internal reflectance R_{fs}. Again, we use the program SUNRAYS to determine the parameter R_{fs}. The specularly reflected near-infrared light rays are absorbed only weakly during the first two passes through the cell and will therefore reach the internal front surface. Light rays with wavelengths in the near-infrared (e.g. wavelength $\lambda = 1000$ nm) propagate towards the inner front surface at an angle of approximately 40° relative to the macroscopic cell normal. For inverted pyramids with 109 nm of SiO_2 the program SUNRAYS finds $R_{fs} = 0.62$.

Checking the analytical generation profile against Monte Carlo ray-tracing

We thus remain with only two unknown parameters, Λ and R_b, that characterize the roughness of the back surface and the back surface reflectance, respectively. All nine parameters, R_f, φ, ϑ_1, ϑ_2, ϑ_r, R_{b1}, R_{br}, R_{f1}, and R_{fr} either are adjusted to sensible values or they depend on Λ and R_b. Both the unknown parameters Λ and R_b are now determined by fitting the experimental reflectance data with Eq. (3.71). We check the accuracy of our analytical generation profiles $g(Z)$ from Eq. (3.70) against the generation profiles we calculate by three-dimensional Monte Carlo ray-tracing. Several solar cell structures were simulated with the program SUNRAYS [26] in order to calculate a reflectance R and the generation profiles $g(Z)$. The wavelength range from 700 to 1200 nm is chosen since here the optical parameters of the back reflector actually influence the reflectance R of the cell. In general, good agreement between analytical and numerical generation profiles is found.

As a typical example, we consider a cell with inverted pyramids of period $p = 11$ μm on the front with 1 μm-wide flat ridges between the pyramids, a 109 nm-thick SiO_2 antireflection coating, 50 μm of cell thickness, and a slightly diffuse reflecting back surface (we choose a Lambertian character $\Lambda = 0.2$ for this example). See Figure 3.8a on p. 75 for the unit cell of this texture. No intermediate oxide is assumed to be between the Si substrate and the Al back surface reflector. The reflectance spectrum R as calculated by SUNRAYS is shown by the filled circles in the insert of Figure 3.10. The two parameters Λ and R_b are then optimized by a least squares fit of the analytical reflectance R according to Eq. (3.71) to these filled circles. All other parameters are chosen as described in the preceding paragraphs. A satisfying fit to the reflectance data calculated by SUNRAYS is achieved for $\Lambda = 0.1$ and $R_b = 0.84$. The randomization of the back surface reflectance is underestimated by our one-dimensional optical model, since the fitted Lambertian character, $\Lambda = 0.1$, is only half of the actual value, $\Lambda = 0.2$. This underestimation is caused by an overestimation of the light randomization by the first internal reflectance. In reality, as in our ray-tracing studies, the first internal reflection does not fully randomize the light directions.

However, these approximations of the analytical optical model do not influence the quality of the generation profile calculated by Eq. (3.70). Figure 3.10 shows the generation profiles $g(Z)$ at wavelengths of 800, 860, 930 and 1000 nm, respectively, as calculated by SUNRAYS (symbols) and as determined from Eq. (3.70) (solid lines). The differences between the solid lines and the symbols is below 4% over the whole thickness range displayed in Figure 3.10. Our model is not applicable to the surface region with the pyramids, which are 7 μm deep in our example. Hence, only the range of depths $Z = 10$ to 50 μm is displayed. The error of our generation profile in the region of the pyramids has no influence on the calculated quantum efficiency, if the base diffusion length is considerably greater than the depth of the surface texture.

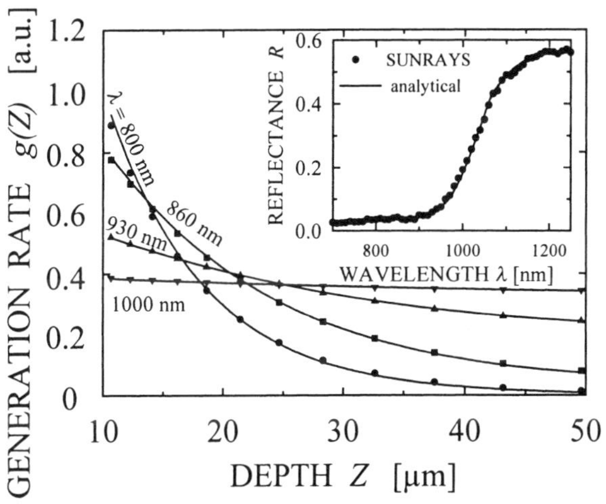

Figure 3.10. The analytical generation profiles (solid lines) show an agreement with those calculated by ray-tracing (symbols). The insert shows the reflectance spectrum as calculated by the program SUNRAYS (symbols) and as calculated by Eq. (3.71) (solid line). Figure reproduced from Ref. [25] with permission. ©1996 IEEE.

Application to Lambertian light trapping

The above model is easily adapted to Lambertian light trapping. By definition, Lambertian light trapping exhibits a full randomization of light propagation for the first pass through the cell. In order to ignore the first two passes in Eq. (3.70) which exhibit incomplete light randomization in our model, we set $T_1 = R_{b1} = T_2 = R_{f1} = 1$. Using the definition of ϑ_r from Eq. (3.69) and the relations $T_f = 1 - R_f$ and $R_{fr} = 1 - T_f / n_s^2$ yields the carrier generation rate $g(Z)$ for Lambertian light trapping schemes that is given by Eq. (A.26) on p. 186. Similarly, the reflectance of a Lambertian light trapping scheme follows from (3.71) and is given by Eq. (A.27) on p. 187. The optical absorption of a Lambertian light trapping scheme as derived from (3.72) equals the absorption given by Eq. (A.13) on p. 183.

3.5.2 Modeling the electronic transport

Figure 3.9 on p. 77 shows schematically a solar cell structure with $W_e << p << W_{bas}$. The situation is well approximated by a one-dimensional transport model. The internal quantum efficiency

$$IQE = \sum_{i=e,scr,bas,sub} IQE_i \tag{3.76}$$

has contributions from the emitter *e*, the space charge region *scr*, the base region *bas*, and the substrate region *sub*. A photon contributes to the quantum efficiency

$$IQE_i = A_i \eta_i / (1 - R) \tag{3.77}$$

of the region $i = e, scr, bas, sub$, if it is actively absorbed in region i with a probability A_i and then collected at the junction with collection efficiency η_{ci}. The factor $(1 - R)$ normalizes the quantum efficiency to the fraction of photons that are not reflected by the cell. The relative contribution A_i from different cell regions i to the total active absorption A is

$$\frac{A_i}{A} = \frac{\int_{\text{region } i} g_i(Z)\,dZ}{\int_{\text{total cell}} g(Z)\,dZ} \tag{3.78}$$

Here,

$$g_i(Z) = \begin{cases} g(Z) & \text{for } Z \text{ in region } i \\ 0 & \text{otherwise} \end{cases} \tag{3.79}$$

equals the optical generation profile $g(Z)$ in the region i and is zero elsewhere. The collection efficiency

$$\eta_{ci} = \frac{j(g_i)}{q \int\limits_{\text{region } i} g_i(Z) dZ} \tag{3.80}$$

depends on the short-circuit current density $j(g_i)$ from the device for the (artificial) generation profile g_i and is normalized to the cumulative generation in region i; the symbol q denotes the elementary charge. Inserting Eqs. (3.78) and (3.80) into Eq. (3.77) yields

$$IQE_i = \frac{A}{1-R} \frac{j(g_i)}{q \int\limits_{\text{total cell}} g(Z) dZ} \tag{3.81}$$

The solution of the diffusion equation with a generation profile $g(Z)$ from the above optical model is tedious but straightforward mathematics. We published the solution in Ref. [25] which yields the current density $j(g_i)$, and thus the quantum efficiency IQE_i. The total quantum efficiency IQE follows from Eq. (3.76).

3.5.3 Application to thin high-efficiency cells

The two cells that we analyze here are at the extremes covered by our quantum efficiency model: Cell A has no substrate at all, while Cell B has a substrate that is several hundred microns thick.

Cell A: A silicon layer with a thickness of $W_f = 47$ µm and a hole concentration of $N_{bas} = 4 \times 10^{16}$ cm^{-3} is prepared by thinning a boron-doped (100)-oriented, float-zone-grown wafer in KOH solution. The cell is 1 cm^2 in area, textured with photolithographically defined inverted pyramids of period p = 13 µm, and it has a phosphorus-diffused emitter. An additional heavy phosphorus diffusion is applied underneath the Ti/Pd/Ag contacts. The front surface is passivated with 109 nm of thermal SiO_2 and the Al rear reflector is separated from the silicon by a SiO_2 layer that grows simultaneously with the front surface oxide. Electrical back contact is made through periodic holes in the SiO_2. With this type of cell structure we reached a confirmed power conversion efficiency of 20.6% [113].

Cell B: The Si layer is fabricated by chemical vapor deposition at 1150°C. The cell has a thickness of W_{bas} = 48 µm and a hole concentration within the base of $N_{bas} = 7\times10^{16}$ cm^{-3}. The boron-doped, (100)-oriented, Czochalski-grown Si substrate has a hole concentration of $N_{sub} = 3.5\times10^{18}$ cm^{-3} and a thickness of W_s = 515 µm. The cell is 4 cm^2 in area and textured with similar inverted pyramids to Cell A. The processing sequence is also similar to Cell A; however, the whole back surface of the substrate is covered with Al without a SiO_2 layer between Si and Al. With this type of cells we reached a confirmed power conversion efficiency of 17.3% [177].

IQE analysis

The IQE analysis aims at the determination of the base diffusion length L_{bas}, the back surface recombination velocity S_b (Cell A), and the substrate diffusion length L_{sub} (Cell B).

The measurement is performed under a light bias of 100 mW cm^{-2}, generated with a halogen lamp. The filled circles in Figure 3.11 show the measured internal quantum efficiency *IQE* and the measured hemispherical reflectance *R* of Cell A. With the above-described model, we perform a least squares fit to the experimental data by varying the back surface reflectance R_b, the Lambertian fraction Λ, the base diffusion length L_{bas}, and the differential back surface recombination S_b (substrate thickness W_{sub} is zero for Cell A). The solid lines represent the theoretical quantum efficiency *IQE* and the theoretical reflectance *R* from the optical model (Eq. (3.71) on p. 78). The data for experiment and theory agree reasonably well. The fitting procedure results in $R_b = 0.92$, which is a reasonable value for an Al reflector. The Lambertian character is $\Lambda = 0.55$, which is also reasonable because Cell A has a rough back surface due to the KOH-etching procedure which we applied to thin the lapped side of the originally thick float-zone wafer. The emitter diffusion length L_e and the emitter surface recombination S_e are also adjusted during the fitting procedure; however, their influence on *IQE* at wavelengths $\lambda > 800$ nm is small. The measured reflectance *R* and the quantum efficiency *IQE* are fitted simultaneously.

Similarly, Figure 3.12 shows good agreement between experiment and theory for Cell B, which has a thick and highly doped Si substrate. The reflectance at 1200 nm is almost entirely due to the front surface. Sub-bandgap light that enters the cell is strongly absorbed by free carriers in the thick and highly doped substrate. Hence, the fitting procedure produces a back surface reflectance $R_b = 0$. For the wavelength region $\lambda > 800$ nm, we remain with only two important fitting parameters, which are the base diffusion length L_{bas} and the substrate diffusion length L_{sub}. The surface recombination S_{sub} at the back surface of the substrate is not important because the thickness W_{sub} of the substrate exceeds the expected substrate diffusion length L_{sub} by more than an order of magnitude.

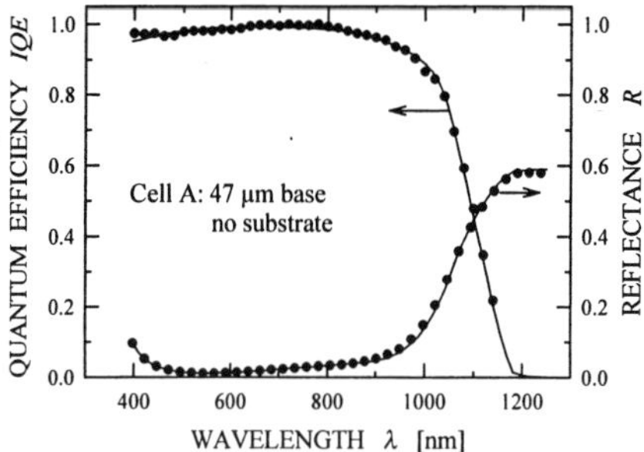

Figure 3.11. Filled circles show measured internal quantum efficiency *IQE* and reflectance *R* of Cell A, which is 47 μm thick, has inverted pyramids on the front, and an Al reflector on the back. Solid lines are fitted theoretical data.

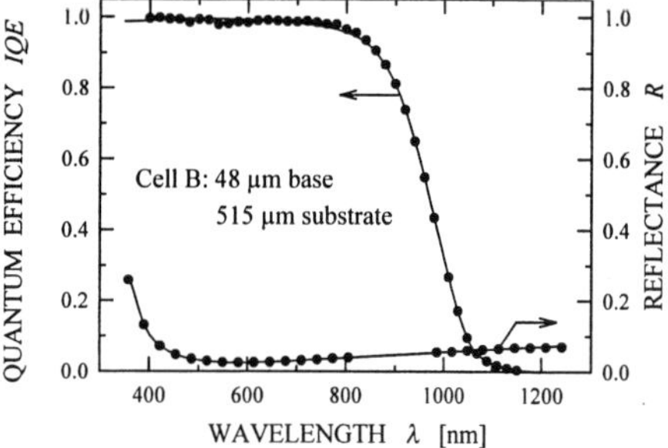

Figure 3.12. Filled circles show measured internal quantum efficiency *IQE* and reflectance *R* of Cell B, which is 48 μm thick, has a substrate of 515 μm thickness, and inverted pyramids on the front. Solid lines are fitted theoretical curves. Figure reproduced from Ref. [25] with permission. ©1996 IEEE.

Figure 3.13 re-plots the IQE data of Figure 3.12 on a semi-logarithmic scale. The quantum efficiency contributions from the emitter, the space charge region, the base, and the substrate are also shown. At all wavelengths between 450 and 1020 nm, more than one region contributes significantly to the total IQE. At a wavelength of $\lambda = 800$ nm, the emitter and the substrate do both contribute approximately 2% to the total IQE. Hence, a wavelength region which is entirely dominated by the base contribution does not exist. The overlapping of contributions from different cell regions becomes even stronger at base thickness values $W_{bas} << 50$ µm. Therefore, the current contributions from all the regions should be considered simultaneously when analyzing thin-film Si solar cells.

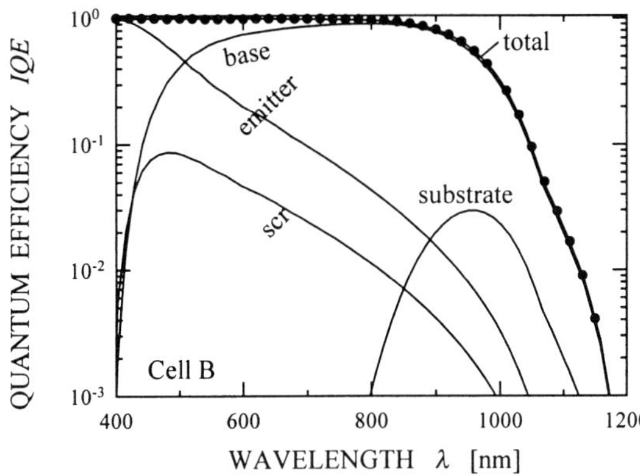

Figure 3.13. The total internal quantum efficiency *IQE* of Cell B from Figure 3.12 is the sum of contributions from the emitter, the space charge region (scr), the base and the substrate. The *IQE* contributions of the different cell regions overlap. Figure from Ref. [25] with permission. ©1996 IEEE.

Parameter confidence plots

We now turn to the question of dezermining the diffusion length L_{bas} and the back surface recombination S_{sub} of Cell A and B. A problem arises from the fact that the measurement can be described equally well with many different pairs for L_{bas} and S_{sub}.

Figure 3.14 shows those regions of the (L_{bas}, S_b)-parameter plane in which the stan-

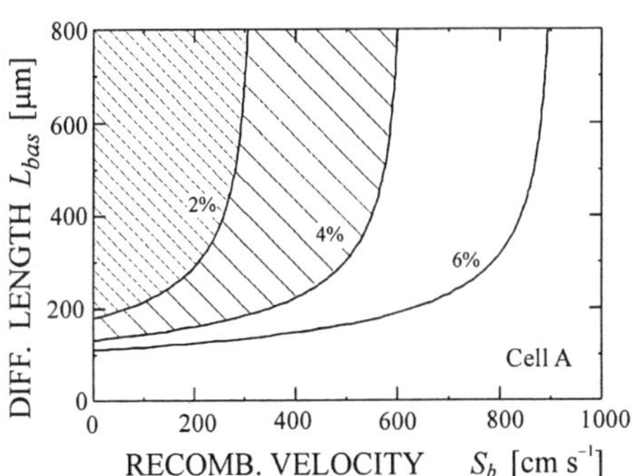

Figure 3.14. Parameter confidence plot for the base diffusion length L_{bas} and the differential back surface recombination velocity S_b of Cell A. With an assumed measurement accuracy of 2%, 4%, and 6%, respectively, all (L_{bas}, S_b)-pairs in the hatched regions are in agreement with the measurement. The conclusions that can be drawn from the IQE analysis therefore depend strongly on the measurement accuracy.

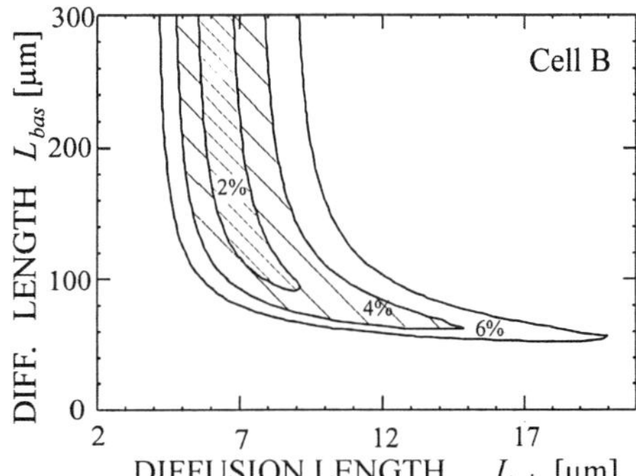

Figure 3.15. Parameter confidence plot for the base diffusion length L_{bas} and the substrate diffusion length L_{sub} of Cell B. With an assumed measurement accuracy of 2%, 4%, and 6%, respectively, all (L_{bas}, L_{sub})-pairs in the hatched regions are in agreement with the measurement. Figure from Ref. [25] with permission. ©1996 IEEE.

dard deviation between calculated and measured IQE is less than 2, 4, and 6%, respectively. We calculate the standard deviation in the region 800 nm $< \lambda <$ 1150 nm, because in this region the current contributions from the emitter are small, and hence the interdependence of emitter and base parameters is also small. An agreement between measurement and fit that is better than the measurement accuracy of 3% is not significant. Hence, we consider all pairs (L_{bas}, S_b) which show a standard deviation smaller than 3% to be consistent with the measurement. As a result, the base diffusion length L_{bas} exceeds 130 μm and the differential back surface recombination velocity S_{sub} is below 500 cm s^{-1}. We call Figure 3.14 a *parameter confidence plot* or for short, a PC-plot. A PC-plot may also contain lines of constant diode saturation current. Such iso-lines give valuable information on the efficiency limiting process if used in conjunction with measurements of the diode saturation current [113]. It is possible to generalize the concept of the PC-plot to more than two variables; however, the graphical presentation of higher dimensional PC-plots is difficult.

Cell B is treated similarly to Cell A. Here, the interdependent parameters are the base diffusion length L_{bas} and the substrate diffusion length L_{sub}. Figure 3.15 shows the PC-plot for the (L_b, L_s)-plane. The parameter pairs that agree with the measurement fall into a banana-shaped area. This shape demonstrates that PC-plots contain more information than just upper and lower bonds for L_{bas} and L_{sub}. Parameter-confidence-plots also reveal interrelations between the parameters of interest. For example, we would get much more precise information on L_{sub}, if we knew from independent experiments that the base diffusion length L_{bas} was 250 μm. However, without such a priori knowledge we can only conclude that Cell B has a diffusion length $L_{bas} >$ 80 μm and a substrate diffusion length 5 μm $< L_s <$ 11 μm. Again, we learn from this discussion that the conclusions to be drawn depend critically on the measurement accuracy. The value of the substrate diffusion length is reasonable. Coulomb-enhanced Auger recombination limits L_{sub} to values < 15 μm [224]. Defects may further reduce the value of the substrate diffusion length L_{sub} to values smaller than the Auger limit. Our model does not include defect-induced recombination in the space charge region or at the base/substrate interface, which would further reduce the apparent substrate diffusion length L_{sub}.

3.6 Differential and actual recombination parameters

Figure 3.16 shows the measured internal quantum efficiency of a cell processed identically to Cell A in the previous section. The thickness is again 45 μm-thick. The back surface has a Si/89 nm SiO_2/Al structure. Holes in the intermediate oxide are 70 μm in diameter and permit the electrical contact from Al to Si. The distance between neighboring holes is 1000 μm

In contrast to the data shown in Figure 3.11 on p. 84, this measurement is performed without bias light. The excess minority carrier concentration at the back of the cell is instead defined by applying a voltage bias using the circuit shown in Appendix C on p. 242. The enhancement of the quantum efficiency at a wavelength around 1000 nm with increasing bias is caused by a back surface recombination velocity S_b that decreases with increasing minority carrier concentration Δn_W [31]. The magnitude of the quantum efficiency enhancement demonstrates the necessity to include the injection level dependence of surface recombination into thin-film cell modeling.

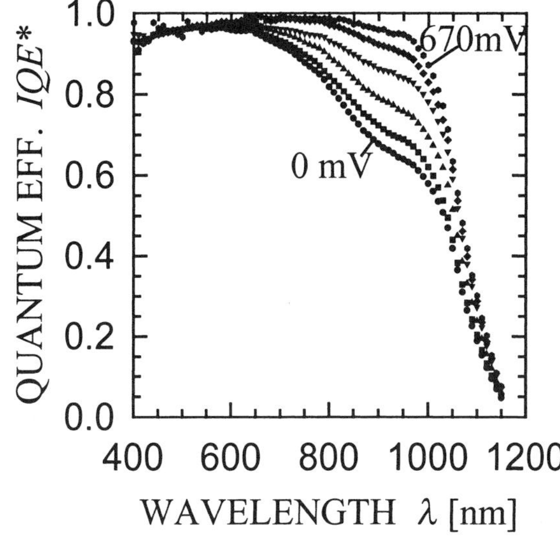

Figure 3.16. Quantum efficiency IQE^* in the dark for a forward bias voltage U_b = 0, 425, 460, 495, 530, and 670 mV.

All these measurements apply modulation techniques that are illustrated in Figure 3.17. A constant bias excitation (light or voltage) defines the injection level Δn_{Wb}. A superimposed small modulated excitation (chopped monochromatic light in our case) generates a variation $d\Delta n_W$ of the injection level. The measurement signal (change in short-circuit current density) is proportional to the change in the surface recombination rate dU_{sur}. The intensity of the modulated excitation is small compared to the bias, thus defining the injection level. We call these types of measurements modulation techniques.

In Figure 3.17, the surface recombination rate U_{sur} depends non-linearly on the excess minority carrier concentration Δn_W. Modulation techniques measure a differential recombination velocity $S_{eff,diff} = dU_{sur}/d\Delta n_W$ in this case and not the actual surface recombination velocity $S_{eff} = U_{sur}/\Delta n_W$, as first reported by the author [30]. Apparently, the necessity to distinguish actual and differential SRV has been overlooked in the past e.g. Refs. [179, 176, 178], resulting in quantitatively wrong interpretation of the measurement. We omit the index *eff* and denote the effective SRV simply by S to simplify the notation. The excess carrier concentration at the edge of the space charge region Δn_{Wb} is simply denoted by n_b. The actual SRV

$$S(n_b) = \frac{U_{sur}(n_b)}{n_b} = \frac{1}{n_b}\int_0^{n_b} S_{diff}(n)\,dn \tag{3.82}$$

follows from an integration of the differential SRV S_{diff} over all injection levels $n < n_b$. The actual SRV S and the differential SRV S_{diff} are equal, if U_{sur} is linear in n_b or, equivalently, if the actual SRV S is constant. The limited experimental sensitivity only allows us to measure at injection levels n_b higher than a minimum level n_{bo}. In this case we start the integration

$$S(n_b) = \frac{S(n_{bo})n_{bo}}{n_b} + \frac{1}{n_b}\int_{n_{bo}}^{n_b} S_{diff}(n)\,dn \tag{3.83}$$

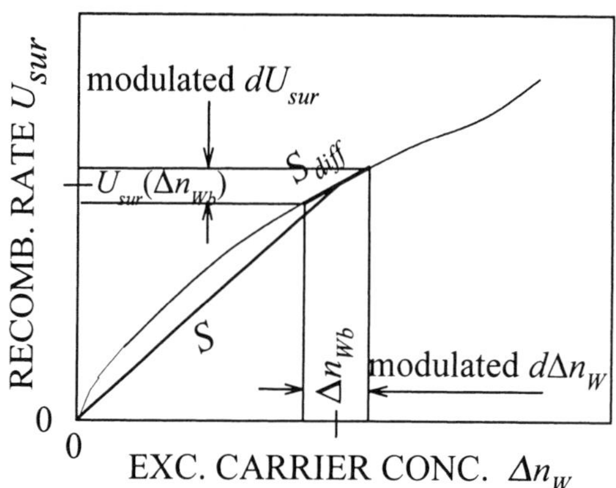

Figure 3.17. For a recombination rate U_{sur} that depends non-linearly on the excess minority carrier concentration Δn_W, the differential surface recombination velocity S_{diff} and the total surface recombination velocity S differ.

at level n_{bo}, leaving an unknown integration constant $S(n_{bo})$. This integration constant represents an error source for the determination of the actual SRV S. The error decreases with increasing injection level n_b and becomes negligible if n_{bo} is sufficiently small to make S_{diff} independent of the injection level. If S_{diff} does not saturate, the S values up to a carrier concentration of about 10 times n_{bo} must be rejected, while the S values at higher injection levels are negligibly affected by the choice of n_{bo} [32].

Differential and actual SRV at the Si/SiO$_2$/Al back surface

We derive the differential back surface recombination velocity S_{diff} marked with triangles in Figure 3.18 by analyzing the IQE^* spectra of Figure 3.16 with the one-dimensional transport model described on p. 73. The excess carrier density n_b is derived from the applied bias voltage [31]. This voltage-biased *IQE* measurement covers six orders of magnitude in the injection level. In order to obtain more data points for a proper integration, we also extract S_{diff} from the *IQE* data that we measure at a single wavelength $\lambda = 880$ nm [31] at various injection levels. These data are denoted by circles and agree with S_{diff} values extracted from the full *IQE* spectra.

Now we are in the position to calculate the injection level dependence of the *actual* SRV $S(n)$ from the measured *differential* SRV S_{diff} via Eq. (3.83). The actual SRV S (squares in Figure 3.18) decreases by two orders of magnitude with the injection level n_b.

The actual SRV S is about a factor of two larger than the differential SRV S_{diff} at medium injection levels. This factor is about equal to the measurement accuracy in this particular case.

Generalization of the concept to light and voltage bias

It is common practice for IQE measurements to apply bias illumination to the short-circuited cell. However, at the maximum-power point the injection level in a one-sun-illuminated cell is even higher than for a short-circuited cell. The strong dependence of the SRV S on the injection level n_b indicates that the IQE measurement should be performed under light *and* voltage bias. The determination of the actual SRV S at the maximum-power point is analogous to the procedure presented here. The only difference is that the we have to increase the bias voltage *and* the bias illumination in small steps until we reach the maximum-power point.

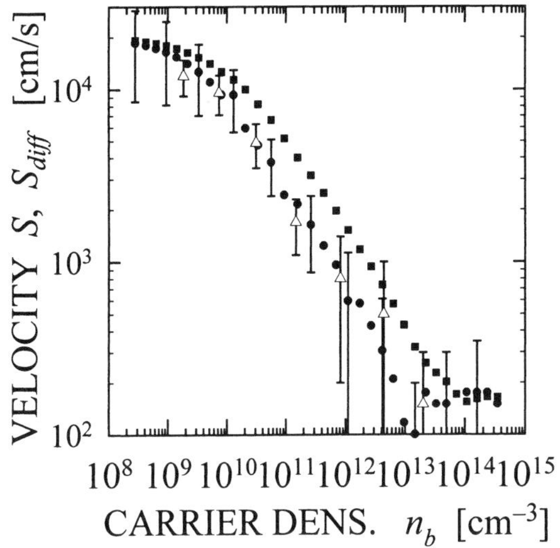

Figure 3.18. Injection level dependence of the recombination velocity at the point-contacted back surface of the Si cell: Differential SRV S_{diff} as deduced from the IQE spectra shown in Figure 3.16 (triangles); SRV S_{diff} from the *IQE* at wavelength λ = 880 nm (circles); and actual SRV S as calculated from Eq. (3.83) (squares).

Applications of the author's concept of a differential surface recombination velocity to oxide-passivated p-type monocrystalline Si wafers, revealed a factor of up to 4 between differential and actual surface recombination velocity for an Si/SiO_2 interface [30-32]. An experimental verification of Eq. (3.82) was recently given by Schmidt [225], who directly compared the small signal differential photoconductance decay measurements with large signal quasi-steady-state photoconductance (QSSPC) measurements [226].

4 Technological approaches to thin-film cells

The driving force for the development of thin-film cells is the reduction of the high amount of costly crystalline Si material consumed by conventional 300 µm-thick wafer cells. The photovoltaic community agreed to consider crystalline silicon (c-Si) cells to be thin if the effective thickness W_{eff} is below 50 µm. The effective thickness is the ratio of cell volume to cell area. This definition is also applicable to devices with spatially varying thickness W (see for example Figure 4.23a on p. 113). Considering that an epitaxy process such as chemical vapor deposition utilizes only a fraction of the Si introduced in the reactor, a thin-film cell should certainly be much thinner than just 50 µm. A thickness below 10 µm is desirable, to keep a clear margin in Si saving.

The feasibility of a high cell efficiency with crystalline Si layers that are thinner than 50 µm was demonstrated experimentally with a confirmed device efficiency of 20.6% by Brendel et al. [113] and 21.5% by Zhao and co-workers [227, 228]. Both cells are 47 µm thick and were fabricated from thinned FZ wafers. The schematics of these cells are shown in Figure 1.1 on p. 1. We discussed the complex cell design in the introduction. In contrast to the 20.6% cell [113], the 21.5% cell [227] received an additional p^+-type boron doping underneath the back contacts in order to reduce the contact recombination. Hebling et al. succeeded in fabricating a 19%-efficient cell 49 µm in thickness that is grown by chemical vapor deposition (CVD) on a SIMOX (separation by implantation of oxygen) wafer [48]. The cell has both contacts on the front surface. All three of these cells were fabricated with highly complex processes using five or more photolithography steps [113, 227, 48]. In addition, all three cell concepts consume a full Si wafer. These processes are not cost-effective. We exclude therefore these high-efficiency devices from further consideration in this chapter.

A large variety of substrates, deposition techniques, and cell designs are currently under investigation fabricating thin-film cells and modules in a manner that is potentially suitable for low-cost large-scale production [229-231]. Most of the developments are still in the laboratory phase resulting in solar cells of few square centimeters in area and with efficiencies ranging from 5 to 14%. As mentioned in the introduction, we group these thin-film approaches into three classes, according to the substrate used for deposition of the crystalline Si film [232, 233]:

- cells on high-temperature substrates (HTS),
- cells on low-temperature substrates (LTS), and
- cells from layer transfer processes (LTP).

This chapter presents a review on these three types of crystalline Si thin-film approaches with a focus on layer transfer processes [234]. Their specific advantages and challenges will be summarized and discussed in the summary of this work on p. 168.

4.1 High-temperature substrate (HTS) approach

Cells on high-temperature substrates (HTS) are typically fabricated at temperatures above 1000°C on substrates such as low-cost Si [235], graphite [47, 236], and ceramics [237]. Some approaches even apply the melting temperature of 1420°C to recrystallize the Si material in order to enhance the grain size.

We discuss the substrates, the active device layer, and the device results of the HTS approach separately.

4.1.1 Substrates

Substrate requirements

The substrate has to withstand the high Si deposition temperatures of 1000 to 1200°C. Due to these high temperatures, it also has to be adapted to the thermal expansion of polycrystalline Si, which is 3.9×10^{-6} K^{-1} at 1000 °C [238]. The substrate has to be sufficiently inert in order not to react with the hot Si. A high purity of the substrate is advantageous to reduce the contamination of the active device layer. Electrical insulation is another important requirement, to permit an integrated series connection. A highly reflecting substrate or a transparent substrate eases the implementation of light trapping schemes (see p. 194 for light trapping with transparent substrates). Last but not least, the substrate should be of low cost.

Table 4.1 lists substrate materials that are currently under test for the HTS approach. The preferred materials contain the elements Si, C, O, and Al as major constituents. Figure 4.1 illustrates the reason: these elements are tolerable in higher concentration than others. Carbon is iso-electronic with Si. Oxygen forms strong Si-O bonds that saturate the dangling bonds. Electronic-grade Si from the Czochralski process, for example, contains oxygen atoms at a concentration of 10^{18} cm^{-3} and still exhibits a high minority carrier lifetime. Aluminum functions as an acceptor.

Mullite

Mullite is a ceramic material that consists of Al_2O_3 and SiO_2. The content of Al_2O_3 controls the thermal expansion coefficient that reaches values of 4 to 6×10^{-6} K^{-1} at 1000°C. A particularly attractive feature of mullite is its white appearance. The hemispherical reflectance is above 90% at wavelengths from 800 nm to 1200 nm [237]. The reflectance values increase with the thickness of the substrate. The light reflectance is due mainly to optical scattering at mullite particles in the volume of the material rather than at the air/mullite interface. The scattering introduces light trapping.

Table 4.1. Substrate materials used for thin-film Si solar cells fabricated with the HTS approach. Some properties relevant for thin-film solar cells are listed.

	Major ingredients	Insulator	Reflector	Costs	Reference
Mullite	SiO_2:Al_2O_3	yes	yes	low	[237]
Graphite	C	no	no	high	[47]
High-T glass	SiO_2:Al_2O_3:BaO:MgO	yes	yes*	high	[238]
Si-infiltrated SiC	SiSiC	no	no	low	[48]
Ribbon Si	Si	no	no	high	[235]

* In combination with an Al mirror evaporated onto the glass substrate.

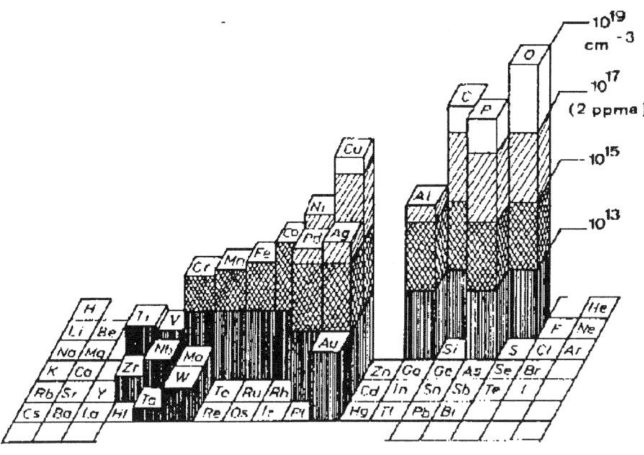

Figure 4.1. Relative tolerable concentration of impurities for solar cells from p-type Si. Reproduced from Ref. [239], Copyright 1981, with permission from Springer-Verlag.

Graphite

Graphite has the required temperature stability and is also available with a thermal expansion coefficient matched to Si. So far, graphite with high chemical purity used for solar cell processing [47, 48, 236]. This material is too expensive for cost-effective production. Cheaper graphite with more impurities requires a diffusion barrier between the active part of the cell and the substrate. Graphite is conductive and therefore requires an insulating barrier layer to permit the fabrication of an integrated series connection.

Glass

Common window glass meets the cost goal but softens above 450°C. It is not suited to the HTS approach. Special glasses with a high-temperature resistance such as Corning 1737 glass and a glass that consists of 64%wt SiO_2, 21%wt Al_2O_3, 7%wt BaO, and 8%wt MgO [238] have been developed. The latter glass has a particularly high transformation temperature of 822 °C and a softening temperature of 874 °C. While we successfully tested this glass to withstand thermal diffusion at 870°C [109, 197] for half an hour, the glass tends to crystallize during high-temperature processing. The sample contamination is significant without using a diffusion barrier. Since this high-temperature glass is not floatable, it probably cannot be produced at low cost.

Quartz glass has the required temperature stability and can be produced by pulling glass ribbons from a melt. While low-cost production is, in principle, possible for quartz, the thermal expansion is not matched to Si. Although glass would be the ideal substrate for efficient light trapping, a low-cost glass for the HTS approach has not yet been developed.

Silicon-infiltrated silicon carbide

Recently, a Si-infiltrated SiC-ceramic (SiSiC) was tested for solar cell applications [48, 240]. Various techniques for the fabrication of this type of ceramic are under investigation. A combination of the sintering of the Si-SiC green tape with the recrystallization of a Si film is particularly attractive since one high-temperature process is avoided [241].

Silicon

Silicon is ideally matched in thermal expansion. As a low-cost substrate for solar module production the Si material has to be much cheaper than a Si wafer. Metallurgical grade Si fulfills the cost requirement but not the purity requirement. Again, diffusion barriers are required. The direct growth of Si ribbons by the silicon-sheet from powder (SSP) technique [242] is an alternative. Silicon powder is recrystallized to form a multicrystalline Si ribbon. Using highly doped Si substrates, electronic passivation of the back surface can be achieved [177].

From the various substrate choices listed in Table 4.1 mullite is a good candidate for thin-film Si solar modules since it is a low-cost material that permits efficient light trapping and integrated series connection. All other materials do not satisfy one of the requirements.

4.1.2 Active layer

Direct growth on substrate

Growth rates of up to 5 μm min^{-1} are feasible with high-temperature deposition of crystalline Si from chlorosilanes [235]. The active Si layer has a thickness of 15 to 40 μm. The layer deposition of crystalline Si at high temperatures onto non-Si substrates yields grain sizes of 1 to 5 μm [243, 244]. The efficiency of devices fabricated from such layers has, so far, been limited to below 6%. Processes conducted on oxidized Si wafers that have negligible impurity concentrations do not yield higher efficiencies. This indicates that the contaminants from the ceramic substrate are not yet a limiting factor in these devices [243]. Grain boundaries with recombination velocities in the range of 10^4 to 10^6 cm s^{-1} limit the efficiency.

Zone melt recrystallization

A substantial increase of the grain size from the micron range to several hundreds of microns is achieved by zone melt recrystallization (ZMR). An overview over this technique, as well as the description of a state of the art recrystallization apparatus, was

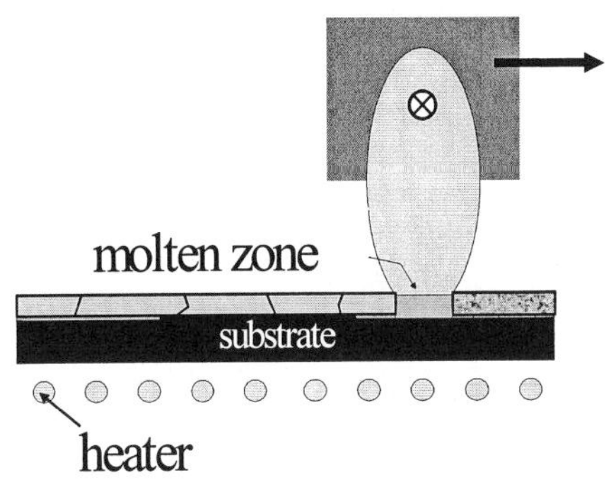

Figure 4.2. Schematic of a zone melt recrystallization apparatus (ZMR): a halogen lamp in one focus of an elliptical mirror melts a thin Si film.

Figure 4.3. Optical microscope image of ZMR silicon on a SiC-coated graphite substrate. The left part is recrystallized while the right part is not.

given by Hebling [245]. A schematic of a recrystallization apparatus is shown in Figure 4.2. A halogen lamp at one focus of an elliptical mirror melts a narrow region of the microcrystalline Si film that has been deposited onto a high-temperature stable substrate. The molten zone crosses the substrate by moving the lamp with a velocity of 1 to 30 mm min^{-1} [48, 237]. Thin films with a thickness of only 1 μm permit a high scanning speed of 180 mm min^{-1} [246]. The scanning speed is crucial for future high-capacity production. Auer et al. observed a proportionality between the thickness of the recrystallized Si layer and the average distance between small-angle grain boundaries [47]. Hence, a compromise between thin layers that permit fast crystallization and thick layers that yield large grains has to be found. The lateral temperature gradient close to the molten zone is reduced by heating the substrate with halogen lamps from underneath the substrate as shown in Figure 4.2.

During zone melt recrystallization a capping layer on the thin Si film hinders the formation of Si droplets. Silicon oxide films have been used for this purpose [244, 199]. After recrystallization, the capping layer is removed by wet chemical etching [244] or by plasma etching [236].

The optical microscope image of a ZMR-silicon on a SiC-coated graphite substrate is shown in Figure 4.3. Grain sizes in the recrystallized left part are up to 5 mm. The right part is not recrystallized. A grain enlargement into the millimeter range was also achieved with mullite substrates [237] and on SiSiC-ceramics [48].

Diffusion barrier

Several micron-thick diffusion barriers from CVD-deposited SiC [47] or plasma-deposited silicon oxide/silicon nitride/silicon oxide (ONO) [48, 240] are deposited onto the substrate prior to the deposition of the Si seed layer in order to reduce the interaction of the molten Si with the substrate. In addition, the diffusion barrier closes the pores of the graphite and ceramic substrates. Thus, cleaning agents cannot fill the pores of the substrate. Depending on the substrate material, such a diffusion barrier might deteriorate or improve the back surface reflectance. An ONO layer on highly reflective mullite reduces the diffusive reflectance [237], while a SiC layer on low-reflectivity graphite improves the back reflectance [47]. During the recrystallization process these layers are heated to the melting point of Si (1420°C), which was observed to lead to a deformation

Figure 4.4. Transmission electron micrograph of a SiC particle that extends beyond the seed layer. This particle causes dislocations that appear as dark lines in the epitaxial layer. (Photo: C. Twardawa, University of Erlangen-Nuremberg)

Figure 4.5. Transmission electron micrograph of a SiC particle that does not penetrate the seed layer. (Photo: C. Twardawa, University of Erlangen-Nuremberg)

of the ONO diffusion barrier. The deformation may even lead to local distortion of the barrier [48]. Cells with locally distorted barriers show markedly reduced performance. Using glow discharge mass spectroscopy (GDMS), it was verified that ONO layers reduce the impurity concentration (Fe, Ti, V, Cr, Mn, Ni, Co Cu) in the active device layer by a factor of about 10^3 relative to the seed layer. The diffusion barrier is hence a vital part of the HTS concept if low-cost and impure substrates are to be used.

We find that the recrystallization of thin Si seed layers on top of a rough SiC-coated graphite causes problems. A high defect concentration in the epitaxial layer is induced by SiC-particles that penetrate the surface of the seed layer as shown in Figure 4.4. Those SiC particles that are underneath the surface of the seed layer do not generate defects (see Figure 4.5). As a rule of thumb the seed layer should be three times as thick as the roughness of the substrate. The interaction of the molten Si with the substrate material was also observed for mullite and SiAlON substrates [244].

4.1.3 Devices

The FORSOL cell on a graphite substrate

An outstanding device result in the area of HTS cells on foreign substrates was reported by the Bavarian cooperative research activity FORSOL [247]. A layer system developed under the leadership of ASE GmbH achieved a record efficiency of 11% for a 1 cm^2 cell using a graphite substrate [47, 236]. A schematic of the device is shown in Figure 4.6. The cells were fabricated by wet chemical processing [47] and by dry plasma processing [236]. Dry processing is advantageous when using porous substrates, such as ceramics or graphite: the accumulation of chemicals in the pores of the substrate, as well as time-consuming drying procedures, are avoided.

Device process: The key processing steps are: i) encapsulation of the graphite with conducting SiC (50 μm); ii) chemical vapor deposition (CVD) of p^+-type μc-Si (40 μm) with a boron concentration of 5×10^{18} cm^{-3} to 2×10^{19} cm^{-3} and deposition of a SiO_x capping layer (2μm); iii) recrystallization at a speed of 1 mm s^{-1} and removal of the capping layer; iv) CVD of a 15 to 30 μm-thick p-type Si layer at 1100°C; v) phosphorus diffu-

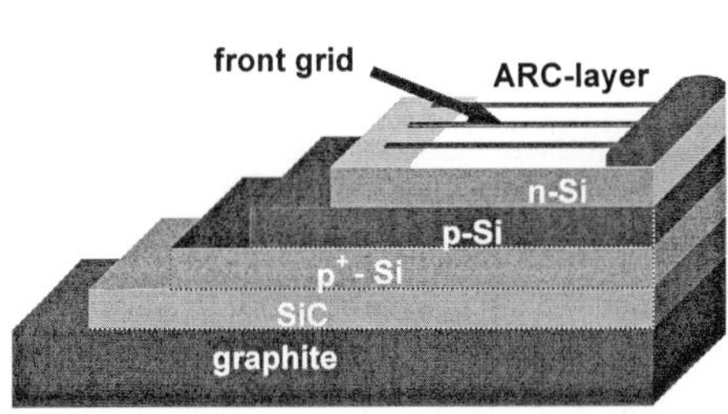

Figure 4.6. Schematic of the thin-film Si solar cell on SiC-coated graphite as developed in the FORSOL research cooperative.

Figure 4.7. Photograph of a thin-film Si cell fabricated on an SiC-coated graphite substrate. The cell area is 1 cm^2.

sion at 820 to 850°C; vi) removal of the phosphorus glass and application of contacts; vii) remote plasma hydrogen passivation; and viii) deposition of a single- or double-layer antireflection coating.

Device results: The resulting device parameters are an open-circuit voltage 570 mV, a short-circuit current density of 25.6 mA cm^{-2}, and a confirmed cell efficiency of 11.0% for the best dry-processed cell. Figure 4.7 shows a phorograph of a photolitographically and dry-processed thin-film Si cell on graphite. A photograph of the 11%-efficient cell is given in Ref. [248]. A voltage U_{oc} = 560 mV, a short-circuit current density of 24.2 mA cm^{-2}, and an unconfirmed cell efficiency of 10.2% were achieved for the best wet-processed cell. Dry processing gives slightly better results. Both process types were, however, applied to Si film material from different epitaxy runs. A direct comparison of dry and wet processing on identical thin-film material is, unfortunately, still lacking. The efficiency of dry and wet chemical processed cells from monocrystalline Si wafers differs by less than 5% relative [249, 250], thus indicating the capability of dry processing to replace wet processing.

Device characterization: We analyze the opto-electronic properties of the 10.2%-efficient thin-film cell. The layer system is 40 µm SiC/30 µm p^+-type seed Si/ 21 µm p-type active Si. The diffused P-doped emitter has a depth of 0.4 µm. Grid contacts are applied by evaporating Ti/Pd/Ag contacts that shadow 6% of the emitter surface. After hydrogen passivation, a double-layer silicon nitride ARC is deposited. The cell size is 1 cm^2. For further details see Ref. [251].

The measured hemispherical reflectance *R* and the measured internal quantum efficiency *IQE* are shown as symbols in Figure 4.8. The double-layer ARC can be improved by adjusting the dielectric layer thickness values to yield a minimum of the reflectance at wavelength 600 nm rather than exhibiting a maximum. The IQE analysis reveals a low carrier collection efficiency in the emitter due to a high surface doping concentration, which causes a so-called dead layer. Auger recombination reduces the diffusion lengths to values smaller than the thickness of the emitter. The *IQE* data contain quantitative information on the minority carrier diffusion length L_b in the base and information on the back surface recombination velocity S_b. For thin-film cells it is, however, important to

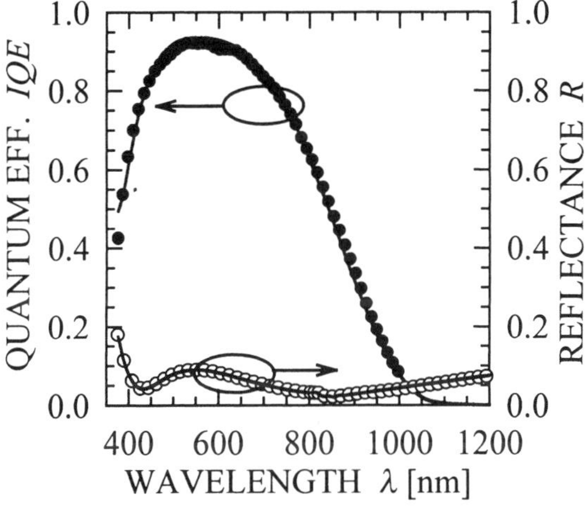

Figure 4.8. Lines: fit of internal quantum efficiency and hemispherical reflectance of the 10.2% wet-processed thin-film cell on SiC-coated graphite substrates. Symbols: measured data.

include the emitter properties in the model when analyzing the *IQE* spectra.

The one-dimensional transport model that we developed for the analysis of thin-film cells [25] is given in Chapter 3 on p. 73. We find an optimum fit to the experimental data for a base diffusion length $L_b = 21$ μm and for a back surface recombination velocity $S_b = 9\times10^3$ cm s^{-1}. This pair of recombination parameters is, however, not uniquely determined by the fit. Using parameter confidence plots (see, for example, Figure 3.15), we deduce a range of parameters that explain the measurement. This range is $L_b > 17$ μm and $S_b < 3\times10^4$ cm s^{-1}.

The effective quantum efficiency diffusion length L_Q is 23 μm (see Figure 3.2 on p. 56 for the definition of L_Q). The length L_Q controls the recombination in the base of polycrystalline solar cells under forward injection in the dark. In contrast, the collection length L_C is related to the short-circuit current. We find $L_C = 13$ μm from the *IQE* data shown in Figure 4.8.

Multiplication of the external quantum efficiency by the solar spectrum and integration yields a short-circuit current density of 24.4 mA cm^{-2}, which is similar to the measured short-circuit current density of 24.2 mA cm^{-2}. Short circuit current mappings (LBIC) of this cell reveal an average distance of $G = 300$ μm for the recombination-active grain boundaries and a typical grain boundary recombination velocity of 10^5 cm s^{-1} [199]. Within the grains, a density of 10^4 to 10^5 cm^{-2} dark spots in the LBIC mapping originate from enhanced recombination at dislocations. A dislocation density of 10^5 cm^{-2} corresponds to an average dislocation distance of 30 μm, which is of the same order of magnitude as the lower bound of the bulk diffusion length $L_b = 17$ μm. Hydrogen passivation reduces grain boundary recombination and recombination at the dislocations in the grains. Both effects contribute a similar amount to the enhancement of the open-circuit voltage that we observe after hydrogen passivation [199].

Devices on mullite substrates

Using an approach similar to the FORSOL cell, a Si cell fabricated by recrystallization of a 10 μm-thick Si film on a mullite substrate with a subsequent deposition of a 49 μm-thick absorber layer was reported recently to have an efficiency of 8.2% and a short-circuit current density of 23.8 mA cm^{-2} [237]. Light trapping induced by the diffuse reflection of the mullite contributed to the short-circuit current density. No barrier layer was used.

4.2 Low-temperature substrate (LTS) approach

Cells on low-temperature substrates (LTS) are deposited at temperatures up to 550°C. Low-cost substrates such as window glass [252], metal [46], or plastics are being used. Using a low deposition temperature T_d keeps the interaction of the substrate with the active device layer small. Thereby, the out-diffusion of contaminants from the substrate into the devices is reduced when compared to high-temperature depositions.

4.2.1 Substrates

At a deposition temperature T_d = 200°C it is possible to use floated Na-containing low-cost window glass [253], while Na-free glass is required for temperatures above 550°C [254]. Floated low-cost glass is planar. Using a rough substrate introduces light trapping and enhances the optical absorption. Chemo-mechanical texturing of the glass substrates has been investigated at the Sanyo Corporation. [255].

The group at the University of Neuchâtel uses glass substrates coated with a textured SnO_2 layer. Efficiency records have been achieved using F-doped tin oxide film fabricated by the Asahi Glass Corporation by atmospheric-pressure CVD in an off-line process [253]. This substrate is currently not available with large areas (m^2) at low cost.

A promising substrate was developed at the "Forschungszentrum-Jülich" [257, 258, 256], where they use sputtered ZnO layers that are Al-doped. A surface texture, as shown in Figure 4.9, is introduced by wet chemical etching [259]. Coating planar glass with a textured SnO_2 or textured ZnO layer is an alternative to using textured glasses. Since transparent and conductive oxides are required for the contacts anyway, texturing is achieved with little extra processing.

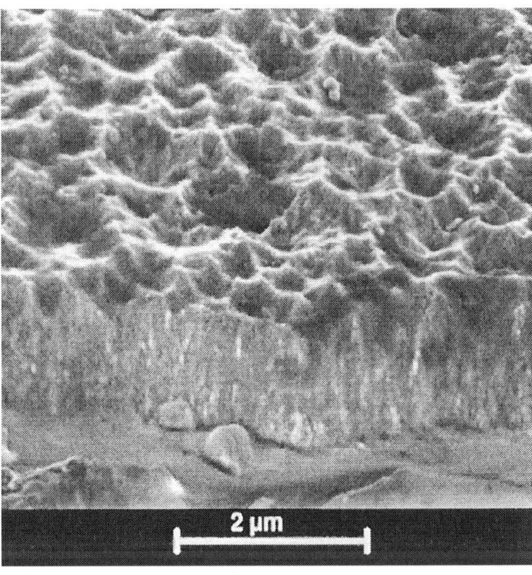

Figure 4.9. SEM micrograph of a HF-etched sputtered Al-doped ZnO film. Reproduced from Ref. [256], Copyright 1999, with permission from Springer-Verlag.

Superstrate and substrate device configurations are possible. In the superstrate configuration the substrate has to be transparent. In the substrate configuration a glass/Ag/TCO stack is used as a back reflector.

4.2.2 Active layer

The electronic quality of the active layer decides the performance of the solar cell. Various approaches to depositing the active layer at low temperatures are under investigation.

Solid-phase crystallization

The Sanyo group deposits amorphous Si from hydrogen-diluted SiH_4 at temperatures of around 600°C in a plasma reactor [46]. A solid-phase crystallization process (SPC) [260] at 600°C for 10 h transforms the initially amorphous layers into microcrystalline Si (μc-Si). The nucleation of crystallites is controlled by the partial doping technique [255]. The amorphous Si is highly P-doped at the bottom of the film. Since P-doping enhances the nucleation rate, all the crystallization nuclei are located at the bottom of the film. Columnar grain growth from the bottom to the top of the film is thus achieved. Grain sizes are typically 1.5 μm [260, 46].

Plasma-enhanced chemical vapor deposition

The Neuchâtel group deposits device grade microcrystalline Si layers at 220°C on a soda-lime glass by plasma-enhanced chemical vapor deposition [261, 253]. A high hydrogen dilution leads to the growth of μc-Si instead of amorphous Si (a-Si). The mechanisms that cause the growth of μc-Si at high H_2 dilution are still under discussion. Etching of strained bonds by atomic hydrogen was suggested [262]. The enhancement of the surface diffusion length of growth precursors due to a saturation of the growth surface with hydrogen is another possible mechanism [263]. Further mechanisms are discussed in Ref. [264].

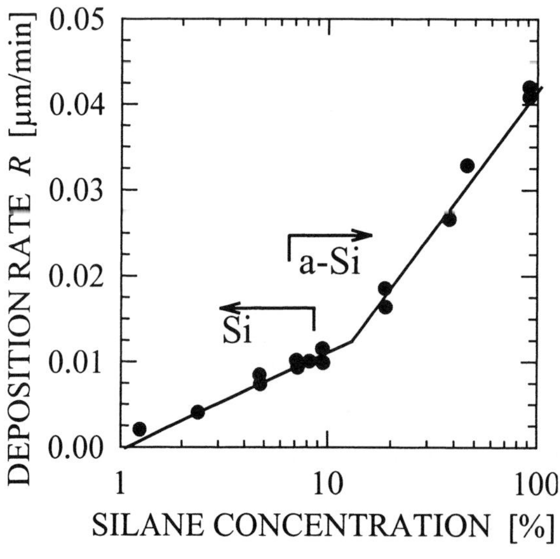

Figure 4.10. Deposition rate as a function of silane concentration $[SiH_4]/([SiH_4] + [H_2])$ in hydrogen dilution. Data from Ref. [265].

Figure 4.10 shows the deposition rate as a function of the silane concentration $[SiH_4]/([SiH_4] + [H_2])$ for Si deposition at 225°C in a very high-frequency (VHF) glow discharge at 70 MHz [265]. The knee at about 7% silane concentration indicates a change in the growth mechanism. Films at silane concentrations < 7% are microcrystalline, while those at concentrations > 7% are amorphous. A high hydrogen dilution produces μc-Si and reduces the growth rate *R* to values below 10^{-2} μm min^{-1}. The growth rate enhances with increasing plasma power. High plasma power results, unfortunately, in films of poor electronic quality. Hence, 10^{-2} μm min^{-1} is a typical growth rate for μc-Si device layers. A 2 μm-thick device thus grows in 3 h, pumping cycles not included.

Very high-frequency plasma excitation

The advantage of using VHF deposition at frequencies > 50 MHz in contrast to conventional 13.6 MHz deposition is a reduction in the Si ion energy. Dutta et al. estimated that, at frequencies above 50 MHz, the peak of the Si ion energy distribution falls below the threshold for defect formation, which lies at 16 eV [266]. In addition, the flux of low-energy Si ions onto the growth surface is increased when using VHF plasma excitation [267]. Figure 4.11 shows an SEM micrograph of the so-called micromorph Si solar cell [268]. The μc-Si is about 3 μm thick and is deposited onto an amorphous solar cell, thus generating a tandem cell. The μc-Si films exhibit a preferential (110) orientation. Depending on the deposition parameters, even a full (110) orientation may be achieved [269]. X-ray diffraction (XRD) analysis reveals a typical grain size of 20 nm with a columnar structure. Transmission electron microscopy (TEM) analysis of the grain size is apparently not available but is urgently required to clarify the structure of the grain boundaries.

Optical properties of μc-Si

An analysis of the optical absorption of μc-Si was performed by Poruba et al., who demonstrated that the natural surface roughness induces an almost Lambertian light trapping [92]. The scattered light intensity from a Lambertian interface is proportional to $\cos(\varphi)$, where φ denotes the scattering angle. This angle dependence stems from a re-

Figure 4.11. Scanning electron microscope micrograph (SEM) of a microcrystalline Si/amorphous Si tandem cell, deposited on an SnO_2-coated substrate (type U). Reproduced from Ref. [354], Copyright 1999, with permission from Springer-Verlag.

duction of the apparent cell area when it is projected in the direction of light scattering. Figure 4.12 compares this ideal Lambertian behavior with the measurement of external light scattering at the surface of a μc-Si. Atomic force microscopy reveals a root-mean-square surface roughness of 15 to 50 nm [92, 269]. Due to the reduced wavelength of light in Si, internal scattering is expected to be stronger than external scattering.

Polishing the surface avoids surface scattering. However, there is still the possibility of internal light scattering, e.g. at inclusions of a-Si in the μc-matrix. Using the constant photocurrent method (CPM), it is possible to determine the amount of internal light scattering by changing the spacing of the electrodes [270]. Measurements with widely spaced electrodes show a current enhancement due to scattering, while narrowly spaced electrodes do not. In this way it was demonstrated that internal light scattering is negligible in device quality μc-Si [92].

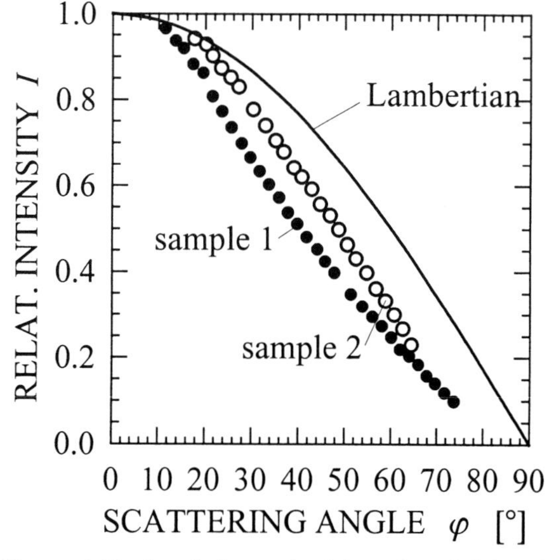

Figure 4.12. Angularly resolved intensity of light with wavelength 632 nm that is reflected from the surface of a 2 μm-thick μc-Si layer. Data from Ref. [92].

Figure 4.13 shows the optical absorption coefficient α_s as determined from CPM data in comparison to the data, for crystalline Si [22]. In the near-infrared at wavelength around 900 nm the optical absorption coefficient is identical to that of c-Si. In the sub-bandgap regime, μc-Si has an enhanced optical absorption due to gap states that are probably located at the grain boundaries. For a small grain diameter of about 20 nm, the atoms at the grain boundary make up 5% of the total number of Si atoms. The dangling bonds at the grain boundary need to be electronically passivated by hydrogen bonds. Hydrogen is incorporated during the film growth at a concentration that is typically

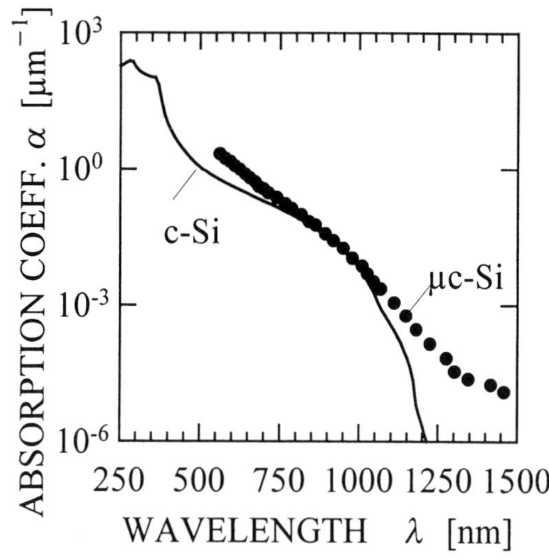

Figure 4.13. Absorption coefficient α as determined from CPM data for a μc-Si sample with a root-mean-square roughness of 21 nm [92] in comparison to c-Si.

5% in device quality material [265]. Non-passivated dangling bonds and strained bonds account for a sub-bandgap absorption that is qualitatively similar to that measured in a-Si. For μc-Si, a dependence of the mobility-lifetime $\mu\tau$-product on the position of the equilibrium Fermi-level was found, which is in agreement with the defect pool model developed for amorphous Si [271]. Hence, the structure of the grain boundary could be similar to the structure of amorphous Si. In the visible range, the absorption coefficient of μc-Si is significantly larger than that of crystalline Si. Again, the origin might be the quasi-amorphous Si layers at the grain boundaries. Amorphous Si has a larger absorption coefficient than c-Si since the lack of long-range periodicity permits direct optical transitions without phonon participation.

Active layer of the STAR cell

A research team at the Kaneka Corp. grows μc-Si layers at 550°C on glass substrates [272]. The only reference to the CVD technique is one to a rather unusual technique described in Ref. [273]. Here, the film growth occurs by repeatedly growing 2 nm of amorphous Si that is then crystallized in a flux of low-energy hydrogen atoms that emerge from a remote electron cyclotron resonance (ECR) source. The deposition rate of the amorphous layer is 6×10^{-3} μm min^{-1}. Including the annealing time, the effective deposition rate is even smaller. A TEM micrograph of an early cell is shown in Figure 4.14 [274]. The Si surface has a texture with a typical scale length of 0.1 μm that naturally evolves during film growth. The grain size in the p^+-type laser-crystallized seed is 300 nm. This layer functions as an electrical back contact. While the seed layer is preferentially (111)-oriented, the μc-Si layer is preferentially (110)-oriented. The μc layer does not grow epitaxially on the seed. A preferential (110) orientation was also observed in the μc-Si layers of the Neuchâtel group. It was argued [275] that this preferential (110) orientation is vital to form electrically inactive grain boundaries. In fact, [100]-tilt type grain boundaries are electrically inactive since they have no dangling bonds [276]. We

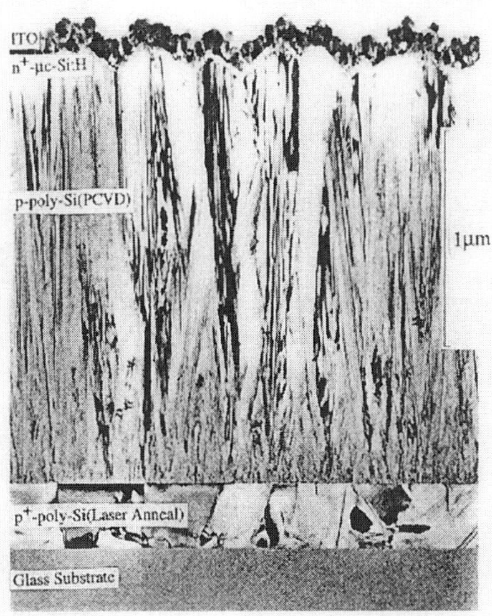

Figure 4.14. Cross-sectional TEM micrograph of a microcrystalline solar cell on a glass substrate. Figure from Ref. [274].

expect the typical grain size to be of the same order of magnitude as the surface texture, hence $G \approx 0.1$ μm. This value is also indicated by the TEM micrograph shown in Figure 4.14. Hence the grain size to thickness ratio is approximately $G/W \approx 0.1$, which is larger than for the Neuchâtel material by a factor of 5. The larger deposition temperature enhances the desorption of ad-atoms from the amorphous substrate, thereby reducing the probability of the formation of stable nucleation centers. Thus, the grain size increases. Please note that the Kaneka research team did not publish what type of Si deposition they use for the above 10%-efficient cells. It is therefore well possible that the grain size of these cells is different from 0.1 μm, the value assumed in this work.

4.2.3 Devices

Devices from solid-phase crystallization

The Sanyo research team fabricates thin-film c-Si solar cells by solid-phase crystallization of amorphous Si on metal substrates. The film sequence is metal/n^+-type Si/ 10 μm n-type poly-Si/10 nm i-type a-Si/70 nm p-type a-Si/ITO. An efficiency of 9.2% was achieved with a 1 cm^2 cell [277, 46]. The short-circuit current density is 25.0 mA cm^{-2}, due to light trapping induced by a textured glass substrate. The open-circuit voltage is 553 mV. The apparent effective diffusion length of a similarly produced cell is $L_Q = 24$ μm, as deduced from surface photovoltage measurements [277].

Devices from plasma-enhanced deposition

One of the μc-Si cells from the Neuchâtel group has a layer sequence glass/textured SnO_2/p^+-type Si/about 3 μm i-type Si/n^+-type Si/ITO. The efficiency for a single-junction and purely microcrystalline cell is 8.5% with an open-circuit voltage of 531 mV and a short-circuit current density of 22.9 mA cm^{-2} under AM1.5G illumination [253].

The group at "Forschungszentrum Jülich" achieved an efficiency of 7% with a μc-Si film just 1 μm in thickness [252]. The short-circuit current density is 17 mA cm^{-2} and the open-circuit voltage is 510 mV. A 3.5 μm-thick cell on the above-mentioned textured ZnO substrate achieves 7.5% efficiency, a short-circuit current density of 25.1 mA cm^{-2}, and an open-circuit voltage of 453 mV. The layer sequence of the latter cell is glass/Ag/textured ZnO/50 nm n-type μSi/3.5 μm i-type/20 nm p-type Si/80 nm ZnO. Both films were deposited with a VHF glow discharge technique in hydrogen-diluted silane. The growth rate is 7 nm min^{-1}.

STAR cell

Higher efficiencies were achieved with the so-called STAR (surface texture with enhanced absorption and back reflector) cell developed at the Kaneka Corporation [254]. The layer system depicted in Figure 4.15 consists of a 300 nm-thick n^+-type microcrystalline layer that is P-doped to a concentration of 10^{20} cm^{-3}. On top of the n^+-type layer a nominally intrinsic microcrystalline Si layer of 1 to 5 μm thickness is deposited. Secondary ion mass spectroscopy (SIMS) reveals, however, a P concentration of 10^{16} cm^{-3} in the absorber layer [278]. Thus, the transport is diffusion-dominated rather than drift-dominated. A 30 nm-thick p^+-type Si layer that is B-doped to a concentration of 10^{20} cm^{-3} finishes the Si deposition. The lateral conductivity of the p^+-type Si is enhanced by an 80 nm-thick indium tin oxide (ITO) and an Ag front metal grid.

For amorphous Si solar cells it is advantageous to have the light passing the p^+-type layer first, because then the holes have, on average, a shorter distance to the p^+-type

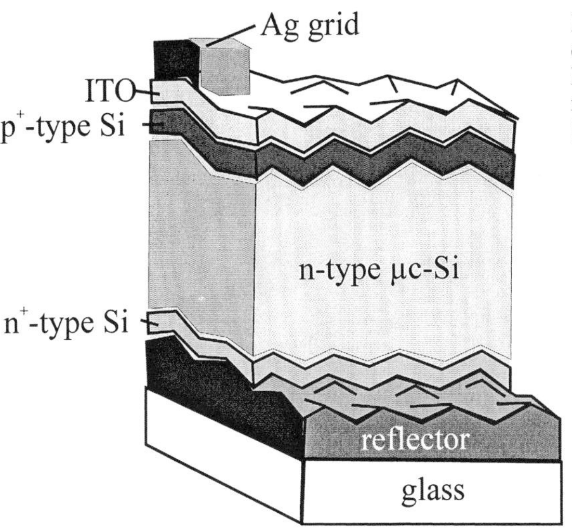

Figure 4.15. Schematic of a microcrystalline STAR cell developed at the Kaneka Corporation. with a total thickness of 1 to 5 μm. Redrawn from Ref. [254].

layer than the electrons to the n^+-type layer. Since the recombination of holes in the intrinsic absorber layer dominates [279], this order of layers improves the cell. If we assume that recombination at the a-Si-like grain boundaries dominates in the STAR cell, a n^+-type layer is advantageous for the same reason as in a-Si cells.

Device results: A solar cell with 10.1% efficiency and a total thickness of 2 μm was recently fabricated with a short-circuit current density of 24.35 mA cm^{-2} and an open-circuit voltage of 539 mV [254]. For a cell of similar quality but a device thickness of 1.5 μm the internal quantum efficiency is shown in Figure 4.16.

Modeling the device optics: It is vital to consider light trapping when analyzing the internal quantum efficiency of the STAR cell. The optical model we developed [25] is explained on p. 76. Following Yamamoto [254], we assume that the microcrystalline cell has the optical constants of monocrystalline Si; this has not yet been checked experimentally. The measured reflectance in Figure 4.16 shows Fresnel interference fringes caused by light reflection at the front and at the back of the Si film. From these fringes we deduce a film thickness of 1.5 μm. Adjusting the back surface roughness, the back surface reflectance, and

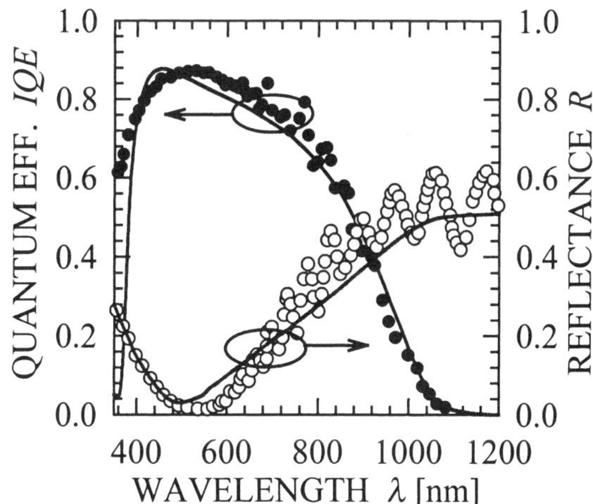

Figure 4.16. Symbols: Measured internal quantum efficiency IQE and hemispherical reflectance R of a 1.5 μm-thick Si cell on glass. Data from Fig. 10 of Ref. [254]. Lines: Fit of quantum efficiency and reflectance with a one-dimensional transport model.

the internal front surface reflectance permits us to fit the experimental reflectance spectrum. The quality of the fit is poor because the optical parameters of the indium tin oxide (ITO) coating have not been modeled. Losses in this layer are approximately accounted for by using a reduced back reflectance value for the calculation. From the optical model we deduce an average path length for the weakly absorbed light of 19 μm (see Eq. (3.73) on p. 78). Hence, the light travels a distance through the cell that is, on average, 13 times greater than the film thickness W_{eff}.

Modeling the device quantum efficiency: We consider diffusive transport in a three-layer system consisting of a 30 nm-thick p^+-type emitter and a 1.5 μm-thick base with n-type doping. To simulate the *IQE* data with the same model that we applied to the FORSOL cell on p. 97, we adjust the recombination parameters and pay particular attention to the base diffusion length L_b and the back surface recombination velocity S_b. The optimum fit of the experimental data shown in Figure 4.16 is obtained for $L_b = 3$ μm. An inspection of the sensitivity of the shape of the IQE curve to the fit parameters shows that the conclusion we can draw is that, $L_b > W_{eff} = 1.5$ μm, values $L_b < W_{eff} = 1.5$ μm lead to significantly worse fits, while values $L_b > W_{eff}$ yield fits of the quality shown in Figure 4.16. The effective collection diffusion length L_C is 16 times the device thickness. See p. 56 for the definition of L_C. It is not possible to deduce the effective quantum efficiency diffusion length L_Q from the data shown in Figure 4.16.

Two-dimensional modeling of the current-voltage curve: The investigation of the current-voltage (*I-U*) curve of the 10.1%-efficient cell [254] yields additional insight into the injection dependence of the recombination parameters. For the semi-logarithmic plot of the current-voltage curve in Figure 4.17 we subtracted the short-circuit current density of 24.3 mA cm^{-2} from the measured current. This type of data representation makes the different voltage dependence of recombination in the space charge region and in the neutral region of the base more apparent than a linear plot.

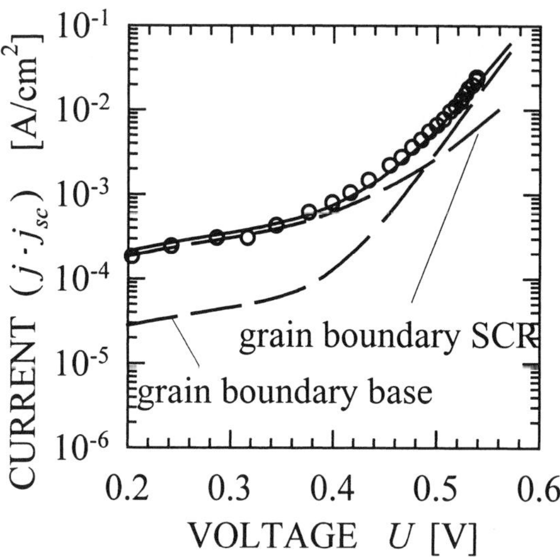

Figure 4.17. Illuminated current-voltage curve of the 10.1%-efficient and 2 μm-thick microcrystalline thin-film Si STAR cell [254] (symbols). Short circuit current j_{sc} = 24.3 mA cm^{-2} subtracted. Capture cross-sections $\sigma_n = \sigma_p = 10^{-15}$ cm^2 and interface state density $N_t = 6\times10^{10}$ cm^{-2} eV^{-1} (lines).

The model we apply to simulate this *I-U*-curve is sketched in Figure 4.18. We neglect the emitter because the thin and highly doped emitter of the STAR cell makes only a small contribution to the diode saturation current density. The carrier-depleted space charge region of thickness W_{SCR} is part of the lightly doped base. We distinguish Shockley-Read-Hall (SRH) recombination in the volume of the space charge region and at the grain boundaries. The grains have a cross-

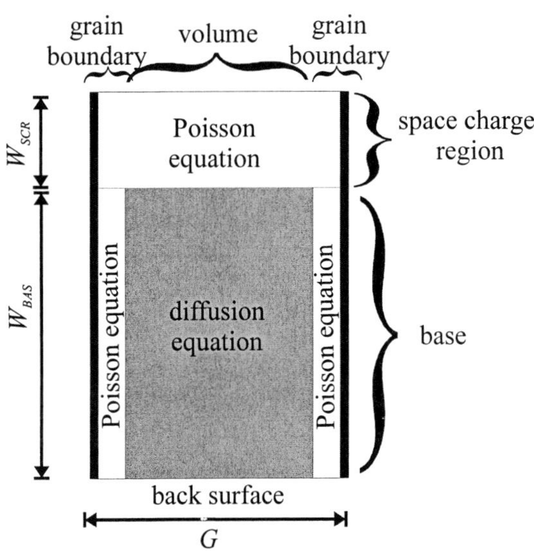

Figure 4.18. Schematic representation of a simplified solar cell model. The lightly doped cell region consists of the space charge region (SCR) and the neutral base. The SCR is fully depleted and the Poisson equation controls the electrostatic potential. Carrier diffusion controls the transport in the neutral base. In addition to recombination in the volume of the SCR and the base, surface recombination at the grain boundaries and at the back of the cell is considered.

sectional size of $G = 0.1$ µm. In Appendix B on p. 237 we describe this analytical model, which calculates the electrostatic potential in the SCR from a solution of the Poisson equation. Our model accounts for the charges in the grain boundary states.

The recombination in the base region of thickness W_{BAS} has two contributions, one from the neutral volume of the grain and one from the grain boundaries. Our analytical and transport model for base solves the two-dimensional diffusion equation for the minority carriers in a neutral base. This model is treated in Appendix B on p. 233.

The grain boundary recombination also has two contributions. One from the defects at the grain boundary and the other from defects in the space charge regions that surround the grain boundary. We characterize the grain boundary by an energy-independent defect density and energy-independent capture cross-sections for electrons and holes. The neutral level is assumed to be near the midgap. Using the extended SRH recombination model we calculate the grain boundary recombination velocity S_{grb}, as explained on p. 228.

The simulation in Figure 4.17 assumes that the carriers recombine solely at the grain boundaries. Capture cross-sections $\sigma_n = \sigma_p = 10^{-15}$ cm^2 and an interface state density $N_t = 6\times10^{10}$ cm^{-2} eV^{-1} yield a reasonable fit to the experimental data. In the base, these parameters correspond to a grain boundary surface recombination velocity $v_{th}\,\sigma_n\,N_t\,E_g/2 = 300$ cm s^{-1} at the surface of the columnar grains which is only slightly enhanced to 450 cm s^{-1} by band banding near the grain boundary. A fit of similar quality is also possible with no grain boundary recombination and an intra-grain diffusion length of $L_b = 7$ µm. Whatever the relative contributions of volume and grain boundary recombination are: we conclude that the intra-grain minority carrier diffusion length is greater than 7 µm and that the grain boundaries are well passivated with a recombination velocity smaller than 450 cm s^{-1}.

4.3 Layer transfer process (LTP) approach

Layer transfer processes (LTP) use a special surface conditioning of the growth substrate that permits the transfer of the device layer from a re-usable growth substrate to a low-cost device carrier. The deposition of dielectric layers [280], ion implantation [281], and porous Si formation [52, 53] are under investigation for surface conditioning. Using a monocrystalline Si wafer as growth substrate permits us to fabricate monocrystalline cells by homo-epitaxy. The layer transfer process is schematically sketched in Figure 1.3 on p. 10.

Layer transfer with GaAs

Layer transfer for solar cell fabrication was invented by McClelland et al. to reduce the fabrication costs of GaAs cells, made of a particularly costly photovoltaic material [50]. With the CLEFT (cleavage of lateral epitaxial films for transfer) process, McClelland et al. grew single-crystalline GaAs films by vapor phase epitaxy on re-usable GaAs substrates. A carbonized photoresist mask with stripe openings is deposited onto the GaAs wafer. The openings function as a seed and lateral overgrowth produces a continuous film that is a single crystal. The film surface is then bonded to a glass and cleaved from the growth substrate. The material quality of the film was found to be comparable to films grown without a mask. A 17%-efficient GaAs cell with a thickness of 10 μm [282] and the fourfold use of the substrate were demonstrated [50]. This early work suggested applying layer transfer processes to other semiconductor materials also.

Layer transfer with Si

Landis proposed the growth of an epitaxial Si film on a textured Si wafer covered with a lattice matched salt [283]; however, the feasibility of this process has never been verified experimentally. More than a decade after the introduction of the CLEFT process a number of Si layer transfer processes are now being developed. We describe these processes in historical order. Strengths and challenges of the various layer transfer processes are briefly discussed on p. 118.

4.3.1 Mitsubishi's VEST process

Deguchi et al. fabricated a Si solar cell by layer transfer using the VEST (via hole etching for the separation of thin films) process that is currently being investigated at Mitsubishi [285, 286]. As shown in Figure 4.19, the process starts with an oxidized monocrystalline Si wafer. The oxidized wafer serves as substrate for the fabrication of 1 to 3 μm-thick and large-grained polycrystalline Si film by chemical vapor deposition (CVD) and zone-melt recrystallization (ZMR). The recrystallized layer is then thickened by CVD and randomly textured by KOH etching. Wet chemical etching through via-holes of 100 to 300 μm diameter and to a distance of 1 to 2 mm etches the SiO_2 layer and detaches the device layer. The via-holes are also used to form an emitter-wrap-through structure [287] that reduces the grid shadowing, simplifies the series interconnection [246], and enhances carrier collection [288]. The use of carrier substrates is not reported by Mitsubishi.

Multicrystalline cells about 80 μm in thickness achieve efficiencies of 16% on areas of 100 cm^2 [289] and 13% on 924 cm^2 [246]. Thin-film cells with thickness values below 50 μm were not reported.

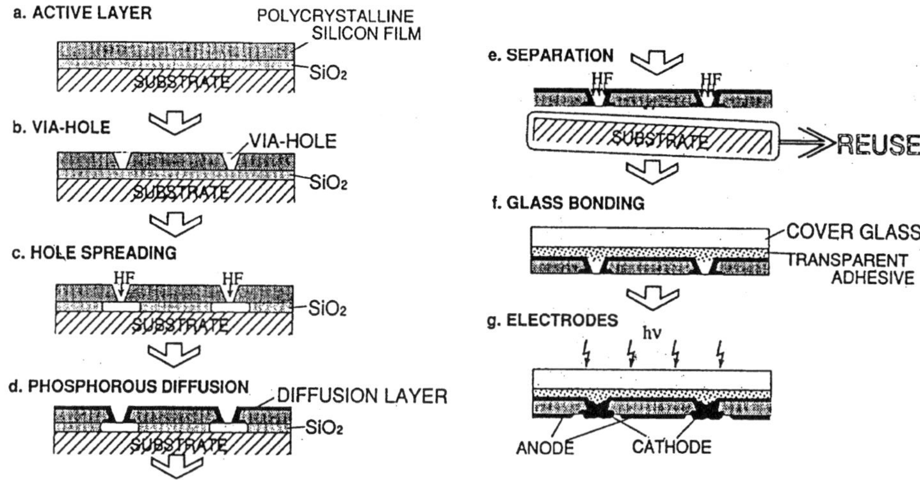

Figure 4.19. VEST process investigated by Mitsubishi. Figure from Ref. [284]

4.3.2 Canon's ELTRAN process

The successful transfer of a monocrystalline Si film using porous Si for surface conditioning was first demonstrated by Yonehara et al. using the ELTRAN (epitaxial layer transfer) process developed at Canon [51]. The top 10 μm of a planar monocrystalline Si wafer are transformed into porous Si by anodic etching in aqueous hydrofluoric acid (surface conditioning). We describe the formation of porous Si on p. 122. The typical pore diameter is 20 to 50 nm.

The porous Si is then oxidized in order to stabilize it against re-organization during high-temperature processing. An HF dip removes the oxide at the outer surface. The etching solution does not penetrate into the depth of the porous Si and thus the oxide prevails at the inner surface.

Annealing the porous Si in hydrogen at temperatures around 1000°C closes the pores at the sample surface [290]. The driving force is a reduction of the surface energy due to reduction of the surface area and an enhancement of the fraction of the (100)-surface which is known to have the smallest surface energy among all surface orientations. We discuss the annealing behavior of porous Si on p. 124. A closed (100) surface, is an ideal starting condition for the CVD epitaxy process.

The epitaxial layer is then bonded to an oxidized carrier wafer and the substrate wafer is sacrificed by grinding it down to the porous Si layer [51]. The oxidized inner surface of the porous Si hardly re-organizes during H_2 annealing and epitaxy. The inner surface of the stabilized porous Si remains large. The porous Si is etched with a selectivity of 10^5 against bulk Si [291]. This high selectivity results in a planar surface of the transferred epitaxial Si layer that is further smoothed by annealing in hydrogen.

Silicon-on-insulator (SOI) wafers fabricated by ELTRAN are available commercially [292]. No solar cell results applying the ELTRAN process have been reported. The reuse of the Si substrate has been published only recently in a modified version of the ELTRAN process that we discuss along with the SCLIPS process [293] on p. 117.

4.3.3 SOITEC's SMART CUT process

The irradiation of GaP with sufficiently high fluences of high-energy protons was observed to cause "flaking". During annealing, the cracks within the stressed region below the implanted layer spread out and split the implanted layer off from the bulk [294].

The Smart Cut (SC) process introduced by Bruel [281, 295] is sketched in Figure 4.20. The "flaking" effect is utilized in a controlled manner to peel of a thin layer of Si from the Si bulk crystal. Again, the substrate is a monocrystalline Si wafer. However, for surface conditioning hydrogen ions are implanted into an oxidized Si wafer to a well-controlled depth using an ion energy of 70 to 200 keV. The Si wafer is then bonded to a carrier wafer. Heating the whole system to temperatures around 500 °C expands the implanted hydrogen and splits off the thin Si layer. The surface of the splitting layer is rough, as shown in the TEM

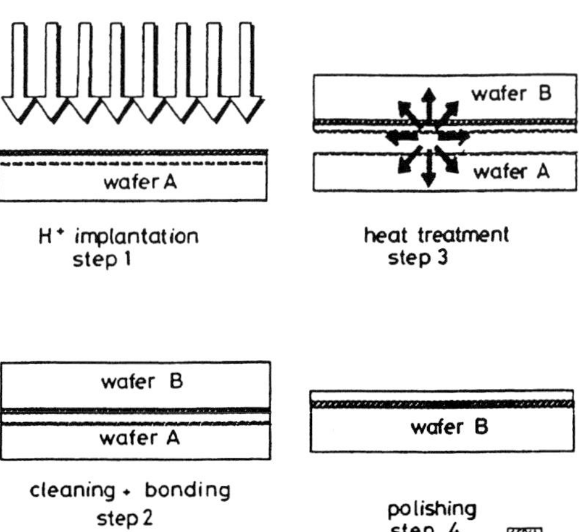

Figure 4.20. The Smart Cut process. Reprinted from Ref. [281].

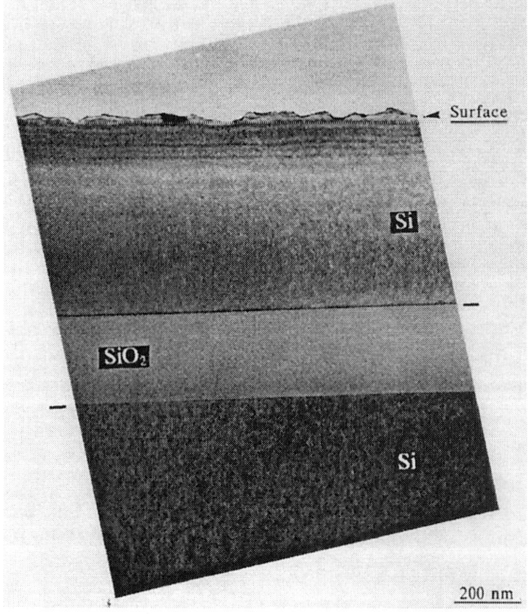

Figure 4.21. Transmission electron micrograph of an SOI sample from the Smart Cut process. Reproduced from Ref. [295], Copyright 1995, with permission from Elsevier Science.

micrograph in Figure 4.21, and requires polishing. Similarly to the ELTRAN process, the SC process aims at the SOI market. The SC process has not yet been applied to solar cell fabrication. The thickness of the transferred layer is defined by the penetration depth of the H^+ ions, which is typically less than 1 µm. This fact, and the necessity of ion implantation, limits the applicability of the SC process to PV.

4.3.4 Sony's SPS process

Tayanaka et al. introduced a layer transfer technique [52] that in this work we call the SPS process for sintered porous Si. The SPS process is now being developed at Sony [52, 296] and at the University of Stuttgart [297]. Similarly to ELTRAN, annealing in hydrogen closes the surface of the porous Si layer system prior to epitaxy. A porous multi-layer system with a low surface porosity and a high porosity at depth permits high-quality epitaxial growth and a subsequent detachment of the epitaxial layer. In contrast to the ELTRAN process, no stabilization of the porous Si by thermal oxidation is used. Thus, voids form in the volume of the low-porosity layer.

Figure 4.22 shows SEM micrographs of the epitaxial layer, the porous layer system, and the substrate. After annealing and epitaxy, the porous Si with low porosity forms voids with a typical diameter of 100 nm. The thin layer with high porosity dissolves. The Si atoms diffuse on the inner surface towards the region of lower porosity, leaving be-

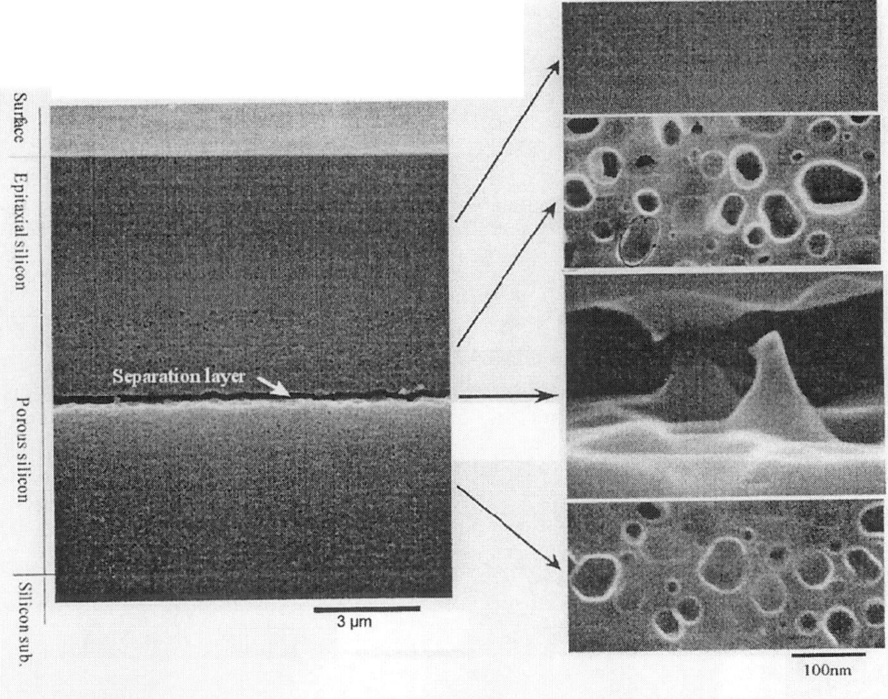

Figure 4.22. Scanning electron micrographs of the epitaxial silicon grown on a sintered multilayer of porous Si. The separation layer forms weak Si bridges that connect the epitaxial layer and the substrate. Reproduced from Ref. [296] with permission of the European Commission.

hind fragile Si bridges that connect the epitaxial layer and the substrate. Our Monte Carlo simulations demonstrated that the only driving force necessary to explain this reorganization is a reduction of surface energy [298]. The cell is then attached to a flexible plastic film. Plastic film and epitaxial layer are detached by applying tensile mechanical stress that breaks the weak Si bridges. A minority carrier lifetime corresponding to a diffusion length of 500 μm was claimed for the SPS process [299].

The sintered low-porosity layer forms the back of the solar cell and the voids in this layer introduce some degree of light trapping. Using photolithography and high-efficiency cell processing, encouraging solar cell results were recently reported with a 12 μm-thick monocrystalline Si cell achieving an efficiency of 12.5% [296] and a 24 μm-thick cell achieving an efficiency of 14.0% [300]. The latter value, achieved with the aid of an antireflection coating on the glass superstrate, is currently (March 2001) the highest efficiency of all thin-film cells, excluding cells from thinned float-zone Si wafers, e.g. Ref. [113]. An open-circuit voltage of 634 mV underlines the high electronic quality of the material. A full analysis of the *IQE* spectrum of this cell is not possible, since the reflectance spectrum is not published. The published internal quantum efficiency [300] reveals an effective quantum efficiency diffusion length $L_Q = 38$ μm and an effective collection diffusion length $L_C = 74$ μm. See Figure 3.2 on p. 56 for the definitions of L_Q and L_C.

4.3.5 ZAE Bayern's PSI process

The porous Si (PSI) process was introduced by Brendel in 1997 [53] and demonstrated for the first time the possibility of fabricating a textured monocrystalline Si film by starting from a textured Si substrate with a porous Si surface layer for detachment. At that time the work on the SPS process was published in a Sony research report in Japanese language only [52]. It was thus the PSI process that introduced the photovoltaic community to the concept of a re-usable Si substrates with a porous Si surface for transfer. The PSI process is under development at the Bavarian Center of Applied Energy Research (ZAE Bayern). Figure 5.1 on p. 121 illustrates the process sequence.

Device fabrication

We use chemical vapor deposition (CVD) on a monocrystalline Si substrate that is textured with randomly positioned upright pyramids. We thus avoid costly photolithography. The textured substrate receives a double layer of porous Si that has low porosity at the top and high porosity at the bottom. The separation layer forms as described for the SPS process. Figure 4.23 shows the SEM micrograph of a film as viewed obliquely on the side from which the substrate was removed. The thin film is textured with random inverted pyramids. This is a novel type of surface texture. The Chemical vapor deposition at 1100°C and at atmospheric pressure tends to smooth out the texture. For a 15 μm-thick film, the bottom-side in Figure 4.23 is almost planar.

Figure 4.24 shows a schematic of a 12.2%-efficient cell that we fabricated from this type of material. This cell has a phosphorus-diffused emitter. The Ti/Pd/Ag metal grid in front is evaporated through a shadow mask. An antireflection coating is evaporated onto the grid. After the cell has been glued to a glass carrier the epitaxial Si film is separated from the substrate. We remove the sintered porous Si by plasma etching. After etching, the effective film thickness is 15.5 ± 0.3 μm, as evaluated from an SEM image that shows a cross-section with a length of 185 μm. The Al reflector at the back is vacuum-evaporated.

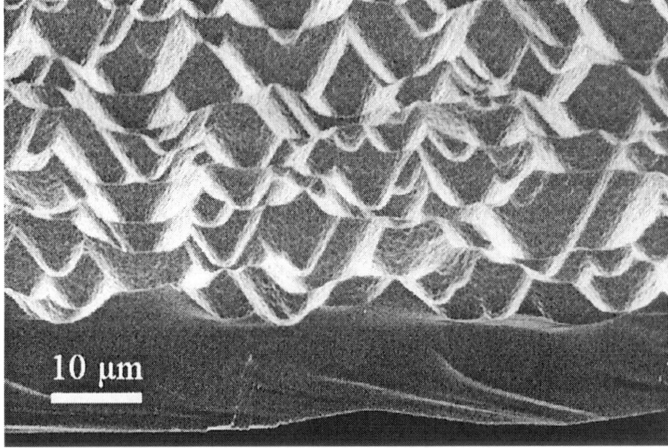

Figure 4.23. Scanning electron micrograph of a thin monocrystalline Si film fabricated by the PSI process on a Si substrate wafer with randomly positioned upright pyramids. We remove the sintered porous Si on the textured surface by plasma etching.

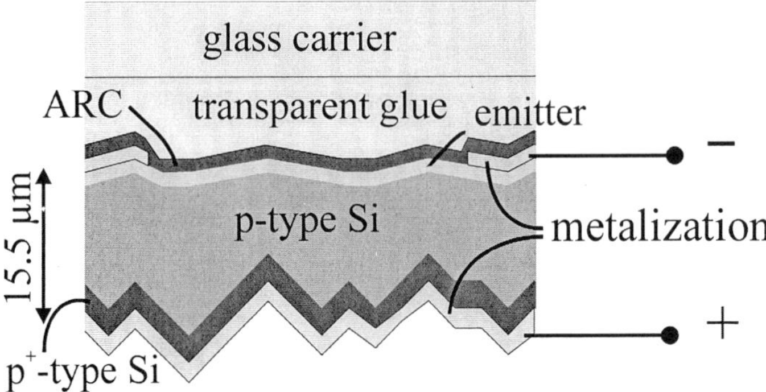

Figure 4.24. Schematic of the PSI cell with the random texture on the back. Compare Figure 4.23a for an SEM image. No photolithography is used in the cell process.

Device results

The current-voltage curve under AM1.5G illumination as measured at the calibration laboratory of FhG-ISE in Freiburg is shown in Figure 4.26. With a confirmed efficiency of 12.2% this is the highest thin-film efficiency among those devices that are processed without using any photolithography.

Device characterization

We analyze the internal quantum efficiency and the hemispherical reflectance to find the dominating loss mechanisms.

Figure 4.25 depicts the measured data (symbols) and the results of the simulations (lines). On p. 73 we explain the model that we use for the analysis. A fit of the reflectance spectrum and the internal quantum efficiency is only possible with a wavelength-dependent back surface reflection. We calculate the back reflectance of a randomly textured Si surface covered with an Al layer by Monte Carlo ray-tracing as described on p. 74. The minimum of the reflectance of the Si/Al interface at around 800 nm (see also Figure A.22 on p. 203) causes the kink at that wavelength in the internal quantum efficiency. The refleciance calculated by ray-tracing has to be multiplied by 0.85 in order to achieve the fit of the reflectance spectrum shown in Figure 4.25. This factor is the

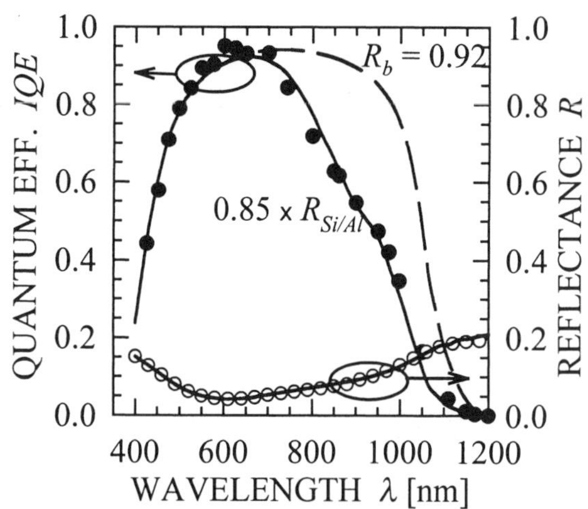

Figure 4.25. Filled circles: measured internal quantum efficiency of a 15.5 µm-thick PSI cell. Open circles: measured hemispherical reflectance. Solid lines: simulation results.

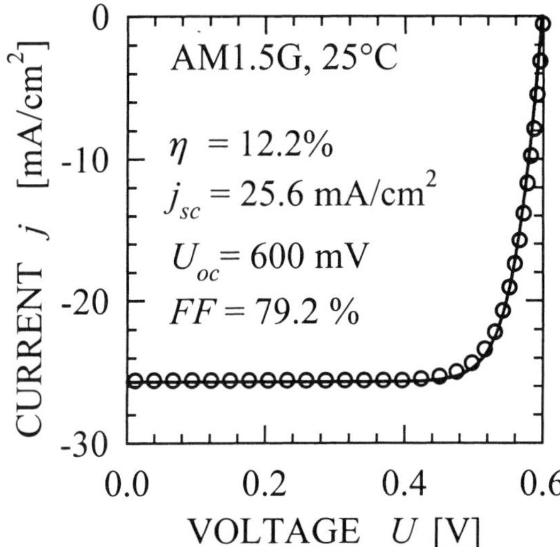

Figure 4.26. Filled circles: Current-voltage curve of an illuminated and 15.5 μm-thick cell. Solid line: Simulation with the parameters extracted from the fits shown in Figure 4.25.

only free parameter we use to fit the reflectance spectrum. The broken line in Figure 4.25 is calculated with a back surface reflectance of 0.92 for all wavelengths. The corresponding short-circuit current improvement is 5 mA cm^{-2}.

The best fit of the measured quantum efficiency is achieved with the bulk diffusion length L_b = 500 μm and the surface recombination velocity S_b = 1000 cm s^{-1}. These parameters also fit the current-voltage curve, as is shown in Figure 4.25. The parameter confidence plots that we introduce on p. 85 of Appendix C show that $L_b \geq$ (20 ± 5) μm and $S_b <$ (5000 ± 2000) cm s^{-1}. The low quantum efficiency at 400 nm originates from a high recombination in the emitter caused by Auger recombination. The emitter accounts for a fraction of 50% to 90% of the total recombination at the maximum-power point. The lower bound of 50% applies if the bulk diffusion length is 20 μm and the upper bound of 90% applies for negligible volume recombination. The emitter recombination not only reduces the voltage by at least 18 mV, it also reduces the short-circuit current by 3 mA cm^{-2}. To summarize, the encouraging efficiency of 12.2% can be improved significantly by optimizing the back reflectance and by improving the emitter. The effective quantum efficiency diffusion length is L_Q = (21 ± 5) μm and the collection length is L_C = (110 ± 20) μm. See p. 55 for the definitions of L_Q and L_C. Our first thin-film cells with a metal grid at the back have a short-circuit current density of 29.0 mA cm^{-2}. Compare also the updating remarks on p. 180.

Re-use of substrate wafer

Recently we re-used the Si substrate for the first time. From the original 4-inch wafer we separated four 4 cm^2 randomly textured cells with an average efficiency of 9.5 ± 0.8%. After forming porous Si for the second time, a second epitaxial layer was deposited. From this layer we separated four 4 cm^2 cells with an average efficiency of 10.2 ± 0.8%. In this first experiment on re-use, the material loss of the Si substrate wafer was 15 μm per generation due to porous Si formation and re-texturing. The loss of substrate material scales with the texture height.

4.3.6 Epilift process of the University of Canberra

The Epilift (EL) process was introduced by Weber et al. [280, 301] and is being developed at the University of Canberra. Figure 4.27 sketches the EL process. The EL is a layer transfer technique that starts with a (100)-oriented monocrystalline Si wafer as the growth substrate. A mesh-patterned silicon oxide layer is processed onto the surface of the wafer by photolithography. The lines of the mesh run in the [110] direction. An epitaxial layer is then grown by liquid-phase epitaxy (LPE). The surface of the epitaxial mesh structure exposes (111)-oriented facets as shown in the SEM micrograph in Figure 4.28. The cross-section of the Si mesh is diamond-shaped. With continuing growth the mesh overhangs the oxide, thus leading to a connection of the mesh and the substrate that is thin (compared to the cross-section). At the end of the growth process, the silicon-containing metal solution escapes through the mesh openings.

Typical film thickness values are 50 to 100 μm. Effective thickness values of only a few microns are claimed also to be possible with the EL process [302]. The Si mesh is detached by wet chemical or electrochemical etching [302]. The EL technique shares the idea of epitaxial lateral overgrowth through line-shaped seeds with the CLEFT process. However, the mesh structure forms a continuous seeding area, and therefore no merging of separately seeded crystals occurs in the EL process.

Transient photoconductance decay measurements revealed minority carrier lifetimes corresponding to diffusion lengths of 100 μm [303]. Efficient light trapping in the Si meshes was demonstrated on the basis of theoretical light trapping studies. No cell efficiencies have yet been reported. Using substrates of various orientations, a wide variety of layer shapes are fabricated [301].

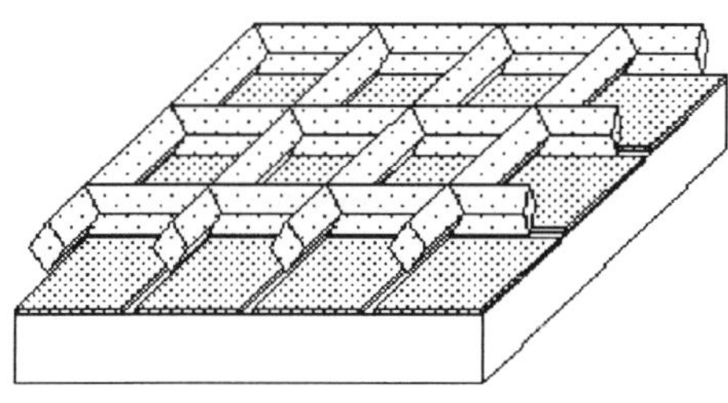

Figure 4.27. Schematic of the epitaxial lift-off process. Epitaxial mesh grown on a Si wafer with a patterned oxide. Reproduced rom Ref. [301], Copyright 1998, with permission from Elsevier Science.

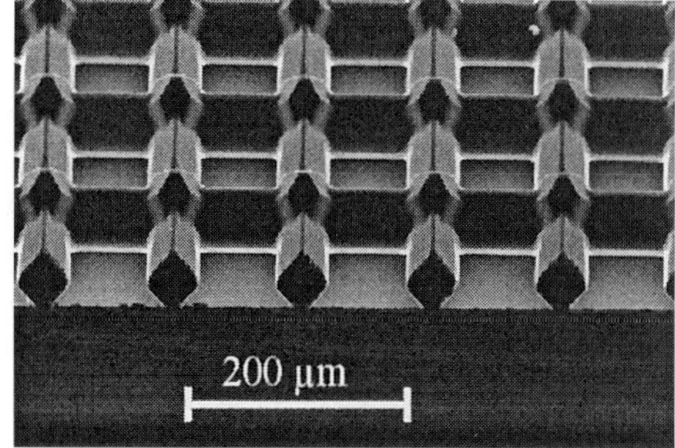

Figure 4.28. SEM micrograph of an epitaxial mesh grown on a monocrystalline Si wafer. Reproduced from Ref. [301], Copyright 1998, with permission from Elsevier Science.

4.3.7 QMS process of the University of Stuttgart

The quasi-monocrystalline Si process (QMS) was introduced by Rinke et al. [304] and is under development at the University of Stuttgart. The QMS process is identical to the SPS process with one difference: the sintered porous Si layer is the device layer and not a seed for epitaxial growth. The advantage of this approach is that no epitaxy is required. Figure 4.29 shows schematically the formation of the separation layer and the formation of the void layer. The voids introduce light trapping into a film that has two planar surfaces. Surface texturing, as required for the PSI process, is thus avoided. Rinke et al. call the void-containing layer quasi-monocrystalline [304]. Considering that the porous Si in the as-etched state is also void-containing and monocrystalline without any sintering, in this work we prefer the term sintered porous Si (SPS).

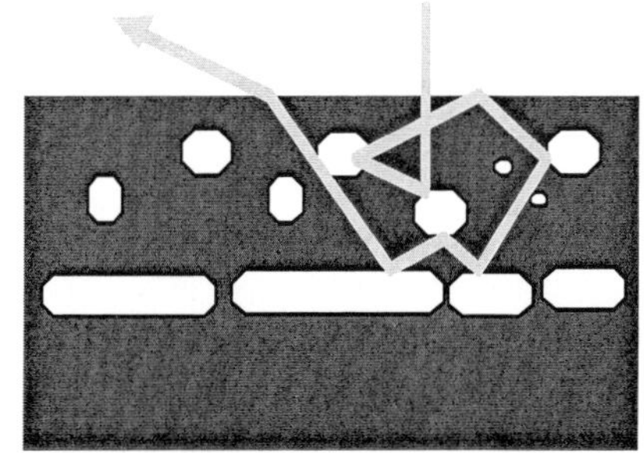

Figure 4.29. Schematic cross-section of a solar cell from the QMS process prior to detachment. The voids of the sintered porous Si introduce light trapping. The separation layer allows detachment.

A short-circuit current density of 11.1 mA cm^{-2} was measured for a 4 μm-thick solar cell with a base from sintered porous Si. The sintered layer was not detached from the substrate [305]. The power conversion efficiency was not reported. The fabrication of 30 sintered porous Si films from a single Si wafer was demonstrated.

4.3.8 Canon's SCLIPS process

A modified version of the ELTRAN process was recently applied to Si solar cells for the first time [306]. This process is named SCLIPS (solar cells by liquid-phase epitaxy over porous Si) [293]. Silicon wafers are electrochemically etched to form a porous Si multi-layer about 10 μm in thickness at the surface. Prior to epitaxy, hydrogen annealing at 950°C is necessary to close the surface of the porous Si. Without that surface closure, the porous Si would dissolve in the melt because the porous Si structures are smaller than the critical radius required for stable grain growth [307]. The epitaxial layer grows by liquid-phase epitaxy using a reactor that rotates the substrates off-axis to enhance the convection in the solvent. Growth rates as high as 1 μm min^{-1} are reported.

A recent version of the SCLIPS process re-uses the Si substrate: the SCLIPS process applies a porous multi-layer system similar to the SPS and the PSI processes. A water jet splits the device layer from the growth substrate [308, 292, 293].

A minority carrier lifetime of 10 μs, corresponding to a diffusion length of 160 μm, was reported [293]. An efficiency of 9.3% was achieved with a small (0.2 cm^2) transferred cell. The thickness of that cell was not reported. Considering the thickness range investigated in Ref. [306], the device thickness is probably in the range of 10 to 50 μm.

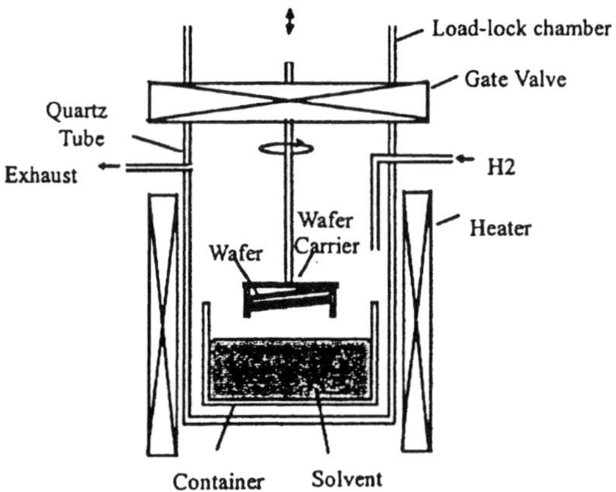

Figure 4.30. Schematic of the LPE system with obliquely mounted wafers for the enhancement of convection and growth rate. From Ref. [306].

Table 6.3 on p. 175 lists the solar cell results discussed in the preceding sections of this chapter. See also the updating remarks on p. 180.

4.3.9 Discussion

Since our first report on the PSI process [53] many transfer processes using porous Si have been reported by various research labs [296, 304, 306, 309, 310]. Layer transfer processes that do not use porous Si are also under investigation [280]. The various approaches are summarized in Table 4.2. This table also lists the key innovations introduced by the various processes. We now discuss the specific strengths and challenges of these transfer approaches with respect to various aspects that are important for low-cost thin-film solar cells.

Re-use of the substrate wafer

Re-using the substrate wafer becomes easier if the process stresses the wafer as little as possible. Hence, low-temperature processing and, in particular, low spatial temperature gradients are beneficial. The PSI process using IAD deposition around 700°C stresses the substrate less intensely than the VEST process, in which a narrow zone of molten Si is pulled across the oxidized Si substrate.

Epitaxy reactor throughput

Epitaxy is clearly the most expensive processing step in all LTPs. To bring the residence time in the epitaxy reactor into the range of a few minutes, the film thickness should be kept in the micron range. Texturing of planar monocrystalline and micron-thick films from the SPS process with random pyramids by wet anisotropic etching with KOH is not possible, because the etch depth is insufficiently controlled, unless photolithography is used. In contrast, the epitaxy on textured substrates with the PSI process permits the fabrication of micron-thick cells without using photolithography.

With deposition times in the minute range, the set-up time of the epitaxial reactor becomes important. For high-throughput large-area epitaxy, in-line atmospheric-pressure

CVD machines, similar to the one developed by Faller et al. [311], are to be preferred over batch-type high-vacuum machines such as the IAD reactor built at ZAE Bayern.

The SCLIPS and the EL processes use liquid-phase epitaxy (LPE), which permits batch processing. The QMS process even avoids epitaxy and could therefore be cost-effective, provided efficient cells can be fabricated despite the large inner void surface that getters impurities and causes internal surface recombination.

Table 4.2. Layer transfer processes for thin-film silicon using monocrystalline Si substrates.

Surface conditioning	Device layer formation	Carrier	Detachment	Key innovation	Process/ References
oxidation	poly-Si by CVD and ZMR	none	via-hole wet etching of SiO_2	separation of Si layer, via hole etching	VEST/ [285]
porous Si, pre-oxidation, H_2 anneal for surface closure	H_2-anneal and CVD	bond to oxidized Si	grinding substrate wafer	transfer of single-crystalline Si using porous Si, closure of porous Si surface during H_2 anneal	ELTRAN/ [51]
H^+ implantation	CVD	bond to oxidized Si	heating expands H_2	re-use of Si substrate, H^+ implantation for formation of separation layer	Smart Cut/ [281]
porous Si	CVD	glue to plastics	stress fractures porous Si	re-use of substrate, thermal formation of separation layer	SPS/ [52]
textured porous Si	IAD/CVD	glue to glass	stress fractures porous Si	transfer of textured films using porous Si	PSI/ [53]
oxidation and patterning	LPE	none	wet chemical etching of Si	mesh-shaped film for light trapping, batch processing	EL/ [280]
porous Si	annealing porous Si	glue to glass	stress fractures porous Si	internal light trapping, no epitaxy process	QMS/ [304]
porous Si	LPE	glue to glass	water jet fractures porous Si	separation by water jet, high-rate LPE	SCLIPS/ [306]

Type of surface conditioning

The most frequently used surface conditioning technique is the formation of porous Si (ELTRAN, SPS, PSI, QMS, SCLIPS). Use of porous Si for growth and separation yields closed monocrystalline films, in contrast to the VEST and the EL processes. The VEST process also has the disadvantage of yielding polycrystalline films. A ZMR process is not necessary if porous Si is used for epitaxy and separation.

At the present state of development of the PSI process at ZAE Bayern we find the technological window for a porous Si layer system that permits high-quality epitaxy and detachment to be rather small. Hence, the application of stress for separation with a well-controlled water jet in the SCLIPS process is of advantage for a high yield.

We consider the SC process to be of no relevance for photovoltaics because ion implantation is more complex and expensive than porous Si formation. In addition, the thickening of the less than 1 μm-thick monocrystalline Si film has to be done either at temperatures below 400°C (implying high defect concentrations) or after transfer to a low-cost substrate (e.g. glass) [295]. Neither of these possibilities permit the use of high deposition temperatures as is required for high-rate epitaxy.

Novel approaches to thin-film crystalline Si cells?

Promising combinations of the LTP and the HTS approaches are also being studied, e.g. transferring a porous Si layer to a high-temperature resistant substrate [312, 310, 313]. Epitaxy and cell processing are done *after* the transfer. This approach reduces the maximum temperature that the ceramic substrate has to withstand from 1420°C in the HTS approach to 1100°C. The process yields monocrystalline Si films on large-area ceramic substrates.

I am sure that quite a number of innovative fabrication sequences will cause the research on thin-film Si cells to remain an exciting endeavor, with numerous novel opportunities for the low-cost fabrication of thin-film crystalline Si solar cells.

5 Waffle cells from the porous Si (PSI) process

This chapter presents the status of the porous Si (PSI) process [53] that we are developing at ZAE Bayern. The key features of the PSI process are epitaxy on textured porous Si substrates and a subsequent transfer without sacrificing the substrate. The major steps are illustrated in Figure 5.1: a) surface texturing of a monocrystalline Si wafer; b) the formation of a porous Si layer system; c) the growth of an epitaxial Si layer; d) fabrication of the cell and attachment of a glass carrier; e) separation of the epitaxial Si from the substrate wafer; and f) the application of a detached back surface reflector.

In this chapter we discuss porous Si fabrication and annealing, epitaxy on porous Si, the fabrication of cells and modules, light trapping in waffle-shaped thin films, and a theoretical estimation of the efficiency of waffle-shaped thin-film solar cells.

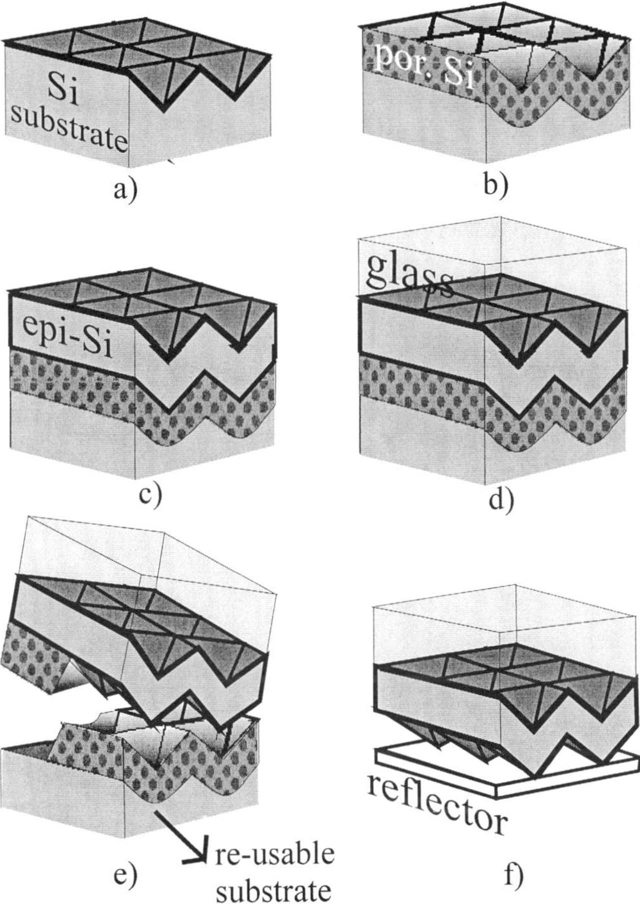

Figure 5.1. Porous silicon (PSI) process: a) surface texturing of a monocrystalline Si wafer; b) formation of a porous Si layer system; c) growth of epitaxial Si layer (epi-Si); d) fabrication of the cell and attachment of a glass carrier; e) separation of the epitaxial Si from the Si wafer; f) application of a detached back surface reflector.

5.1 Epitaxy on porous Si

The quality of epitaxy films that grow on porous Si depends critically on the microstructure and the pre-treatment of the porous Si.

5.1.1 Porous Si

The structure of the porous Si is determined by the electrochemical formation process and by the unavoidable annealing of the porous Si prior to and during the epitaxy process.

5.1.1.1 Electrochemical formation

Porous Si forms by anodic etching in aqueous hydrofluoric acid (HF:H_2O). The details of the etching mechanism are still debated, as discussed in recent papers on porous Si formation [315, 316]. The dissolution of Si results from a sequence of chemical reactions [314] shown in Figure 5.2. The reaction starts with a hydrogen-terminated Si surface. After the substitutions Si-H → Si-OH → Si-F the weakened and strongly polarized Si-Si bonds are cracked by non-dissociated H_2O and HF. The Si surface remains H-terminated. The Si complex in solution is then further hydrolyzed, thus leading to the production of gaseous H_2.

Holes and F^- ions are required for Si dissolution. The p-type Si interface to the aqueous HF solution is a Schottky barrier. Holes are rare in the depletion region of the diode. Hence, at low forward bias (minus to the Si bulk, plus to the solution) the etch rate is controlled by the hole concentration. Imagine that the Si surface is not planar, as depicted in Figure 5.3a. Holes supplied from the bulk silicon follow the field lines and are primarily delivered to the valleys of the electrolyte/Si interface. Hence, etching proceeds faster in the valleys than on the hills where holes are depleted. Pores propagate into the Si volume. The situation is different at higher current densities. The hole concentration at the surface increases exponentially with the voltage across the Si depletion region. Hence, for a large voltage the etch rate is no longer limited by the supply of holes. Instead, the low mobility of the F^- ions leads to a depletion of F^- in the valleys, as depicted in Figure 5.3b. In consequence, the hills are etched at a higher rate than the valleys. The Si surface is flattened. This is the current regime of electropolishing.

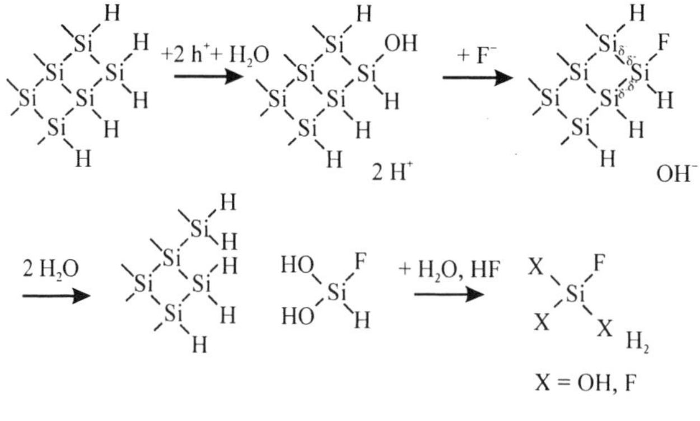

Figure 5.2. Reaction model for the dissolution of Si during anodic etching in aqueous hydrofluoric acid after [314].

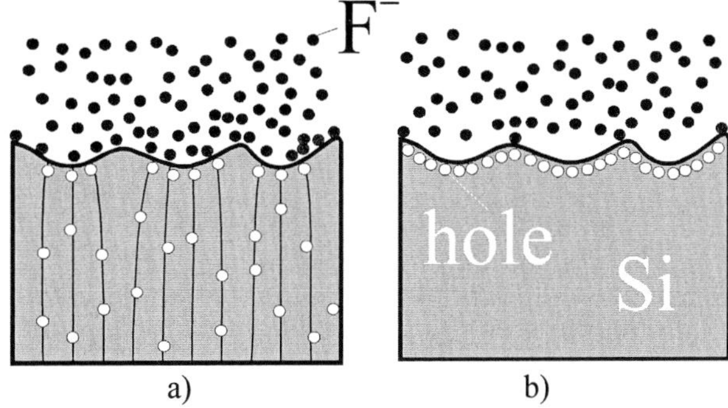

Figure 5.3. a) Low current densities cause a depletion of holes at the hills. The surface roughness increases and pores grow. b) At high current densities the F^- ions are the rare species. This is the regime of electropolishing.

A fascinating aspect of porous Si is the simplicity of its fabrication. We use a double tank reactor as shown in Figure 5.4. A 4-inch p-type Si wafer is immersed in HF that is diluted in water. Ethanol is added as a surfactant that helps the H_2 gas bubbles to escape from the porous Si. A solution for the fabrication of porous layers with 20% porosity into p^+-type wafers is 49% wt. HF:H_2O:ethanol = 1:1:1. Typical current densities are 5 to 200 mA cm^{-2}. The above discussion of the dissolution shows that the morphology of porous Si depends heavily on the composition of the etching solution, the doping concentration of the semiconductor, homogeneity of doping, and the electrical potential at the Si surface. For etching n-type Si, additional holes are supplied by illuminating the semiconductor. Experimental details for the fabrication of porous Si are given in Ref. [317].

Sample preparation for the PSI process starts with a p^+-type, 10^{19} cm^{-3} boron-doped, (100)-oriented monocrystalline Si wafer of 100 mm in diameter that receives a surface texture. We use periodic textures defined by photolithography as well as random textures that avoid photolithography. In both cases Si etching is done with diluted KOH. The porous Si layer with a thickness of 1 μm consists of two layers with different porosity values. The top layer has a porosity of 20%. Low porosity improves the quality of the

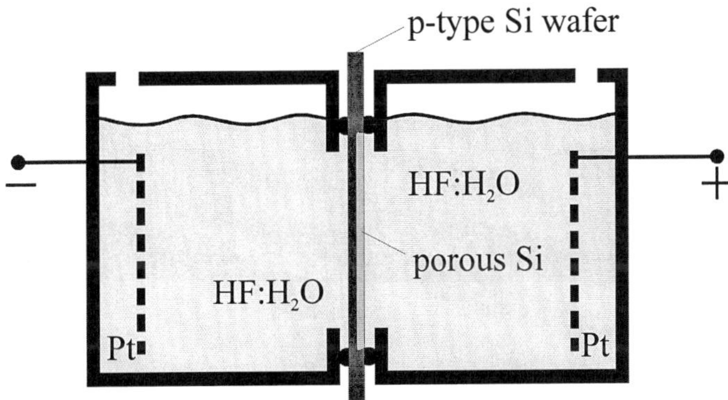

Figure 5.4. Double tank reactor for fabricating porous Si layers on 4-inch Si wafers. A current is applied to the Pt grids.

epitaxial layer [318]. The bottom layer has a porosity preferably exceeding 50%. Higher porosity reduces the mechanical strength of the layer, which is a precondition for the separation of the epitaxial layer from the substrate. The porous sample is rinsed in deionized water and dried in air.

5.1.1.2 Annealing

The PSI process uses epitaxy on porous Si. Since epitaxy requires elevated temperatures (we use 700 °C for ion-assisted deposition and 1100°C for chemical vapor deposition), annealing of the porous Si inevitably occurs during loading or heating of the reactor. As a consequence the layer system re-organizes.

In a recent paper we investigated the annealing of porous Si theoretically and experimentally [319]. We developed a Monte Carlo annealing model that is based on thermally activated jumps of surface atoms to neighboring positions. The model measures the annealing time in units of jump trials per atom (tpa) [319]. Figure 5.5a shows a model representation of the cross-section of a sample with 20% porosity. After annealing, the simulation in Figure 5.5b shows the formation of voids in the porous layer and a closure of the Si surface. This is qualitatively what we observe in our experiments: Figure 5.5c and d show cross-sectional scanning electron microscope (SEM) micrographs of a sample with 20% porosity prior to annealing and after annealing at 1200°C for 30 min in H_2. The surface closure during annealing was first observed by the researchers developing the ELTRAN process [290] (see p. 109) and is also used in the PSI process [320] (see p. 113).

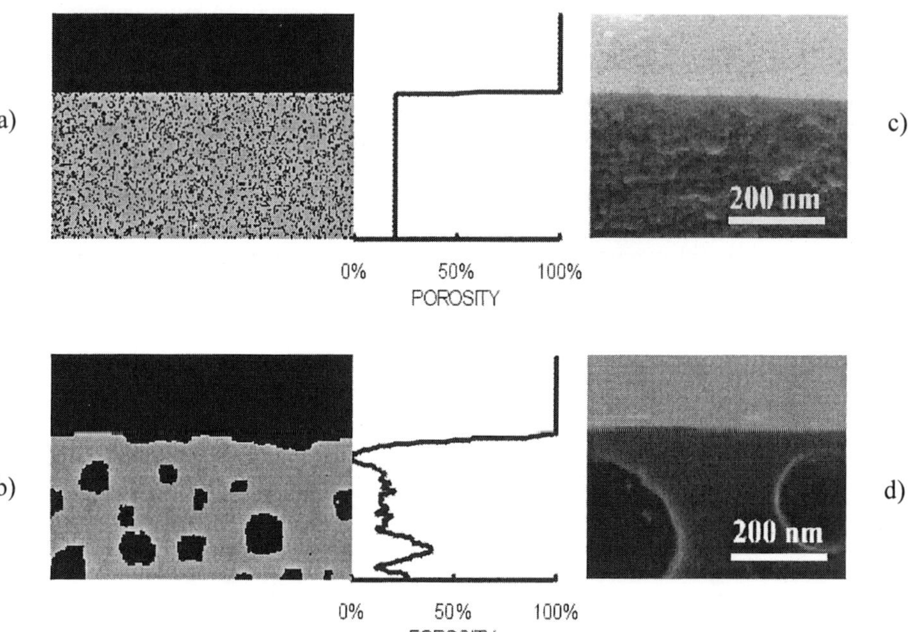

Figure 5.5. a) Representation of a sample as etched with 20% porosity and the corresponding porosity depth profile. b) Sample after simulated annealing for 10^5 tpa at a temperature of 1100°C. c) Oblique cross-sectional SEM micrograph view of a sample with 20% porosity prior to annealing, and d) after 0.5 h annealing at 1100°C. Figure from Ref. [319].

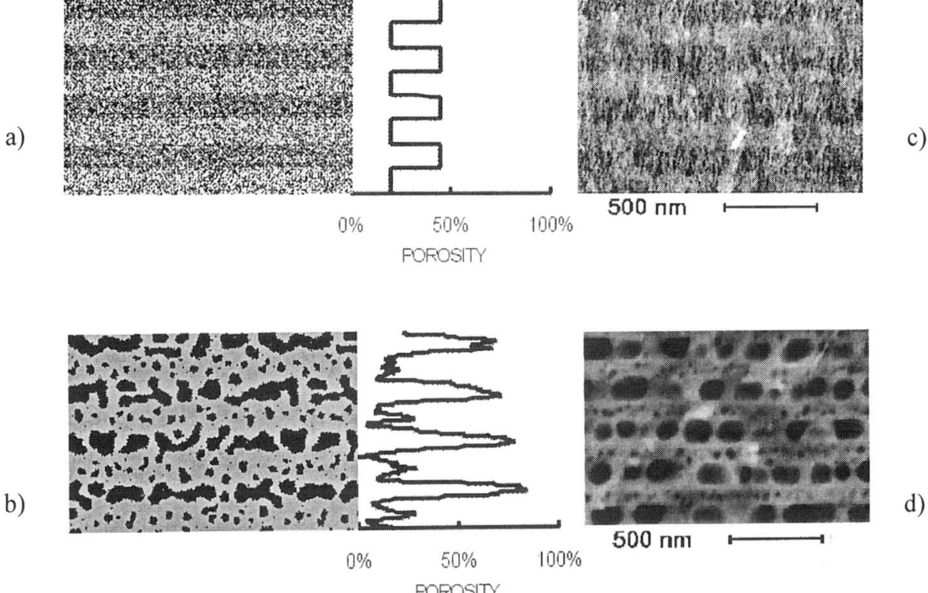

Figure 5.6. a) Representation of the cross-section of a sample with alternating porosity values of 20% and 45% in the as-etched state, and b) after annealing for 10^6 tpa at 1100°C. c) Cross-sectional SEM micrograph of a porous multi-layer with alternating porosity values of 20% and 50%. d) SEM micrograph after annealing at 1100 °C for 0.5 h.

In a second simulation we investigate the annealing of a multi-layer system consisting of layers with alternating porosity values of 20% and 45%. The simulated annealing preserves the multi-layer structure as shown in Figure 5.6a and b. Large voids form in the high-porosity layers with a height equal to the layer thickness. Small voids line up along the middle of the low-porosity layers. Figure 5.6c shows the SEM micrograph of an as-etched multi-layer with alternating porosity values of 20% and 45%. Figure 5.6d shows that annealing at 1100°C for 0.5 h drastically increases the pore size. In accordance with our simulation, we observe the formation of large voids with a height equal to the thickness of the high-porosity layer, while small voids form in the low-porosity layer. Increasing the porosity of the high-porosity layer leads to the formation of elongated voids with increased lateral extension. For porosity values exceeding 50% only weak bridges connecting adjacent low-porosity layers remain, similar to those observed in the SPS process [52] (see Figure 4.22 on p. 111).

Our Monte Carlo model for the annealing of porous Si [319] thus qualitatively explains the experimental findings. The only driving force implemented is a reduction of the surface energy. The ambient hydrogen used in our experiments is, in principle, not necessary for the observed effects. Hydrogen mightbe necessary, however, in practice to evaporate surface oxides and to enhance the material transport.

5.1.2 Ion-assisted deposition (IAD)

The ion-assisted Si deposition technique was developed by Oelting et al. to facilitate low-temperature epitaxy at high growth rates [321]. A schematic of an IAD reactor is shown in Figure 5.7. The concept of the IAD technique is to enhance the surface mobility of the Si atoms by ion bombardment rather than by an elevated sample temperature. The Si ions are accelerated towards the substrate. Their kinetic energy depends on the acceleration voltage U_{acc} and is 10 to 50 eV. This energy is transferred into electron and phonon excitations. These excitations would only occur at much higher temperatures without the ion bombardment. The bombardment thus generates an enhanced "virtual" temperature at the Si surface [321, 322] and enhances the surface mobility [323].

The epitaxy reactor we built at ZAE Bayern is an ultra-high vacuum chamber with a base pressure of 2×10^{-10} mbar. Si atoms are evaporated by a 15 kW electron gun. The ionizing voltage U_{ion} = 100 V is chosen as to optimize the ionization rate for the formation of Si^+ ions. A fraction of 1 to 5% of the Si atoms is ionized. The evaporated Si is refilled with a mechanism that pushes Si cubes with a volume 4 cm^3 into the Si source. The substrate is carried by a molybdenum sample holder. The substrate is a Si wafer 10 cm in diameter. A graphite heater heats the Si wafer to temperatures ranging from T_d = 500 °C to 1000 °C. We measure the wafer temperature with a pyrometer.

5.1.2.1 IAD on monocrystalline Si wafers

In this section, we review IAD film growth for solar cells on (100)-oriented monocrystalline Si wafers. Studying epitaxy on monocrystalline Si is a good starting point for the investigation of IAD on porous Si.

Oelting et al. demonstrated the fabrication of the first monocrystalline thin-film cells by IAD. Those cells have a Ga-doped base and an Sb-doped emitter [321, 324]. Prior to

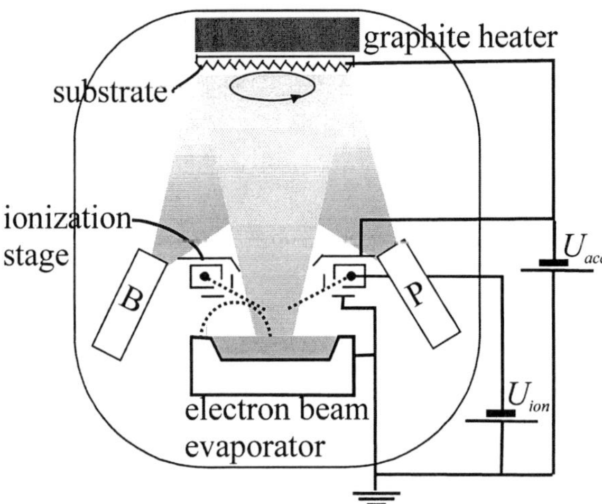

Figure 5.7. Schematic diagram of ion-assisted Si deposition. A fraction of about 5% of the evaporating Si atom is ionized in the ionization stage and accelerated with a voltage U_{acc} towards the substrate. The ionization voltage U_{ion} is chosen to maximize the ionization cross-section.

epitaxy, the Si wafers were cleaned by a standard RCA procedure [325] and the residual oxide was evaporated at 850°C in ultra-high vacuum. The emitter grew with an acceleration voltage U_{ion} = 500 V to enhance the incorporation of Sb by secondary ion implantation [326].

Reducing the distance of the Si source and the substrate to 25 cm enhanced the deposition rate to 0.3 μm min^{-1} [327]. At this rate, a 5 μm-tick layer is deposited in only 17 min. We will show that such layer thickness values are sufficient for effective thin film cells (see Figure 5.33 on p. 152), provided a surface texture is applied to enhance the optical absorption.

Introducing a p^+-type back surface field, replacing the epitaxial Sb-doped emitter by a phosphorus-diffused emitter, and passivating the front surface with a thermal silicon oxide grown at 1000°C raised the efficiency of a 6 μm-thick cell to 8.9% [327]. Evaluating the quantum efficiency with our model (presented on p. 73), this cell was found to have a diffusion length L exceeding the device thickness of 6 μm. The same device modifications allowed an efficiency of 9.9% using a base that was twice as thick (12.5 μm) [328]. High-temperature annealing during diffusion and oxidation improves the open-circuit voltage, an effect that was speculated to be caused by an annealing of structural defects and by a phosphorus gettering [327, 329].

Oberbeck et al. introduced B-doping of IAD layers by using high-temperature B-effusion cells that have recently become available commercially [330]. The comparison of SIMS and Hall data shows that the hole concentration equals the doping concentration. The mobility values of Si bulk material [331] are reached in the acceptor concentration range of 10^{17} cm^{-3} to 5×10^{18} cm^{-3}. Replacing the diffused emitter by a P-doped epitaxial emitter is attractive, since the deposition time for the emitter is much shorter than the time required for a thermal diffusion. An effective diffusion length of 17 μm was reported for IAD material with an epitaxial P-doped emitter [332]. However, since the device thickness was only 7 μm, the effective diffusion length is possibly enhanced by current generated in the underlying p^+-type Si substrate that has a resistivity of 0.01 Ω cm. As discussed in Appendix C on p. 245, we measure a diffusion length of 6

Table 5.1. Literature results for thin crystalline Si solar cells fabricated by IAD. The abbreviation b-Ga-IAD means a base that is Ga-doped and is grown by IAD, while e-P-DIF denotes an emitter that is P-diffused.

Efficiency η [%]	Voltage V_{oc} [mV]	Thickness W_f [μm]	Growth rate R [μm min^{-1}]	Growth temperature T [°C]	Remark	Reference
6.4	500	7	< 0.3	500-700	b-Ga-IAD e-Sb-IAD n^+/p	[324]
8.9	580	6	< 0.3	550-700	b-Ga-IAD e-P-DIF $n^+/p/p^+$	[327]
9.9	600	12.5	0.2	600	b-Ga-IAD e-P-DIF $n^+/p/p^+$	[328]
	555	7	0.07-0.18	710	b-B-IAD e-P-IAD $n^+/p/p^+$	[332]

μm in this type of material. The base diffusion length reported by Oberbeck [332] is, however, certainly larger than the base thickness of 7 μm.

As a general tendency, higher diffusion lengths were found in films deposited at higher temperatures and higher rates [332]. The higher deposition temperature decreases the dislocation density, while a higher deposition rate reduces the contamination concentration in the film. Solar cell data on (100)-oriented monocrystalline Si wafers are listed in Table 5.1.

5.1.2.2 IAD on porous Si

Silicon wafers with a porous surface layer are treated by RCA cleaning and a 10 s HF dip. The RCA cleaning weakens the porous Si layer mechanically. Prior to epitaxy, we heat the sample to 850°C for 10 min to remove the native oxide from the PSL surface. An epitaxial, Ga-doped Si film of thickness $W_f = 5.8$ μm grows by the ion-assisted deposition (IAD) technique at 700°C. The growth rate is typically 0.1 μm min^{-1}. The maximum deposition rate is 0.3 μm min^{-1} and is limited only by the power of the electron gun and the geometry of the reactor. A layer with a thickness of 10 μm is deposited in 30 to 100 min. Figure 5.8 shows scanning electron microscope images of a free-standing Si waffle fabricated by the PSI process. The pyramidal texture is regular and is a replica of the initial surface texture of the substrate wafer. The inclination of the pyramidal facets is 54.7° to the microscopic cell surface. The effective film thickness measured normal to the macroscopic cell surface is $W_{eff} = 10$ μm.

With the PSI process we transfer thin textured Si films with a diameter of up to 100 mm. Figure 5.9 shows a Si waffle with a diameter of 85 mm, a film thickness of $W_{eff} = 7.8$ μm, and a texture period of 13 μm. The thickness of the film is $W_f = \cos(54.7°)\times 7.8$ μm = 4 μm when measured perpendicular to the facets. The waffle is illuminated by the sun

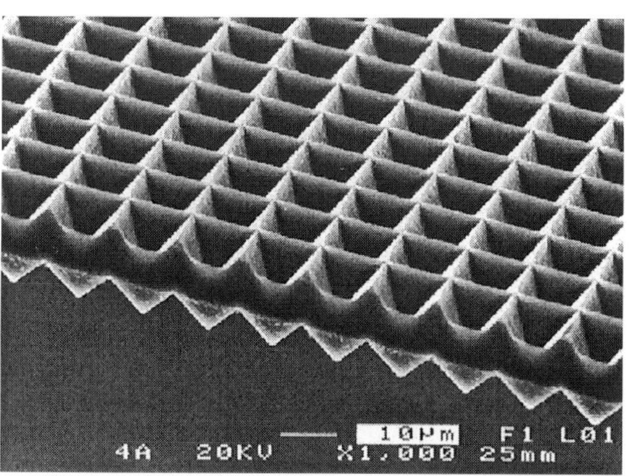

a)

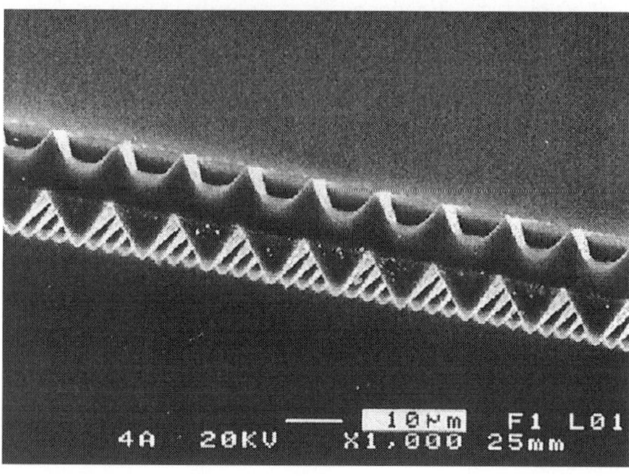

b)

Figure 5.8. Scanning electron image of a free-standing waffle-shaped monocrystalline Si film produced by the perforated-Si process. The facet inclination is $\alpha = 54.7°$, the texture period is $p = 13$ μm, and the film thickness is $W_{eff} = 10$ μm: a) view from top and b) oblique view at the cross-section from the bottom. Figure from Ref. [53] with permission of the European Commission.

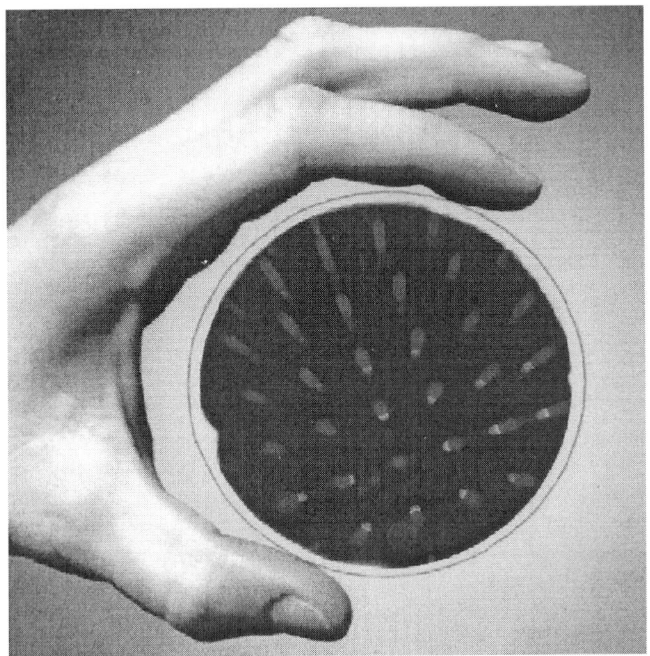

Figure 5.9. Photograph of a Si waffle with a diameter of 85 mm, a Si film thickness of W_{eff} = 7.8 μm, and a texture period of 13 μm. The waffle is illuminated by the sun from the back. The bright spots on the wafer originate from a diffraction pattern of red light that is partially transmitted through the wafer. Sample fabricated at ZAE Bayern. Photo by courtesy of Robert Bosch GmbH, Stuttgart, Germany.

from the back. Red light with an optical absorption length of 3 μm is partially transmitted through the wafer. The transmittance is modulated by the texture period p = 13 μm. Therefore, the Si film functions as a transmitting optical grid and causes constructive interference in certain directions. In Figure 5.9 these directions show up as bright spots on the wafer.

Hall-effect measurements in van der Pauw geometry [333] at 293 K yield a hole mobility of 275 $cm^2\ V^{-1}\ s^{-1}$ and a carrier concentration of $2\times10^{16}\ cm^{-3}$ for a Ga-doped, planar epitaxial film transferred to a glass substrate. The measured mobility is 75% of the literature value for Ga-doped monocrystalline Si [334].

Figure 5.10 shows the X-ray diffraction spectrum measured with the $Cu_{K\alpha}$ lines for a Si waffle on glass in comparison to the spectrum of the monocrystalline Si substrate. The step width of the spectrum is 0.020°. Note the logarithmic intensity scale. All peaks are at the same angles to within 0.01°. Only the large (400) peak originates from Si. This peak is at 69.14° ± 0.01°, corresponding to a lattice constant of 542.991 pm ± 0.6 pm. The literature value for the Si lattice constant is 543.095 pm [79]. All other peaks are more than two orders of magnitude smaller and are artifacts of the X-ray apparatus. Since the X-ray diffraction spectrum of the epitaxial film has peaks at the same angles as the (100)-oriented substrate, all crystalline parts of the epitaxial film are also (100)-oriented. The X-ray diffraction spectrum does not strictly prove monocrystalinity because the epitaxial film could consist of many (100)-oriented crystallites being rotated by various angles around the [100] direction. In order to exclude this possibility, we recorded Laue images of the epitaxial film and the substrate. The Laue images of the epitaxial films show spots at the same positions as for the monocrystalline substrate. We also observe that the thin epitaxial film fractures preferentially in two directions which are oriented perpendicular to each other. This behavior is similar to that of a monocrystalline (100)-oriented Si wafer that preferentially breaks along the <110>

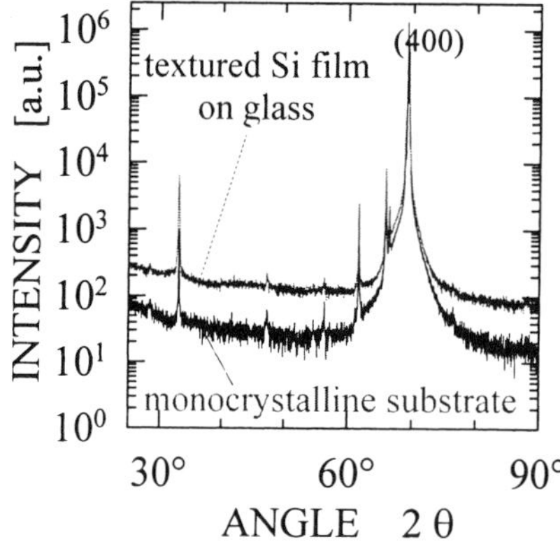

Figure 5.10. X-ray diffraction spectra of the epitaxial Si waffle shown in Figure 5.8 and of the monocrystalline substrate wafer. Note the logarithmic intensity scale.

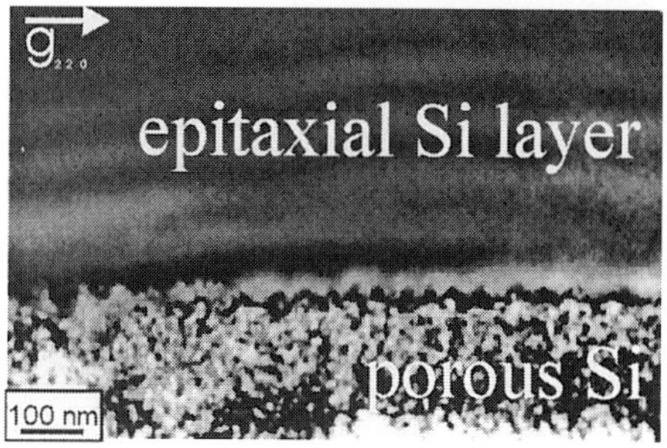

Figure 5.11. Transmission electron micrograph of the interface from the epitaxial layer (top) to the porous Si (bottom). Epitaxy by ion-assisted deposition. Photo by J. Krinke, University of Erlangen-Nuremberg [336].

directions. The higher background intensity of the epitaxial film is due to the amorphous glass substrate. From the X-ray spectrum, the Laue image, and the fracture behavior we conclude that the IAD technique permits epitaxial growth of monocrystalline Si films on porous Si substrates at temperatures as low as 700°C [53].

Figure 5.11 shows a transmission electron micrograph of the interface between the epitaxial IAD layer (top) and the porous Si (bottom). The substrate orientation is (100). The epitaxial film is monocrystalline and has the orientation of the substrate. No defects are visible in the epitaxial layer. Wet chemical etching yields an etch-pit density of 10^3 to 10^4 cm^{-2}. The reproducibility is currently poor. The defect density is often as high as 10^7 cm^{-2}.

We investigated the dependence of the stacking fault density on the deposition temperature, the energy of the Si^+ ions during epitaxy, and the crystallographic orientation of the substrate [335]. The latter is particularly important for growing Si films with a waffle shape that has (111)-oriented facets.

Figure 5.12 shows a schematic cross-section of our samples. We investigate the epitaxial layers on the {111} and {001} facets. Sample preparation consists of separation

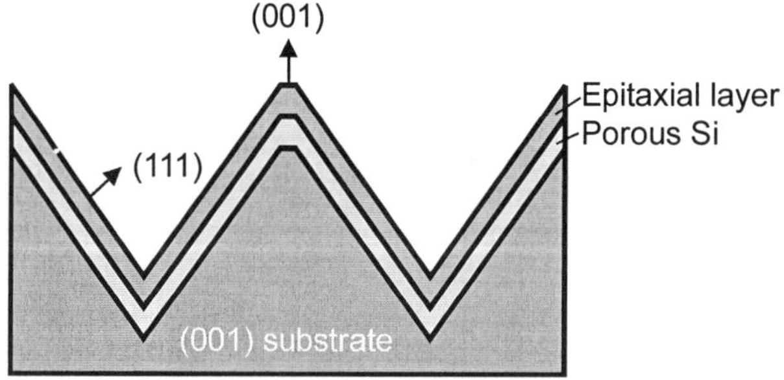

Figure 5.12. Schematic cross-section of the samples investigated. Light gray: porous silicon layer. Epitaxial growth takes place on the inverted pyramids on {111} lattice planes and between the pyramids on {001} lattice planes. Figure after Ref. [335].

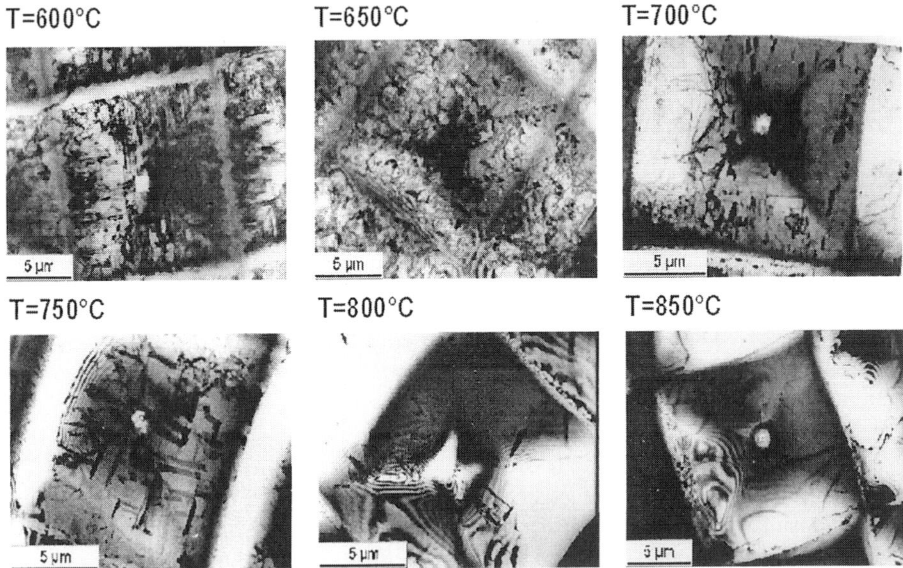

Figure 5.13. Transmission electron micrographs showing stacking faults on the {111} facets of the inverted pyramids [335]. The space between the pyramids is defect-free. The stacking fault density is greatly reduced by increasing the deposition temperature T from 600°C to 850°C.

from the substrate and 3 kV argon ion-beam etching, which leads to electron transparency.

Figure 5.13 illustrates the TEM results as a function of the deposition temperature T_d in the range from 600°C to 850°C. The major defects are stacking faults on the {111} lattice planes of the pyramids. The (100)-oriented regions between the inverted pyramids are free of microstructure defects. The density of stacking faults depends on the deposition temperature. The higher the deposition temperature, the lower the stacking fault density.

Table 5.2. The density of stacking faults on {111} lattice planes for various deposition temperatures [335].

Deposition temperature T_d [°C]	Density of stacking faults N_S [cm^{-2}]
600	9×10^9
650	7×10^8
700	5×10^8
750	5×10^8
800	2×10^8
850	4×10^7

5.1.3 Chemical vapor deposition (CVD)

Figure 5.14 shows a schematic of an optically heated cold-wall reactor that we use for Si layer deposition. The gases enter the reaction chamber at atmospheric pressure. A rotating graphite carrier holds 24 Si wafers with a diameter of 100 mm each. Lamps heat the samples to a temperature of 1100°C. The Si source is $SiHCl_3$ and BH_2 is added for p-type doping. In contrast to the IAD technique, Si epitaxy by high-temperature CVD is well established in microelectronics. Another advantage of chemical vapor deposition is illustrated in Figure 5.15. The CVD is one step of the Siemens process commonly applied to fabricate electronic-grade Si material. Hence using CVD from $SiHCl_3$ saves one melting and crystallization step when compared to all other cell processes that start from solid Si, e.g. Si wafers. For these reasons we consider the high-temperature substrate (HTS) class and layer transfer processes (LTP) using high-temperature deposition to be economically more attractive than the low-temperature substrate (LTS) approach.

5.1.3.1 CVD on monocrystalline Si

Epitaxy by chemical vapor deposition on monocrystalline Si is a standard technique [337]. Transmission electron microscopy of epitaxial films grown on planar monocrystalline float-zone Si wafers show no structural defects. Secco etching of these epitaxial layers reveals a defect density of less than 100 cm^{-2}.

We fabricated thin-film Si cells by CVD epitaxy on a p^+-type Si substrate (without any porous Si). With a layer thickness of 48 μm we achieve a power conversion efficiency of 17.3% and an open-circuit voltage of 655 mV [177].

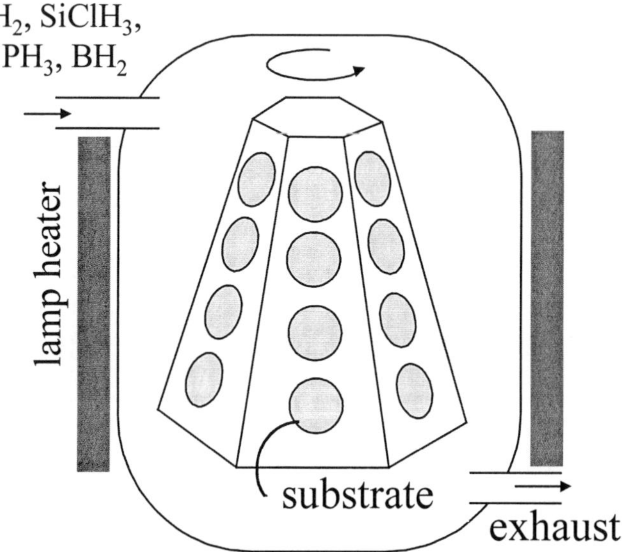

Figure 5.14. Schematic of a atmospheric-pressure CVD reactor for high-temperature Si epitaxy. A rotating graphite susceptor carries 24 Si wafers of 100 cm^2 diameter.

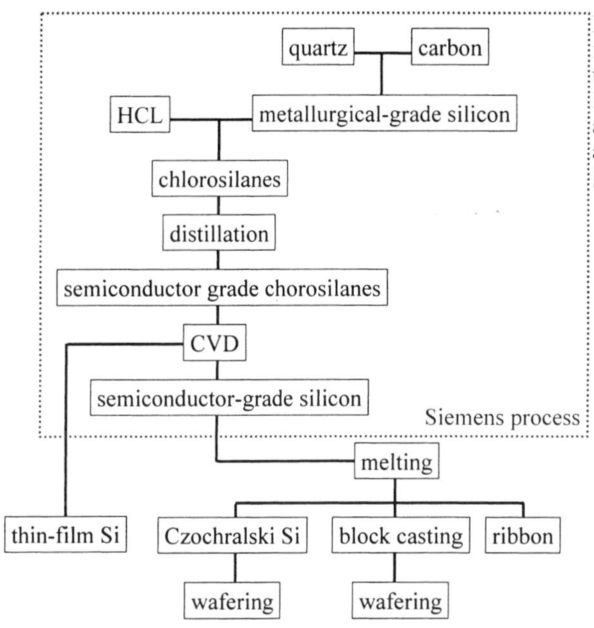

Figure 5.15. Thin-film deposition from chlorosilanes has the advantage of utilizing the well established Siemens process while avoiding one crystallization step and wafering. Redrawn from Ref. [339].

5.1.3.2 CVD on porous Si

As for the ELTRAN [290] and the SPS process [296], we anneal the porous Si in hydrogen at 1100°C for up to 30 min. This closes the pores at the surface and weakens the separation layer (see Figure 4.22 on p. 111). On planar substrates we find a defect density that is smaller than 100 cm^{-2} by Secco etching. A low porosity of 20% gives fewer defects in the epitaxial film than starting with a high porosity of 60% [318, 338].

Figure 5.16 shows Si films that we grow on a p^+-type Si substrate that is textured with

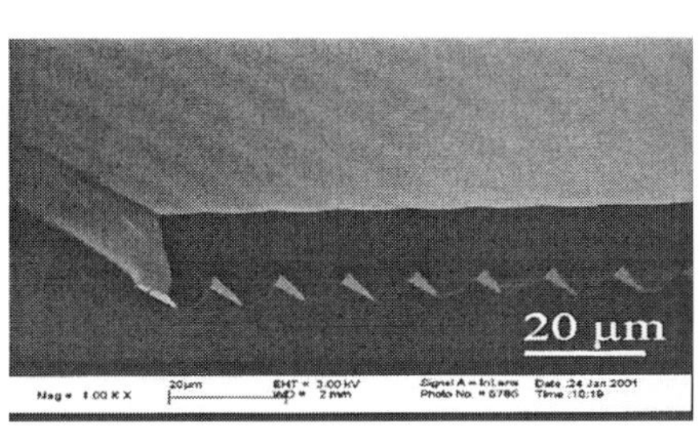

Figure 5.16. Thin monocrystalline Si film fabricated by the PSI process using CVD on a Si substrate that is textured with photolithographically defined V-grooves.

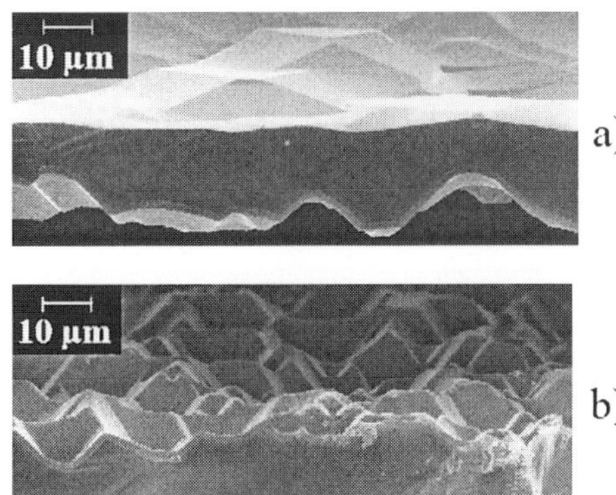

Figure 5.17. Scanning electron micrographs of detached free-standing silicon films fabricated by the PSI process: a) a 15 µm-thick CVD film deposited on a randomly textured substrate viewed from the top; b) the same film viewed from the substrate side, showing random inverted pyramids.

photolithographically defined V-grooves. In contrast to the IAD technique, the Si layer grows non-conformally. The surface of the epitaxial Si layer is almost planar.

Figure 5.17a shows a film grown on a Si substrate that is textured with randomly positioned upright pyramids. Again, we observe a surface close to planar with facets that have an inclination of 8° typically. Figure 5.17b shows the same film as viewed obliquely on the textured side. This side has random inverted pyramids, a novel type of surface texture that has a light trapping performance similar to that of the regular waffle shown in Figure 5.8 on p. 128 [320]. The defect density is 10^2 to 10^3 cm^{-2} in these randomly textured waffles fabricated by the PSI process, as revealed by Secco etching. A dislocation density of less than 10^5 cm^{-2} can be tolerated for thin-film solar cell applications.

5.2 Module concepts

Thin-film solar cells require a cost-effective technique for the series connection of individual cells to modules. Using the PSI process, we demonstrate novel concepts to fabricate a series connection of waffle-shaped monocrystalline Si cells from the PSI process. Different techniques are introduced for IAD-grown films [340] and for CVD-grown films [341].

5.2.1 Integrated series connection of IAD-grown films

The novel process is named Shadex (for shadow epitaxy) and applies Si epitaxy through shadow masks as sketched in Figure 5.18. The principle is best explained along with possible applications: the fabrication of an integrated series connection and the fabrication of a parallel multi-junction solar cell.

Concept for fabricating an internal series connection

A re-usable monocrystalline Si wafer is textured with random or regular pyramids. The surface of the pyramids is transformed into porous Si. On top of the porous Si we deposit a p^+-type epitaxial Si layer through a shadow mask by ion-assisted deposition (IAD) [324]. The mask consists of a set of stretched wires that are positioned by cold

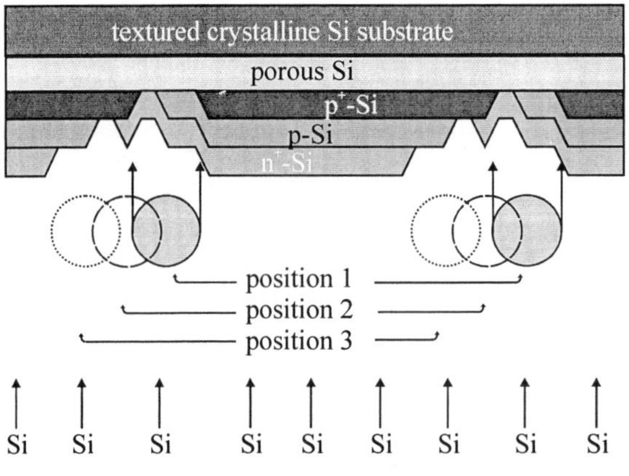

Figure 5.18. Shadow epitaxy by ion-assisted deposition fabricates an integrated series connection during film deposition. The pyramidal texture of the substrate is not shown.

supports in the reactor. Deformation of the mask due to thermal expansion is thereby avoided. The purpose of the wires is to avoid epitaxy underneath them. After deposition of the p^+-type layer, we shift the mask from position 1 to position 2. Then, a p-type Si layer grows. Finally, with the mask in position 3, an n^+-type emitter grows and overlaps with the neighboring p^+-type layer. A metal grid may be evaporated onto the n^+-type Si to enhance the lateral conductivity with the mask still in position 3. We then glue the epitaxial, monocrystalline, and textured layers to a glass and separate the stack of layers from the substrate at the sintered porous Si. Removal of the residual porous Si finishes the integrated series connection.

Experimental proof of concept for series connection

We grow the epitaxial layers by IAD at 700°C at a rate of 0.1 µm min^{-1}. The mask consists of six Ta-wires 0.4 mm in diameter that are stretched 2 mm in front of the sample with a separation of 5 mm between neighboring wires. We determine the minimum width of trenches from the thickness profile of an epitaxial film in the vicinity of the wire. Figure 5.19 shows a cross-sectional scanning electron micrograph of the edge region of a 8 µm-thick Si film grown on a monocrystalline Si wafer. The position of the wire is indicated, but neither the wire nor the distance of the wire from the substrate is drawn to scale. The film thickness decays from 8 µm to a thickness close to zero within a distance of approximately 30 µm. This length results from the geometry of the IAD machine. The distance between the Si melt and the wire is 25 cm, and that from the wire to the substrate is about 2 mm. From the area of Si evaporation is 0.5 cm in diameter, we estimate a profile width of 2 mm × 0.5 cm /25 cm = 40 µm, in good agreement with the measurement. Hence, narrow interconnection lines with a width of less than 200 µm are feasible.

Figure 5.18 shows the layer sequence deposited for the integrated series connection: (i) a 1 µm p^+-type Si layer with the mask in position 1, (ii) a 8 µm-thick layer of p-type Si with the mask in position 2 (that is, 2/3 of the wire diameter to the left to position 1), and (iii) a 1 µm-thick n^+-type layer with the mask in position 3 (that is, 2/3 of the wire diameter to the left of position 2). The n^+-type layer is deposited 4/3 of the wire diameter

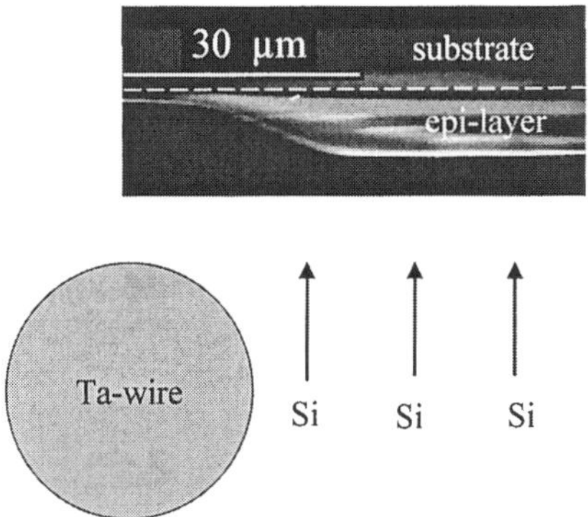

Figure 5.19. Cross-sectional scanning electron micrograph of a 8 µm-thick Si film deposited by ion-assisted deposition onto a monocrystalline Si wafer. A Ta-wire of 400 µm diameter (not drawn to scale) spans 2 mm in front of the Si substrate.

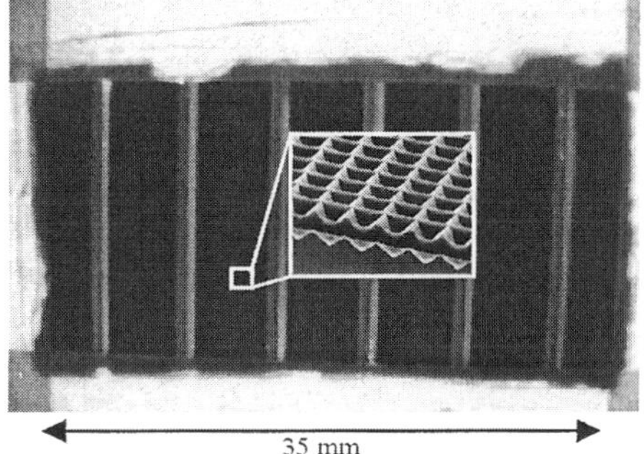

Figure 5.20. Photograph of a mini-module consisting of six series-connected cells. The insert shows the waffle shape of the cells with a texture period of 13 µm.

to the left of the p^+-type layer. The n^+-type layer overlaps the p^+-type layer of the adjacent cell by 1/3 of the wire diameter.

Figure 5.20 shows a photograph of the 10 μm-thick monocrystalline Si *module* attached to glass when illuminated from the back. The interconnection lines are about 1 mm wide. At its thinnest region the interconnection line consists of a 1 μm-thick n^+-type Si layer. Safe detachment of such a thin layer from the porous Si is, surprisingly, no problem. The thin regions transmit more light and appear brighter than the rest of the module in Figure 5.20. The insert shows a scanning electron micrograph of the waffle shape with a texture period of 13 μm. The deposition time required for the complete module (without any metallization) is 30 min to 2h, depending on the deposition rate. The first mini-module already proves the principle: the open-circuit voltage of the module is 1.6 V under one sun's illumination. This voltage is six times the voltage of 266 mV measured for an identically fabricated single cell. The low open-circuit voltage is primarily due to an inappropriate doping profile. The dopants Ga and Sb are evaporated onto the substrate from opposite directions both at an angle of 45° relative to the substrate's normal. Since the facet angle of the waffle texture is 54.7° (see Figure 5.8 on p. 128) two out of four pyramidal facets have either no donors or no acceptors. Thus the junction is missing on half of the cell surface. In addition, the stacking fault density of IAD layers grown at 700°C can be as large as 10^8 cm^{-2} on (111)-oriented pyramidal facets [335].

We achieve open-circuit voltages of 530 mV without any means of bulk or surface passivation from IAD diodes grown on planar (100)-oriented porous substrates. On planar substrates, neither the doping problem nor the high stacking fault densities are present. A modification of our IAD reactor geometry at ZAE Bayern is under way to permit homogeneous doping of waffle-shaped cells.

5.2.2 Parallel junction design with IAD-grown films

The parallel multi-layer junction solar cell design was suggested for cells with a minority carrier lifetime too short for sufficient carrier collection with one or two junctions [116]. Theoretical investigations demonstrated that the parallel multi-junction design is particularly advantageous for cells with poor light trapping and poor surface passivation [223].

Concept for fabricating a parallel multi-junction

Figure 5.21 shows the schematic of a multi-layer cell. A conductive substrate, e.g. a metal, is covered with a p-type seeding layer that makes a low-resistivity contact to the metal. On top of this p-type layer we grow a structured n-type emitter through a wire mask that is held in position 1. Then the mask is shifted to position 2 to deposit an interrupted p-type layer. Finally, the wire mask is removed and a closed n-type layer grows. The advantage of the closed n-type layer is that contact to the n-type Si can be made with a grid that needs no alignment. The example shown in Figure 5.21 is a parallel junction device with three junctions. It is obvious how to fabricate devices with more than three junctions. Series-connected parallel multi-junction solar cells may also be fabricated with the Shadex process.

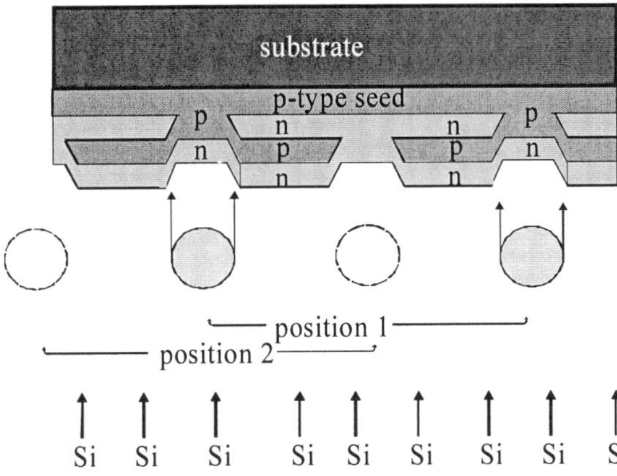

Figure 5.21. The epitaxy of Si through a wire mask permits the realization of a multi-layer cell. In the case of a conducting metal substrate, the final layer may be deposited without a mask. This facilitates the application of a contact grid without alignment relative to the p-type and n-type via lines.

Experimental proof of concept of a parallel multi-junction

In the experiment, we used a p^+-type, 0.01 Ω cm, B-doped, and (100)-oriented monocrystalline Si wafer. The wire mask consists of Ta wires of 1 mm diameter. The gap between the wires is 4 mm. Hence, the mask has a period of 5 mm. Following Figure 5.20, the layer system deposited onto the p^+-type Si wafer is: (i) a 1 µm-thick p^+-type Si layer without a wire mask, (ii) a 3 µm-thick p-type layer again deposited without a mask, (iii) a 0.6 µm-thick n^+-type layer with the mask in position 1, (iv) a 3 µm-thick p-type layer with the mask in position 2, which is located 2.5 mm to the left of position 1, and finally, (v) we deposit a full 0.6 µm-thick n^+-type layer without a mask. An Al grid is evaporated onto the front surface and an Au grid onto the back surface of the device.

Figure 5.22 shows a light beam-induced current mapping of the 1×1 cm^2 solar cell that we record under illumination with light of wavelength 750 nm (optical absorption length about 7 µm) and a spot diameter of 70 µm. Bright areas correspond to large currents, while the black areas correspond to zero current. The shadow of the grid fingers, as well as the structure due to the Ta wires, is clearly visible. We define the square shape of the cell with its rounded corners by an additional mask that is placed in front of the porous substrate. The cloudy shadow close to the bus bar is due to the contacting needle. The cell exhibits three different layer systems, as can be seen in the schematic of Figure 5.21: (i) a region with a thin n-type emitter (for short, the np-region),

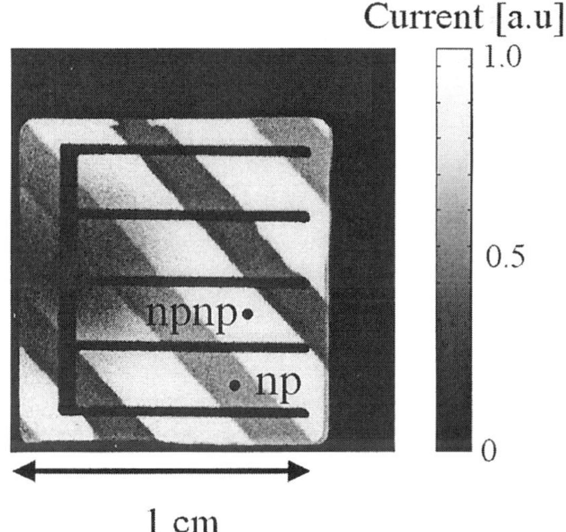

Figure 5.22. Light beam-induced current mapping of a multi-layer solar cell. The cell is illuminated with light of wavelength 750 nm. The current is higher in the npnp-region than in the np-region.

(ii) a region with two electrically connected emitters that are separated by 3 μm of p-type Si (for short, the npnp-region), and (iii) a region with a thick n-type emitter. We only investigate regions (i) and (ii), the np- and npnp-region. The npnp-region shows a larger current density than the np-region, as we would expect from the cell design.

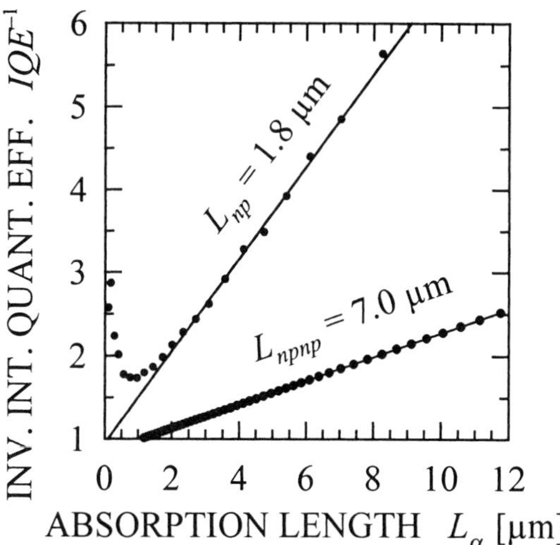

Figure 5.23. Local inverse internal quantum efficiency at the points denoted in Figure 5.22 by "np" and "npnp". The npnp-region has an apparent effective diffusion length that is more than three times greater than the one in the np-region.

In order to quantify the improvement in carrier collection, we measure the external quantum efficiency locally at the points denoted by np and npnp in Figure 5.22. From the measured external quantum efficiency and the measured reflectance of a bare Si wafer we determine the local internal quantum efficiency. The plot of the inverse internal quantum efficiency versus the optical absorption length shown in Figure 5.23 is used to deduce an apparent diffusion length [36]. In the np-region this length is 1.8 μm. In the npnp-region the apparent effective diffusion length is 6.9 μm.

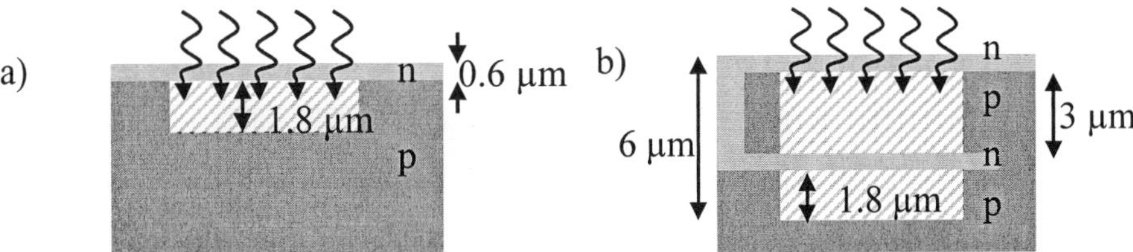

Figure 5.24. Schematic cross-section a) through the np-region and b) through the npnp-region of the multi-layer structure shown in Figure 5.21. The thickness of all n-type layers is 0.6 μm, and that of the p-type layer between the two n-type layers is 3 μm.

Figure 5.24a shows the layer system in the np-region. Carriers are collected from the shaded region, which extends 1.8 μm down from the junction. Now we estimate the depth from which the carriers are collected in the npnp-region, which is shown schematically in Figure 5.24b. The width of the p-type layer between the two n-type layers is 3 μm and is smaller than twice the diffusion length (1.8 μm). Hence, all carriers generated in this region are expected to reach the upper or the lower n-type layer. In addition, carriers generated down to 1.8 μm underneath the lower n-type layer are also collected. The thickness of the two junctions is 0.6 μm. Their space charge region is 0.4 μm wide. Therefore, we expect good carrier collection down to a depth of (2 × 0.6 μm + 3 μm +

0.4 µm + 1.8 µm) = 6.4 µm. This value agrees well with the measured apparent diffusion length of 6.9 µm. The Shadex technique thus enables the fabrication of a multi-junction cell with little extra processing effort.

5.2.3 Integrated series connection of CVD-grown films

In this section we report our novel approach to fabricating an integrated series connection from CVD-grown Si layers, using the porous Si process [341]. The process sequence is illustrated in Figure 5.26. a) As for the fabrication of a single solar cell, we use epitaxy on a porous Si layer. The P-doped emitter is formed by thermal diffusion and an Al grid is vacuum-evaporated. b) We then interrupt the emitter by plasma etching using

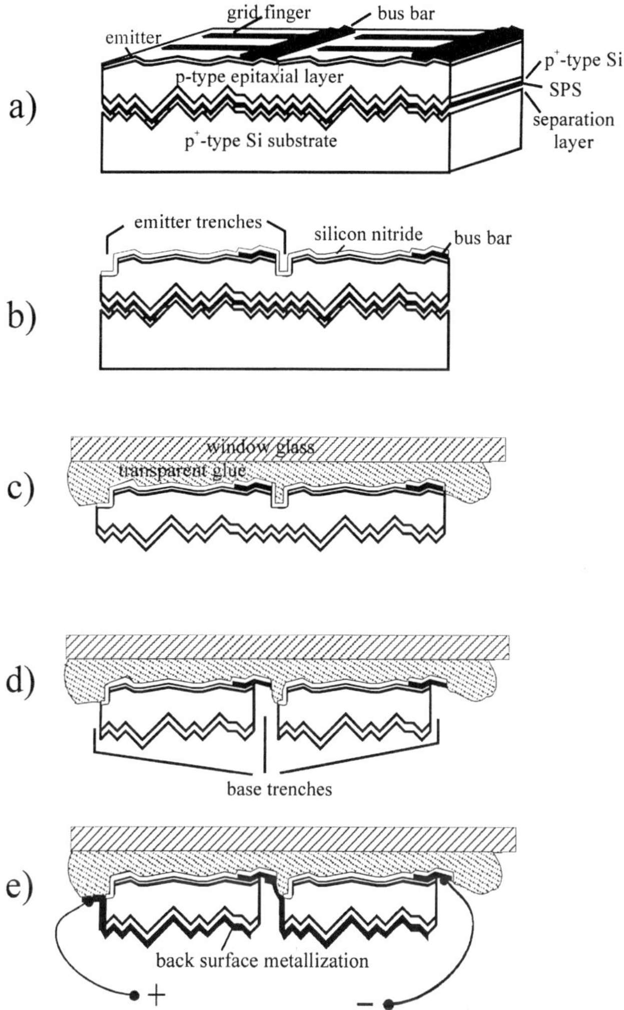

Figure 5.25. Process sequence for the fabrication of an integrated series connection from CVD-grown epitaxial Si layers.

SF_6 and Ar gases. The depth of the trench is 3 μm. Ni foils are used for masking. c) The glass carrier is glued to the cell and the residual sintered porous Si is removed from the back of the cell by plasma etching. d) Using plasma etching at a rate of 1.5 μm min^{-1} and masking by Ni foils, we separate the base of the cells. The plasma etches neither the Al nor the transparent glue. Thus, the timing of the process is not critical. e) Metallization of the back surface through a shadow mask completes the series connection.

This technique for series connection is only feasible because, using a layer transfer process, the thin-film cell is accessible from both sides. The other part of the trick is to use an etching technique that stops at the metallization.

With this process, we have fabricated a small module that has an area of 25 cm^2 and consists of five series-connected cells with an area of 5×1 cm^2 each. The film thickness is W_{eff} = 15 μm. The independently confirmed efficiency is 10.6% under air mass 1.5 illumination at 1000 W m^{-2} [341]. The open-circuit voltage is 3035 mV (607 mV per cell) and the short-circuit current is 115.1 mA (23 mA cm^{-2} per cell). For more details see Ref. [341].

5.3 Optical absorption in Si waffles

This section analyzes the optical absorption in waffle-shaped Si films from the PSI process by theoretical ray-tracing and by spectrally resolved optical reflectance measurements. We concentrate on periodic waffle structures of the type shown in Figure 5.8 on p. 128. Waffles with a random texture have a similar level of light trapping [320]. To determine the photogeneration in a waffle-shaped solar cell, we use our Monte Carlo ray-tracing code SUNRAYS [26].

5.3.1 Hemispherical reflectance measurement

The solid line in Figure 5.26 shows the hemispherical reflectance measured for the encapsulated Si waffle shown in Figure 5.8 on p. 128. A detached Al mirror is held behind the sample. For comparison, the filled circles show the hemispherical reflectance simulated by ray-tracing. The simulation reproduces the measurement without adjustment of model parameters [53]. Small deviations between the measurement and the simulation are explained qualitatively by the micro-roughness of the pyramidal facets, which we do not include in the simulation. We describe the impact of micro-roughness on the reflectance spectrum in Ref. [27].

For our sample with a thickness W_f = 5.8 μm (corresponding to $W_{eff} = W_f/\cos(54.7°)$ = 10 μm) and a waffle period p = 13 μm, we calculate a maximum short-circuit current j_{sc}^* = 36.5 mA cm^{-2} ± 0.5 mA cm^{-2} from the simulated absorption (filled triangles) under illumination with an AM1.5G spectrum of 1000 W m^{-2} integrated intensity.

The simulated absorption of a planar Si film (open circles) of thickness W_f = 5.8 μm is very different from the measurement. Note the strong reduction of the front surface reflection by the texture without antireflection coatings. For the planar film, the ray-tracing simulation yields an optical absorption (open triangles) that corresponds to a maximum short-circuit current density j_{sc}^* = 22.5 mA cm^{-2} ± 0.5 mA cm^{-2}. Hence, light trapping enhances the photogeneration by a factor of 1.6. The enhancement factor for Lambertian light trapping (Model L) at W_{eff} = 10 μm over no light trapping (Model O) at W_{eff} =

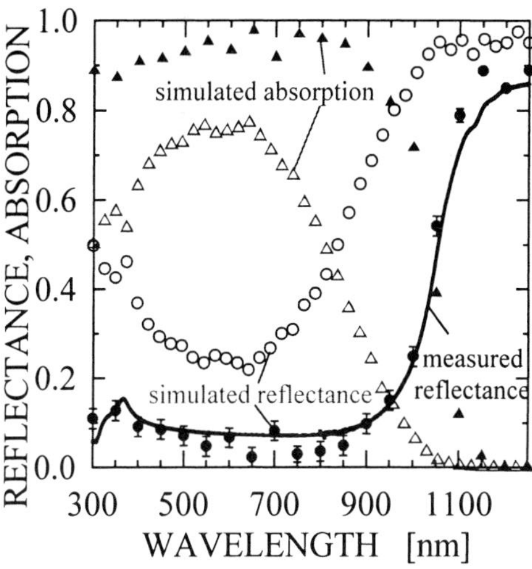

Figure 5.26. Measured hemispherical reflectance of encapsulated waffle structure (solid line); calculated reflectance (circles) and absorption spectra (triangles) for the waffle (full symbols) and a planar Si film of thickness W_f = 5.8 μm. The simulated absorption spectrum corresponds to a maximum short-circuit current j_{sc}^* = 36.5 mA cm^{-2} for the waffle. The error bars are due to the statistics of the Monte Carlo simulation. The Si layers are assumed to have a hole concentration of 2×10^{17} cm^{-3}.

5.8 μm deduced from Figure 2.8 on p. 20 is also 1.6. This enhancement factor underlines the large impact of the texture on the optical absorption of the thin film.

5.3.2 Optical design parameters

Having established that ray-tracing reproduces the hemispherical reflectance measurement, we apply the Monte Carlo simulation to analyze the impact of various optical design features on the photogeneration j_{sc}^*.

The waffle texture that we analyze is shown in Figure 5.27. We assume a structured cover-glass and a transparent glue that both have the optical constants of amorphous SiO_2. The grids on both sides of the cell are neglected. We perform a ray-tracing study that starts with a very optimistic baseline structure (*Case A*) and step by step includes more realistic, and even pessimistic optical features (*Cases B to J*).

Case A, the baseline cell, has a texture period p = 8 μm and an effective film thickness W_{eff} = 6.9 μm corresponding to a film thickness $W_f = W_{eff} \times \cos(54.7°)$ = 4 μm. The reflector has unity reflectance, and the illumination impinges normally onto the module. Free carrier absorption in the Si film is also neglected. We assume a 2 mm-thick cover-glass with the optical constants of amorphous SiO_2 [219]. The front surface of the glass is textured with symmetric grooves with a facet angle $\beta = 45°$. The waffle film has facets inclined by α = 54.7° to the macroscopic cell surface. Optical constants of intrinsic crystalline Si are used [22]. An antireflection coating (ARC) of 60 nm thickness with refractive index 2.3 is assumed to cover both sides of the textured Si film. This baseline structure has a maximum short-circuit current density j^* of 38.4 mA cm^{-2}, as listed in Table 5.3. Only (0.3 ± 0.05)% of the rays of wavelength λ = 600 nm do not enter the Si film, thus demonstrating the high efficiency of the textured front cover-glass in combination with a single-layer ARC at the glass/Si interface. The average path length of a ray of light of wavelength λ = 1000 nm is (55 ± 1) W_{eff}.

Case B replaces the ideal reflector by an Al mirror. The optical constants of Al are compiled in Ref. [219]. Table 5.3 also gives the relative change in the current density j_{sc}^* for each feature when compared to the baseline *Case A* without that feature. The Al

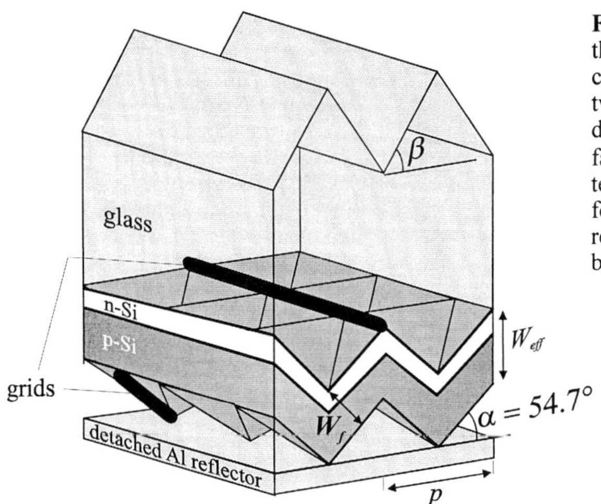

Figure 5.27. Schematic drawing of the optical design of a PSI waffle cell: The waffle is sandwiched between a textured cover-glass and a detached back surface reflector. The facet angle is fixed at $\alpha = 54.7°$. The texture period p requires optimization for maximum cell efficiency. The relative thicknesses of emitter and base are not to scale.

reflector reduces the current j_{sc}^* by 0.5%. For light with wavelength $\lambda = 1250$ nm absorption in the Si film is negligible and the simulated hemispherical reflectance is 96.1% for the encapsulated waffle. The reflectance of the air/Al interface is 96.9% under isotropic illumination. Therefore, the rays interact on average $N = \ln(0.961)/\ln(0.969) = 1.2$ times with the Al mirror.

Case C is identical to Case A, but without the two ARCs. The surface reflectance increases at $\lambda = 600$ nm from 0.3% with an ARC to (3 ± 0.1)% when omitting the ARC. The photogeneration reduces by 2%.

Case D includes parasitic free carrier absorption [77]. The current density j_{sc}^* reduces by 0.8%.

Case E considers a cell with a non-textured front glass ($\beta = 0$ in Figure 5.27), which reduces the j_{sc}^* by 1.8%. This value is of the order of magnitude of the reflectance of the flat air/glass interface (3.5%). Grooves in the cover-glass reduce this reflectance [88].

Case F considers the movement of the sun during the day and the year, giving a reduction in j_{sc}^* of 2%. This low sensitivity of the waffle-shaped texture to the illumination geometry is an important feature for good outdoor performance of future modules.

In practice, the presence of air within a solar cell module is unfavorable. *Case G* replaces the air in front of the reflector by an encapsulant with the optical constants of SiO_2, resulting in j_{sc}^* reduction by 0.8%.

Case *H* considers an Al reflector adjacent to the back surface of the Si film instead of a detached back surface reflector. The photogenerated current decreases by 9.9%. At a wavelength $\lambda = 1250$ nm the hemispherical reflection is 40.8%, while the reflectance of the planar Si/ARC/Al system is 93.7%. Hence, we estimate an average number of $N = \ln(0.408)/\ln(0.937) = 13.8$ interactions with the back surface of the Si film.

Case I assumes an unstructured Si film ($\alpha = 0°$) with the effective thickness W_{eff} of the waffle. The reduction in photogeneration j_{sc}^* is 19.5%.

Finally, *Case J* considers a waffle-shaped texture that is likely to be realized in practice. *Case J* is a combination of some of the previous cases: an Al reflector (*Case B*), a doped Si film (*Case D*), a non-textured cover-glass (*Case E*), the direction of illumination averaged over a year (*Case F*), and a glass encapsulant between the reflector and the Si film (*Case G*). The maximum short-circuit current density j_{sc}^* is 34.9 mA cm^{-2}, a value that is 9.1% smaller than for *Case A*. Only this practical *Case J* will be considered in section 5.4 on p. 144 for the estimation of the power conversion efficiency of waffle cells.

Table 5.3. Parasitic absorption and other optical effects reduce the maximum short-circuit current density j_{sc}^* as calculated by ray-tracing for the structure shown in Figure 5.27, with period p = 8 μm and thickness W_f = 4 μm. The statistical error is 0.1 mA cm^{-2}.

Case	Cell structure	j_{sc}^* [mA cm^{-2}]	change [%]
A	Baseline cell	38.4	0
B	Al reflector	38.2	–0.5
C	no ARCs	37.6	–2.0
D	free carrier absorption	38.1	–0.8
E	flat air/glass interface ($\beta = 0°$)	37.7	–1.8
F	year's average	37.6	–2.0
G	Glass between reflector and Si film	38.1	–0.8
H	Al reflector adjacent to Si	34.6	–9.9
I	flat Si film ($\alpha = 0°$)	30.9	–19.5
J	B + E + F + G + H	34.9	–9.1

Comparison of the waffle with Lambertian light trapping

Note that the 34.9 mA cm^{-2} corresponfs to 90% of the photogeneration of the Lambertian light trapping scheme and 86% of the photogeneration of an ideal light trapping scheme. We studied the distribution of the path lengths in the waffle cell under isotropic illumination in Ref. [342]. The average path length and the standard deviation of the path length distribution approach the respective parameters of a Lambertian light trapping scheme to within 10%. As for an ideal Lambertian scheme, we find a path length distribution that decays exponentially for long path lengths (see Eq. (A.21) on p. 185).

Relative significance of optical design features

To summarize, a photogeneration j_{sc}^* ranging from 35 to 38 mA cm^{-2} is possible with a waffle-shaped Si film of W_{eff} = 6.9 μm and W_f = 4 μm thickness, depending on the details of the optical design. Our case study reveals that the detached back surface reflector and the waffle shape are primarily responsible for the high optical absorption. All other factors are less important by a factor of at least 5. The detached back surface reflector reduces the number of interactions with the metal from 13.8 to 1.2 interactions.

5.3.3 Detached back reflector

Figure 5.28 shows the detached back reflector. Assuming that the direction of light propagation is sufficiently randomized by the front surface, only a fraction $1/n_S^2$ of the isotropically impinging light escapes into the air behind the cell, where n_s is the silicon index of refraction. The effective back reflectance is $R_b = 1 - 1/n_s^2 = 0.92$. This escaping fraction $1/n_s^2$ is then reflected by a metal reflector of reflectance R_m and the total reflectance is $R_b = 1 - 1/n_s^2 + (1/n_s^2)\,R_m$. This is equivalent to

$$\frac{1-R_b}{1-R_m}=\frac{1}{n_s^2} \tag{5.1}$$

We assume here that the interface of Si and air has an ideal antireflection coating. A generalization of this relation to a non-ideal coating and to a non-unity refractive index between Si and the reflector is given in Appendix A on p. 206. The interpretation is as follows: for fully randomized light the intensity of light, or equivalently the path length of light, is proportional to the square of the refraction index n_s^2 [95], as discussed in section 2.1.3 on p. 16. A reduction of the index of refraction in front of the metal from the value n_s for Si (ordinary non-detached reflector) to unity (detached reflector) reduces the optical loss by a factor $(1/n_s)^2$. For an air/Al reflectance of 0.92 we calculate an effective back reflectance of $R_b = 1 - (1 - 0.92)/n_s^2 = 0.994$. A ray-tracing simulation of the textured Si/air/Al system determines an effective back reflectance of 0.997. This value holds for isotropic illumination at $\lambda = 1250$ nm. Thus, high reflectance values are feasible with a simple Al mirror.

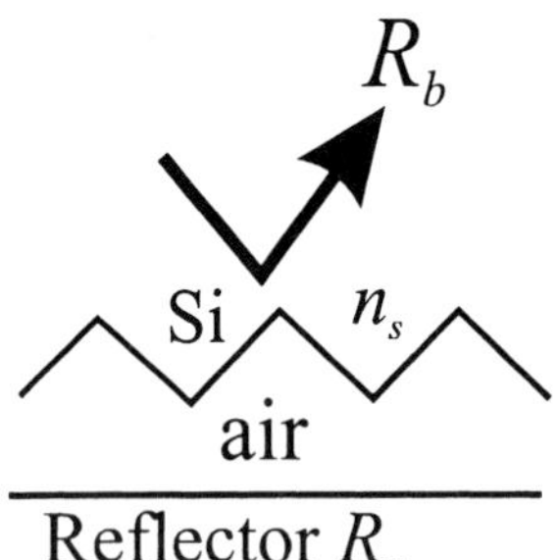

Figure 5.28. Definition of the hemispherical reflectance R_b for light isotropically incident on the inner surface of a pyramidally textured Si/air interface. A planar metal reflector with reflectance R_m is placed below the semi-infinite Si sample.

5.4 Efficiency potential

In this section, we calculate the efficiency potential of waffle-shaped Si solar cells. We aim at the determination of the optimum texture period p and film thickness W_f. The results of this optimization clearly depend on the material quality assumed. We characterize the quality of the thin film by a minority carrier diffusion length L and a surface recombination velocity S. In contrast to our modeling in Chapter 1, we now account for the finite carrier mobility in Si.

The importance of judging the efficiency potential at an *optimum* thickness was pointed out in previous work [343, 28, 223]. Therefore, all our simulations determine the cell efficiency at optimum film thickness W_f.

In order to reduce the tremendous amount of numerical computing, we develop an approximate analytical model for the minority carrier transport in the waffle-shaped Si solar cells. The comparison of the analytical and the more comprehensive numerical model demonstrates agreement to within 10% for all parameters of the illuminated current-voltage curve [342], an accuracy that is sufficient for an efficiency estimate. The analytical modeling described below is sufficiently general to be applicable not only to waffle cells, but also to other conformal films on textured glass substrates, such as those discussed in Ref. [108].

5.4.1 Optimization of the period-to-thickness ratio

Figure 5.27 on p. 142 shows the type of solar cells we plan to fabricate from the Si films described above. The waffle-shaped Si film a few microns in thickness is sandwiched between a front cover-glass and a detached Al back surface reflector. The front glass is textured to reduce the reflectance at the air/glass interface. The PSI process, as described in the section 4.3.5 on p. 113, adjusts the facet angle to $\alpha = 54.7°$. Only the film thickness W_f and the texture period p are free parameters for the optimization of the cell performance.

In general, the thickness of a cell has to be less than the minority carrier diffusion length. However, for the waffle in Figure 5.27 it is not clear whether W_f or W_{eff} is the appropriate length with which to compare the diffusion length L. A clear definition of a "geometrical collection thickness" W_C that is, in general, different from W_f and W_{eff}, is thus needed. We give such a definition in section 5.4.1.1.

The texture period p influences the cell performance in two ways: first, the capability of the cell structure to collect light-generated minority carriers depends on the texture period. Second, the optical absorption changes with the texture period. We consider both effects to determine the optimum period of waffle cells in section 5.4.1.2 on p. 147.

5.4.1.1 Geometrical collection thickness

We want the geometrical collection thickness W_C to monitor the average distance a minority carrier has to travel to the junction. The thickness W_C should characterize the geometry of the cell structure and hence should become independent of the minority carrier diffusion length L and the surface recombination velocity S. Let us consider a solar cell of arbitrary geometry under short-circuit conditions. The cell surface consists of regions with the collecting junctions and other regions which are ideally passivated. We consider a cell with diffusion lengths $L >> W_{eff}$. Hence, the diffusion equation for the minority carriers is well approximated by

$$D\,\nabla^2 n(\boldsymbol{x}) = -g(\boldsymbol{x}) \tag{5.2}$$

wherein the term describing the volume recombination has been omitted. The symbol D denotes the diffusion coefficient. This equation yields a minority carrier concentration profile $n(\boldsymbol{x})$ that is independent of L. The quantum efficiency

$$IQE = \frac{\int g(\boldsymbol{x})\,dV - \int n(\boldsymbol{x})/\tau\ dV}{\int g(\boldsymbol{x})\,dV} \tag{5.3}$$

is the volume integral over the generation rate $g(\boldsymbol{x})$ and the integral over the volume recombination rate $n(\boldsymbol{x})/\tau$. Here, τ denotes the minority carrier lifetime. With the relation $L^2 = D\,\tau$ we find

$$IQE = 1 - \gamma / L^2 \ \text{ for } L >> W_{eff} \text{ and } S = 0 \tag{5.4}$$

with

$$\gamma = \frac{D\int n(x)dV}{\int g(x)dV} \tag{5.5}$$

For a *planar* cell and spatially homogeneous generation rate $g(x)$ the coefficient

$$\gamma = \frac{W_{eff}^{\ 2}}{3} \tag{5.6}$$

depends only on the thickness $W_{eff} = W_f$ of the cell.

Definition of collection thickness W_C

We define the geometrical collection thickness of the waffle cell as the thickness W_C a planar Si cell requires to show the internal quantum efficiency of the waffle cell (in the limit $S \rightarrow 0$, $L \rightarrow \infty$, and for spatially homogeneous generation). Hence, the definition of the geometrical collection thickness is

$$W_C = \sqrt{3\frac{D\int n(x)dV}{\int g(x)dV}} \tag{5.7}$$

Calculation of collection thickness

We calculate the right-hand side of Eq. (5.7) with the numerical device simulator DESSIS [344] and check that the same value W_C is calculated for various large values of the diffusion length L. Figure 5.29 shows the geometrical collection thickness W_C re-

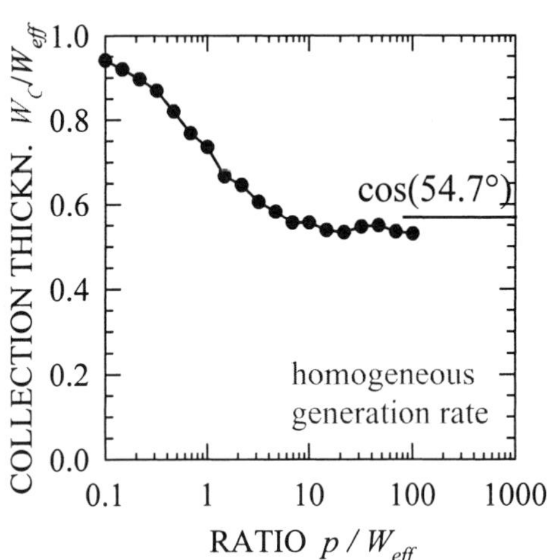

Figure 5.29. Geometrical collection thickness ratio W_C/W_{eff} as a function of period p/W_{eff}. The value W_C/W_{eff} approaches cos(54.7°) for large periods.

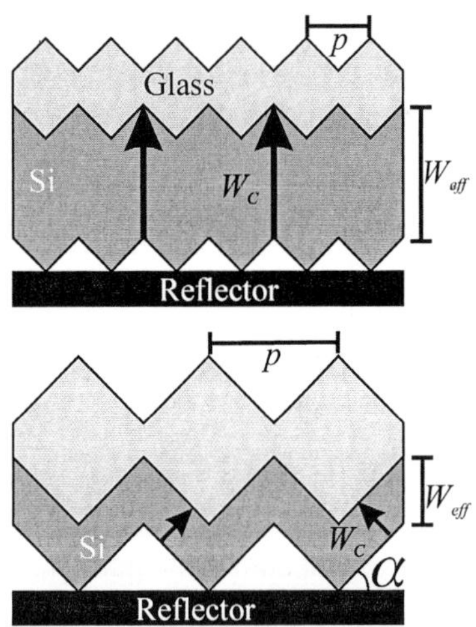

Figure 5.30. The arrows indicate the minority carrier path in waffles with small and large periods. For small periods (top) the path length is equal to W_{eff}, and for large periods (bottom) the path length is $\cos(\alpha) \times W_{eff}$.

sulting from a two-dimensional (2D) transport simulation of the waffle cell. Hence, we approximate the pyramidal texture of the waffle cell by a V-grooved texture. We consider spatially homogeneous generation. The geometrical collection thickness is given in units of the effective waffle thickness W_{eff} = 10 μm. We find a geometrical collection thickness W_C / W_{eff} that decreases from a value close to unity for very small p/W_{eff} ratios to a value close to 0.55 for large p/W_{eff} ratios.

Figure 5.30 gives a simple explanation for the period dependence of the geometrical collection thickness W_C shown in Figure 5.29. For $p << W_{eff}$ the situation is similar to a textured thick Si wafer cell. The geometrical collection thickness approaches the physical thickness $W_C = W_{eff}$. In the limit of large periods $p >> W_{eff}$ the ridges and tips of the waffle cell play a minor role. Carriers have to diffuse a distance $W_C = W_{eff} \cos(\alpha) = W_f$. For the waffle cell, $\cos(\alpha) = 0.58$ is in fairly good agreement with the numerically determined geometrical collection thickness for large periods. The transition from high to low geometrical collection thickness is soft for our 2D simulations. We find the transition to be sharper for three-dimensional simulations. Periods $p > W_{eff}$ show a better carrier collection than periods $p < W_{eff}$, since the effective geometrical collection thickness W_C saturates at a small value.

5.4.1.2 Optical thickness

Ray-tracing studies show that the absorption depends on the texture period p [107, 109, 345]. Here, we study this dependence for the waffle-shaped texture using the three-dimensional ray-tracing program SUNRAYS [26]. We express the calculated photogeneration as an *optical* thickness W_O.

Definition of optical thickness W_O

By definition, the optical thickness W_O is the thickness a planar cell with zero front and zero back surface reflectance (Model N on p. 16) must have to absorb the same photon flux as the textured cell.

Calculation of optical thickness

Figure 5.31 shows the numerically determined optical thickness W_O of the waffle cell with thickness W_{eff} = 10 μm for a period ranging from 1 μm to 10^4 μm. Illumination by an AM1.5G spectrum is assumed. The optical thickness W_O shows a strong dependence on the period. For periods $p < W_{eff}$, the optical thickness saturates at values around 11 W_{eff}. For large $p >> W_{eff}$, the optical thickness decreases to values close to 2 W_{eff}. Geometrical resonance effects are observed at particular p/W_{eff} ratios, e.g. the dip at $p/W_{eff} = 1$. Similar resonance effects in wafer cells have been calculated before [2]. However, these effects do not have a strong influence on the current densities. At p = 10 μm the maximum short-circuit current density is 37.9 mA cm^{-2}, while it is 38.3 mA cm^{-2} at p = 14 μm.

For weakly absorbed light at 1250 nm that causes a spatially homogeneous carrier generation profile, the optical thickness is W_O / W_{eff} = 44 independently of the texture period p.

Figure 5.32 gives the explanation of the dependence of the optical thickness W_O on the texture period. For textures with a small period $p < W_{eff}$ total internal reflection is likely because many light rays interact with more than one facet orientation. In contrast, for large texture periods $p >> W_{eff}$ most of the rays "see" a planar sheet of Si that does not permit total internal reflection. Only light rays entering the Si film near the intersections of adjacent facets have a chance to be trapped. With a larger period an increasing

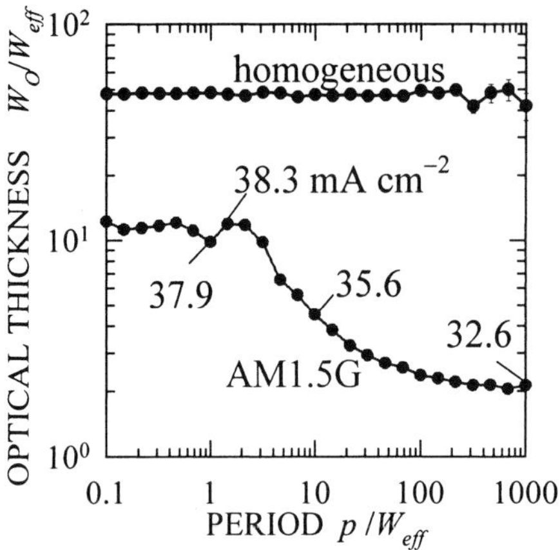

Figure 5.31. Optical thickness W_O of the waffle cell with detached Al back surface reflector and a thickness W_{eff} = 10 μm under isotropically incident AM15G illumination and for a spatially homogeneous generation rate. Maximum short-circuit j_{sc}^* current densities are given for some periods.

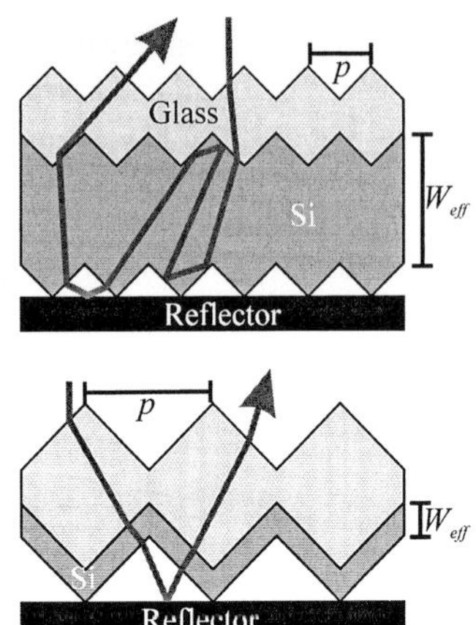

Figure 5.32. Light propagation in waffles of small and large texture period. For textures with small period $p << W_{eff}$ (top), light trapping due to total internal reflection is probable. For textures with large periods (bottom), most of the rays traverse the Si layer twice without being trapped.

fraction of the rays does not enter the Si near these intersections. The majority of rays traverses the film only twice. We expect an optical thickness close to $2W_{eff}$.

Optimum period-to-thickness ratio is about unity

In conclusion, for a large optical thickness W_O a period $p < W_{eff}$ is required. For a small geometrical collection thickness W_C a period $p > W_{eff}$ is required. Therefore, the optimum value of the period is a compromise. Simulations show that $p \approx W_{eff}$ is a good choice. Our next step is the determination of the absolute thickness value W_{eff} for arbitrary minority carrier diffusion length L and surface recombination velocity S.

5.4.2 Modeling

Cell structure

We assume a Si cell with a P-doped emitter of donor concentration 5×10^{18} cm^{-3} and thickness W_e = 0.5 μm. The base is B-doped with an acceptor concentration of 10^{18} cm^{-3} and a thickness $W_b = W_f - W_e$. We assume an equal thickness for emitter and base if the total film thickness W_f is less than 1 μm. The simulated cell has a silicon nitride antireflection coating (ARC) with an index of refraction of 2.3 and a thickness of 66 nm on both surfaces of the Si film. The material between the back surface of the cell and the aluminum reflector is air.

Reducing the dimensionality

Table 5.4 compares the physical situation of the waffle cell (first column) with a two-dimensional (2D) model and a one-dimensional (1D) model, both of which we use in this section. The waffle cell is a three-dimensional (3D) object that requires 3D modeling. However, such 3D modeling takes several hours for a single point the current-voltage curve (e.g. with the device simulator DESSIS [344] on a personal computer).

Table 5.4. Three models for simulating waffle cells. A real solar cell has a three-dimensional generation profile and three-dimensional minority carrier transport. The 2D model uses two-dimensional generation profiles and two-dimensional transport equations. The 1D model uses homogeneous photogeneration and a one-dimensional transport model.

	Waffle	2D model	1D model
Geometry			
Carrier generation	three-dimensional	two-dimensional	homogeneous
Transport	three-dimensional	two-dimensional	one-dimensional

Consequently, a 3D device optimization is currently not feasible. Simplified models have to be developed.

Modeling the optics

The 2D model approximates the waffle shape by V-grooves with the same facet inclination angle α. We generate a 2D generation profile $g(x,z)$ by ray-tracing. From section 2.1.5 on p. 19 we know that light trapping in two-dimensional textures is less efficient than in three-dimensional textures [16]. The photogeneration current density

$$j^*_{2D} = \frac{q}{A}\int_V g_{2D}(x,z)\, dx\, dy\, dz \tag{5.8}$$

is therefore smaller than

$$j^*_{sc} = j^*_{3D} = \frac{q}{A}\int_V g_{3D}(x,y,z)\, dx\, dy\, dz \tag{5.9}$$

For this reason, we scale the 2D generation profile to $\frac{j^*_{3D}}{j^*_{2D}} g_{2D}(x,z)$ to have the photo-generated current density of the 3D waffle cell in its 2D representation. In Eq. (5.9), the photogenerated current density j^*_{3D} is determined by a scaling law explained in Ref. [342] from a single path length distribution $f(l)$ that was calculated by ray-tracing for a wavelength of $\lambda = 1000$ nm and a film thickness $W_f = 4$ µm. The scaling law exploits the similarity of light propagation in textures that are identical except for a scaling factor.

The 1D model further simplifies the analysis by considering only one-dimensional transport. The grooves from the 2D model are now approximated by slabs of Si with film thickness W_f at an inclination angle α. We assume a spatially homogeneous generation rate $g = j^*_{3D} / W_{eff}$. The assumption of a homogeneous generation rate permits a simple analytical solution of the transport equations.

Modeling the transport

For 2D transport simulation we use the DESSIS simulation software [344] in the 2D simulation mode. Contacts to the emitter and the base are provided all over the cell surfaces. Thereby, resistive losses due to lateral current flow in the emitter and the base are neglected. The minority carrier lifetimes of the Shockley-Read-Hall recombination model are adjusted to yield the specified low-level injection minority carrier diffusion lengths L. In order to reduce the number of parameters, we assume equal diffusion length L in the emitter and the base. There is one exception to this equality: Auger recombination limits the diffusion length in the highly doped emitter to 8 μm. The less strongly doped base has a Auger-limited diffusion length of 80 μm. For diffusion lengths in the range 8 μm < $L \leq$ 80 μm the specified L value only applies to the base, while the emitter value is 8.0 μm. Diffusion lengths L > 80 μm are not considered. We also assume the same surface recombination velocity S on both surfaces of the cell. Mobility values and bandgap-narrowing parameters of crystalline Si are taken from Ref. [343]. The simulation of a current-voltage curve requires a couple of minutes with the 2D model.

For the 1D model all material parameters are the same as for the 2D model. We assume one-dimensional transport perpendicular to the pyramidal facets. The current-voltage curve

$$j(U) = \frac{1}{\cos(\alpha)}\left[\left(j_{oe} + j_{ob}\right)\left(\exp(U/U_t) - 1\right) + j_{scr}(U)\right] - j_{sc} \tag{5.10}$$

depends on the applied voltage U, the dark saturation current density j_{oe} and j_{ob} of the emitter and the base, the recombination current $j_{scr}(U)$ in the space charge region, the short-circuit current density j_{sc}, and the thermal voltage U_t. The factor $1/\cos(\alpha)$ describes the increase of the Si volume, the surface, and the junction area with facet angle α. The saturation current densities

$$j_{ox} = \frac{qn_i^2 D_x}{N_x L} \frac{\frac{SL}{D_x}\cosh(W_x/L) + \sinh(W_x/L)}{\frac{SL}{D_x}\sinh(W_x/L) + \cosh(W_x/L)} \tag{5.11}$$

depend on the diffusion coefficient D_x, the doping concentration N_x, the thickness W_x and the effective intrinsic carrier concentration n_i [346]. Here the index $x = e, b$ stands for the emitter region ($x = e$) or the base region ($x = b$) of the cell. Hence, the relation $W_f = W_e + W_b$ holds for the total cell thickness.

A model of the space charge recombination current density $j_{scr}(U)$ in the asymmetrical junction was given by Choo [347]. We assume a single midgap recombination center with equal capture rates for electrons and holes. The capture rate is $1/\tau = D_b/L^2$ rather than D_e/L^2, as most of the space charge region falls in the base region. Tunneling-assisted recombination is neglected for simplicity. The minority carrier generation rate is assumed to be spatially homogeneous. This assumption simplifies the expression

$$j_{sc} = \frac{W_e \eta_{ce} + W_b \eta_{cb}}{W_e + W_b} j_{3D}^{*} \tag{5.12}$$

for the short-circuit current density. Here,

$$\eta_{cx} = \frac{L}{W_x} \frac{\frac{SL}{D_x}\left(\cosh(W_x/L)-1\right)+\sinh(W_x/L)}{\frac{SL}{D_x}\sinh(W_x/L)+\cosh(W_x/L)} \tag{5.13}$$

denotes the collection efficiency in the emitter ($x = e$) and in the base ($x = b$). We do not consider current collection from the space charge region. Hence, we slightly underestimate cell efficiencies because collection efficiency is close to unity in the space charge region. The collection efficiency η_{cx} depends on L, S, W_x, and the diffusion coefficient D_x [24].

Optimization strategy

The strategy for the optimization of the cell geometry proceeds in two steps: in the previous section 5.4.1 on p. 145, we concluded qualitatively that the optimum period-to-thickness ratio is unity. A more quantitative analysis yields $p = 2W_f$ to be the optimum period [342]. Assuming the facet angle to be fixed at 54.7°, the next step is to optimize the film thickness W_f for arbitrary minority carrier diffusion length L and surface recombination velocity S using the 1D model. This optimization requires just a few seconds with the 1D model for every pair of values (S, L), although many current-voltage curves have to be simulated.

5.4.3 Optimization of the thickness

An optimum film thickness W_f exists because the photogeneration increases with thickness, while the collection efficiency decreases due to the finite length of the diffusion length L. The simulation optimizes the film thickness W_f for maximum cell efficiency at constant $p/W_f = 2$ ratio for every pair of values L and S. The cell efficiency, the open-circuit voltage and the short-circuit current density are deduced from the current-voltage curve given by Eq. (5.10).

Simulated efficiencies and optimum thickness values

Figure 5.33a shows the efficiencies (solid lines) at optimum cell thickness W_f (broken lines) for a large range of parameters S and L. At a diffusion length L = 10 μm, we calculate an energy conversion efficiency of 18% at an optimum cell thickness of 3 μm, for a surface recombination velocity $S = 1.3\times10^3$ cm s^{-1} (point A in Figure 5.33). At the maximum-power point, the recombination in the space charge region dominates in the hatched region of small diffusion lengths $L < 10$ μm. In this region, the current-voltage curve is not uniquely described by a minority carrier diffusion length L, because different combinations of energy levels and capture cross-sections yield the same low-injection diffusion length L, but different recombination current densities j_{scr} in the space charge region. In the non-hatched region of large diffusion lengths L, the recombination in the emitter and the base dominates over recombination in the space charge region.

Simulated short-circuit currents and open-circuit voltages

Figure 5.33b shows lines of constant short-circuit current density j_{sc} (solid lines) and open-circuit voltage V_{oc} (broken lines) for the efficiencies and cell thickness values shown in Figure 5.33. At point *A*, the efficiency of 18% is achieved with a short-circuit current density of j_{sc} = 33.2 mA cm^{-2} and an open-circuit voltage of 652 mV.

Figure 5.34 shows the optimum thickness W_f for surfaces passivated to S = 100 cm s^{-1}. The optimum thickness increases approximately linearly with the diffusion length L and follows the relation W_f = 1 µm + L/5. A film thickness deviating from the optimum value only slightly deteriorates the solar cell efficiency. For example, for a diffusion length L = 10 µm, the energy conversion efficiency decreases to 95% of the

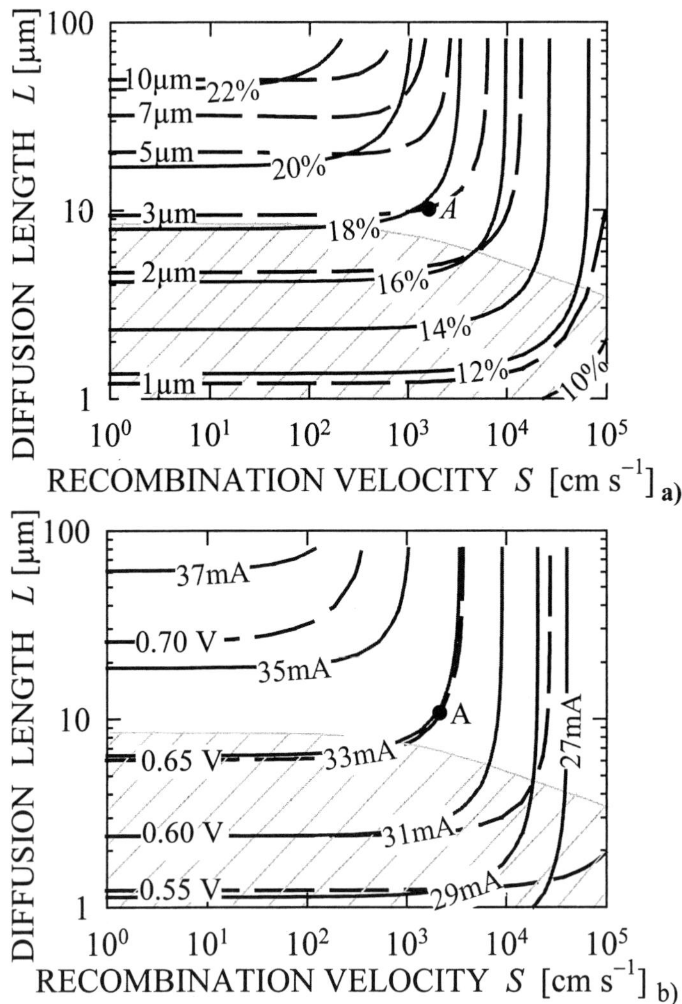

Figure 5.33. a) Theoretical energy conversion efficiency η (solid lines) and optimum film thickness W_f (broken lines) for cells with surface recombination velocities S and minority carrier diffusion length L; b) Short circuit current j_{sc} (solid lines) and open-circuit voltage V_{oc} (broken lines). The simulation assumes the waffle texture of Figure 5.8. Recombination in the space charge region dominates in the hatched region.

optimum (that is from 18% to 17%) if the cell thickness is reduced from W_f = 3 μm to 1 μm. Hence, waffle cells with a planar cover-glass, a detached back surface reflector, and glass between the metal and the silicon film can achieve efficiencies above 17% with a surface passivation S = 100 cm s^{-1} and volume diffusion length L = 10 μm. An efficiency of 18% is theoretically feasible with cell thickness values as low as W_f = 3 μm, a diffusion length of 10 μm, and a surface passivation S = 1000 cm s^{-1}.

Simulations with the 2D model agree with the much simpler 1D simulations to within 10% for all parameters of the current-voltage curve [342].

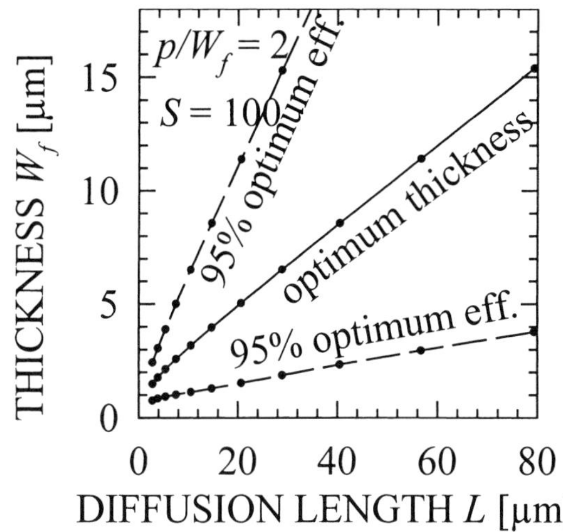

Figure 5.34. Optimum cell thickness W_f as a function of minority carrier diffusion length L for surface recombination velocity S = 100 cm s^{-1}. For L > 4 μm the optimum thickness is approximately $W_f = L/5$.

5.4.4 Optimization of the facet angle

The facet orientation of the waffle cell from the PSI process is fixed to 54.7° relative to the macroscopic cell surface by the orientation of the crystallographic (111) planes. Other facet angles may become possible with a different fabrication technology, e.g. by mechanical grinding or by reactive ion-etching.

Here, we study the light trapping behavior of a Si waffle as a function of facet angle α at constant film thickness W_f. The optical absorption increases with an increase in Si material available and with a reduction in optical reflection. The disadvantage of large facet angles α is a decrease in the cell voltage due to an enhancement of the volume and the surface recombination. In order to emphasize this disadvantage, we simulate a cell with the SRV S = 10^7 cm s^{-1}, equal to the thermal velocity of electrons. At the same time, we choose a large ratio p/W_f = 10 to make light trapping less efficient than for smaller $p/W_f \approx 1$. Our simulations show that for the range of parameters investigated (α ranging from 0° to 80° and diffusion lengths L ranging from 0.1 to 10 μm), the efficiency calculated at optimum cell thickness always increases with increasing facet angle, due to optical absorption that increases with the facet angle α [28]. Facet angles as large as technologically possible yield the best performance, at the price of enhanced Si consumption.

The increase of the efficiency with the photogeneration demonstrates that the efficiency is current-limited rather than voltage-limited. Extremely thin cells (sub-micron thickness, no light trapping) are always current-limited, since the photogeneration becomes approximately proportional to the Si volume, which increases by a factor of 1/cos(α). The dark saturation current density also increases with 1/cos(α) and thus the open-circuit voltage $V_{oc} = U_t \ln(j_{sc}/j_o + 1)$ is α-independent for very thin films, while the current increases with the facet angle α.

5.4.5 Space application

In contrast to terrestrial solar cells, space cells utilize low-doped Si material in order to reduce the impact of defect generation on minority carrier lifetime. High-energy electron or proton radiation enhances the minority carrier recombination rate in Si cells [348]. The enhancement in the recombination rate is proportional to the change

$$\Delta\left(\frac{1}{L^2}\right) = K_L \, \Phi \tag{5.14}$$

in diffusion length L [349]. Here Φ is the particle dose and K_L the damage coefficient for minority carrier diffusion length. This reduction in diffusion length will finally lead to a cell failure. Thin cells permit short diffusion lengths, and therefore space cells are designed to be as thin as mechanically and optically possible. Being thinner than conventional Si cells, the waffle cell is particularly attractive for space applications.

In order to demonstrate the beneficial effect of the waffle geometry, we compare the performance of a waffle cell with a conventional thin-film Si space cell as described by Yamaguchi et al. [349] that has a planar geometry, is 50 μm thick, and has a base doping concentration of 10^{15} cm^{-3}. In our simulation, the front and back surface recombination velocities are fixed to 1000 cm s^{-1}. The doping concentration of the conventional cell and that of the waffle cell are equal. The optical generation profile is calculated for nor-

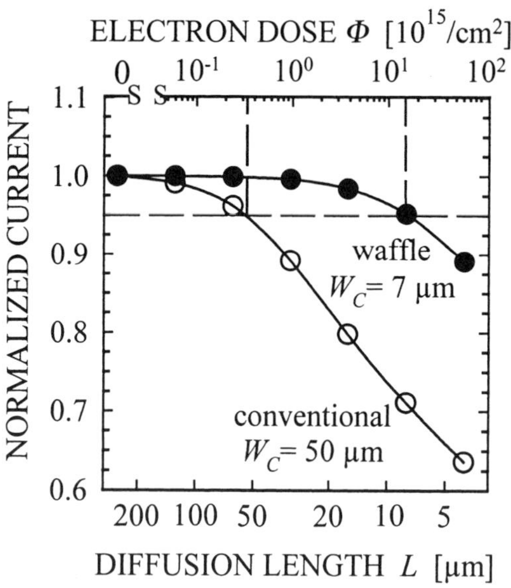

Figure 5.35. Decrease of normalized short-circuit current under 1 MeV electron irradiation. The efficiency of the waffle cell decreases more slowly than the efficiency of the conventional space cell.

mal incidence of an AM0 spectrum using the ray-tracing program SUNRAYS (see p. 74). The 1 MeV electron dose is translated into a diffusion length by Eq. (5.14) with $K_L = 10^{-10}$ [349].

Figure 5.35 shows the simulation results calculated with the 2D simulator DESSIS [344] for a W_{eff} = 10 μm waffle cell. The efficiency of the waffle cell decreases less rapidly. The dose that decreases the current from the waffle to 95% of the initial value is more than an order of magnitude smaller than for the conventional space cell. The ratio L_C / W_C of the critical diffusion length L_C = 7.9 μm that degrades the waffle current to 95% to the geometrical collection thickness W_C = 7 μm is L_C / W_C = 1.1 for the waffle. For the conventional cell, we find L_C = 58 μm, W_C = 50 μm, and L_C / W_C = 1.2. These two ratios are similar. In contrast, when dividing by W_{eff} instead of W_C, we find L_C / W_{eff} = 0.8 for the waffle and L_C / W_{eff} = 1.2 for the conventional cell. Obviously, these two values disagree. This confirms that the ratio of L_C / W_C is important for the degradation rather than the ratio L_C / W_{eff}.

6 Summary and conclusions

In this chapter we summarize solutions to the problems raised in the introduction (Chapter 1). The main objectives of this work are:

- To study the relative importance of the various physical loss mechanisms in thin-film crystalline Si solar cells and to determine their limiting power conversion efficiency as a function of device thickness.
- To discuss device analysis techniques that, in combination with comprehensive analytical device modeling, reveal the limitations of experimental thin-film crystalline Si cells.
- To review the previous experimental work on thin-film crystalline Si cells.
- To report on layer transfer processes that are a novel approach to crystalline Si solar cell fabrication with the potential of overcoming some of the previous technological limitations.

6.1 Physical limitations to power conversion

The photogeneration of electron-hole pairs in thin-film cells is limited by the small optical path length of the light in the cell. Texturing the surface enhances this path length due to multiple total internal reflection at the surfaces (see Figure 2.4 on p. 15).

6.1.1 Light trapping

Maximum path length enhancement

The maximum path length is realized in a thin-film cell that has a maximum radiation energy content. For surface textures sufficiently coarse to obey geometrical optics, the optical theorem of the conservation of the étendue applies. Similarly to Liouville's theorem in classical mechanics, the phase space volume of a bundle of light rays is constant. For homogeneous illumination, any phase space volume of the cell has either the radiance of the illuminating light or the radiance zero. The maximum energy content corresponds to the case where all the phase space has the illuminating radiance. In section A.2.1 on p. 187 we show that for arbitrary cell shapes with reflecting surfaces, the average path length never exceeds $\bar{l}_{max} = 4\ n_s^2\ W_{eff}$ under isotropic illumination. Here, the effective film thickness W_{eff} is the ratio of cell volume to cell area, and n_s is the wavelength-dependent index of refraction in Si. We consider isotropic illumination, which is more relevant in practice than perpendicular illumination because of the sun's daily and yearly movement and because of a large fraction of sunlight is diffuse.

Maximum photogeneration

An optimum light trapping scheme has, by definition, the maximum photogeneration that is physically possible. Such a scheme has to have the optimum optical absorption at *all* wavelengths. An optimum absorption under isotropic illumination is achieved when all rays, independently of their position and their angle of incidence, have a path length that equals $\bar{l}_{max} = 4\ n_s^2\ W_{eff}$. In this case, the standard deviation σ_l of the path length distribution is zero. Such an optimum light trapping scheme is currently not known.

The maximum photogeneration in a crystalline Si cell with optimum light trapping is 35.9 mA cm^{-2} for an effective film thickness of $W_{eff} = 1$ μm, and 41.0 mA cm^{-2} for $W_{eff} =$ 10 μm when photogeneration is expressed as a short-circuit current density (see Figure 2.8 on p. 20), assuming that every absorbed photon creates a single electron-hole pair. Comparing these figures with the short-circuit current density $j_{sc} = 42.2$ mA cm^{-2} of the 400 μm-thick world record crystalline Si cell [196], we can state that a 1 to 10 μm-thick cell can, in principle, absorb a sufficient number of solar photons to permit excellent device efficiencies.

In the framework of wave optics it is possible to have path lengths exceeding the limit $\bar{l}_{max} = 4\ n_s^2\ W_{eff}$ for isotropically illuminated cells. We proved this by studying a Si film that is much thinner than the wavelength. The film is sandwiched between an ideal metal and a cover material with an index of refraction larger than that of Si (see section 2.1.7 on p. 22). The electric field in the Si film increases with the index of the transparent cover. Choosing a large enough value for this index, the geometric optics limit $\bar{l}_{max} = 4$ $n_s^2\ W_{eff}$ is surpassed.

This example demonstrates that we should study sub-micron structures and sub-micron film thickness values to improve light trapping beyond the limit set by geometrical optics. Dealing with sub-micron structures is, however, likely to enhance the surface area of the device drastically, which in turn, increase interface recombination losses.

Lambertian light trapping

By definition, a Lambertian light trapping scheme causes isotropic light propagation after a single interaction of the solar light with the cell surface. Such a scheme has the optimum average path length $\bar{l}_{max} = 4\ n_s^2\ W_{eff}$. In thin cells, the path length distribution is exponential (see Eq. (A.21) on p. 185) and has a minimum standard deviation $\sigma_l = \bar{l}_{max}$. Lambertian light trapping is thus *not* an optimum light trapping although it has the maximum average path length.

We derived an analytical expression for the optical absorption of a Lambertian light trapping scheme (see Eq. (A.13) on p. 183). The maximum short-circuit currents are 8% below that of an optimum light trapping scheme for $W_{eff} = 1$ μm and only 3% below the optimum for $W_{eff} = 10$ μm (see Table 2.1 on p. 20). The performance difference of optimum and Lambertian light trapping is thus less than 8% for all device thickness values above 1 μm.

Lambertian light trapping rather than the optimum light trapping is often preferred as a benchmark for optical device performance, since we have at least an idea, how to realize Lambertian schemes: the surface has to fully randomize the direction of the transmitted and the reflected light. While it is theoretically possible to do better than with a Lambertian scheme for non-isotropic illumination, this has not yet been demonstrated experimentally.

6.1.2 Generalized detailed balance model for thin-film cells

The physical losses can be divided into two groups: (i) intrinsic losses, e.g. radiative recombination losses or Auger recombination losses, that cannot be avoided; (ii) extrinsic losses, e.g. defect recombination, that could, in principle, be avoided by an improved technology. We also consider losses due to "design errors" such as omitting light concentration. By adding one after the other of these physical loss mechanisms to a thin-film cell that has the optical constants of crystalline Si, we determine their relative significance. For our study we develop a cell model that has the following properties:

- The thin-film device has the dielectric function of crystalline Si. The common assumption of an abrupt onset of the absorption at the electronic bandgap energy is too crude for thin-film cells (see Figure 1.2 on p. 4). We include phonon-assisted subbandgap absorption in our model. The smallest phonon-assisted band-to-band excitation we consider has an energy of 0.89 eV, corresponding to a wavelength of 1400 nm.
- We consider illumination with a spectral intensity distribution of the global air mass 1.5 spectrum. Maximum light concentration and one-sun illumination are considered.
- Our model also describes cells without thermalization losses by assuming an optimum quantum yield. Optimum quantum yield means that as many electron-hole pairs are generated per absorbed photon as is energetically possible (see Eq. (2.27) on p. 28). To model these cells we apply a modified version of the author's detailed balance model for carrier-multiplying cells [66, 18].
- We consider three levels of light trapping: optimum light trapping, Lambertian light trapping, and no light trapping (a single light path of length W_{eff}). In each case the absorption increases with the effective thickness W_{eff}, giving, in most cases, higher efficiencies for thicker cells. For large thickness values, the efficiency can decrease with thickness, since the recombination current increases with the cell volume and thereby the small gain in absorption is overcompensated. We thus have to analyze the impact of the various losses as a function of the device thickness.
- Photon recycling, the re-absorption of luminescence radiation, is included. This is accomplished by choosing a detailed balance model with an optical absorption of thermal radiation in the dark that equals the thermal emission (see Eqs. (2.16) and (2.20) on p. 25).
- Considering very thin cells without Auger recombination and without defects does, under maximum light concentration, lead to device voltages that exceed the electronic excitation energy of the smallest band-to-band transition (0.89 eV in our case). For these cases we include the possibility of laser action by assuming a *negative* absorption coefficient whenever the splitting of the quasi-Fermi levels exceeds the electronic band-to-band transition energy (see Eq. (2.21) on p. 26). This effect is, however, of no relevance in real Si solar cells, due to strong Auger recombination.
- We assume an infinite carrier mobility. In homogeneously doped devices the carrier concentration is thus spatially constant, which greatly simplifies the analysis. In practice, electrically thin cells are required to make this assumption approximately valid. Special cell designs, such as the parallel multi-junction design [116] and the Encapsulated-V design [27], yield electrically thin cells (see Figure 2.13 on p. 25) in which the distance that the minority carriers have to diffuse is kept much smaller than the minority carrier diffusion length.

Maximum efficiency of an idealized thin-film crystalline Si cell

The maximum efficiency of a 1 µm-thick crystalline Si cell with optimum geometrical light trapping under isotropic illumination and no defect recombination is 26.7%. The limiting efficiency is 28.6% for a 10 µm-thick device. For comparison, the Carnot efficiency of a thermodynamic machine working between the temperature 5780 K of the sun and 300 K is η_{Car} = 94.8%.

Hypothetical thin-film cells operating at the Carnot limit

We construct a hypothetical thin-film cell that has the optical absorption coefficient of crystalline Si and that works at the Carnot efficiency. Such a Carnot cell has to avoid all irreversibilities. Thus it should absorb all solar photons, avoid carrier relaxation to the band edges, and avoid recombination via defects.

Since our hypothetical thin-film Carnot cell absorbs solar radiation with the optical constant of crystalline Si, luminescence, the inverse process to photogeneration, inevitably occurs. Non-absorption of sub-bandgap light is also inevitable. Since we cannot avoid these losses even in a gedanken experiment, all we can do is to exclude them by *defining* the power conversion efficiency η_{net} as the electrical output power divided by the *net* input power (incident solar power minus luminescence power emitted minus non-absorbed power). There is a justification for this definition: under maximum light concentration the Carnot cell redirects the luminescence light and the light that was not absorbed back to the sun. The redirected power is thus not consumed and is available for *later* conversion. The efficiency η_{net} is, admittedly, of no practical use.

The hypothetical Carnot cell described above has an efficiency η_{net} = 94.5%, a value that is almost equal to η_{Car} = 94.8% if operated at or close to open-circuit conditions (see section 2.2.1 on p. 27). The small difference between η_{net} and η_{Car} originates from the fact that the air mass 1.5 spectrum is not exactly a black body spectrum. The Carnot efficiency value is reached for *all thickness values* and for *all levels of light trapping*.

At open-circuit, the electron-hole gas of the cell is in chemical equilibrium with the sun's radiation. No or little current is flowing. The open-circuit voltage is a fraction $\eta_{net} \approx \eta_{Car}$ of the bandgap energy. The hypothetical cell converts radiation *energy* to electrical energy with the Carnot efficiency but it has a zero or close to zero *power* conversion efficiency. Close to zero power conversion means that the machine works slowly – that is, adiabatically. The commonly used term "energy conversion efficiency" is, strictly speaking, a misnomer and is replaced in this work by "power conversion efficiency".

6.1.3 Significance of intrinsic and extrinsic loss mechanisms

The step-by-step reduction of the power conversion efficiency due to the introduction of intrinsic and extrinsic losses is illustrated in Figure 6.1a for a 1 µm-thick cell with Lambertian light trapping.

Luminescence: Starting from the Carnot cell, we first introduce the loss due to luminescence radiation. The definition of the efficiency η_{abs} is output power per absorbed power (incident solar power minus non-absorbed power). With this change in definition the efficiency reduces from the Carnot value to η_{abs} = 87.5% (see section 2.2.2 on p. 29) for a 1 µm-thick device. The efficiency η_{abs} decreases with increasing film thickness since the luminescence intensity increases with the film thickness.

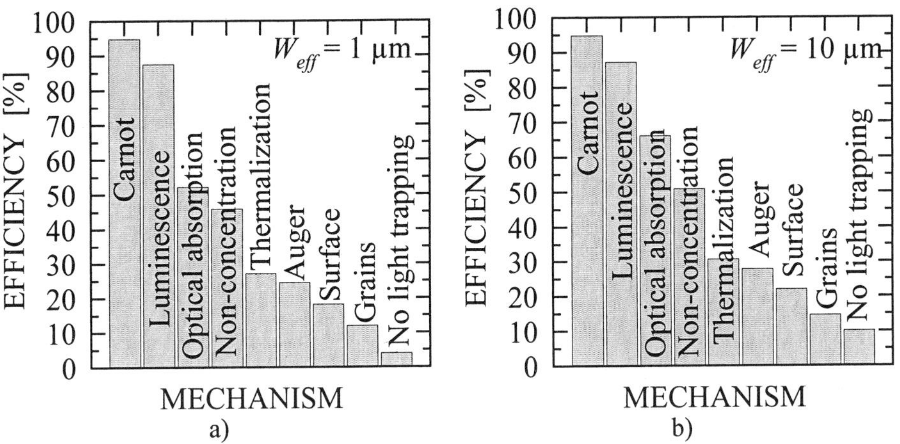

Figure 6.1. Limitations of the power conversion efficiency of thin-film crystalline Si solar cells when adding one loss mechanism after the other in the order from left to right: a) cell thickness W_{eff} = 1 µm; b) cell thickness W_{eff} = 10 µm.

Optical absorption: The non-absorption of sub-bandgap photons and the incomplete absorption of above-bandgap photons are both significant losses, even for Lambertian light trapping. The common definition of the efficiency η is electrical output power per incident solar radiation power and thus includes optical losses. A 1 µm-thick Lambertian cell under fully concentrated air mass 1.5 solar illumination without thermalization losses, without Auger-recombination, and without defect recombination has a conversion efficiency of 52.2% (see section 2.2.3 on p. 30).

Non-concentration: Cell operation with non-concentrated solar light and without restricting the optical acceptance angle lets the cell see the dark sky under a solid angle that is 10^5 times larger than the solid angle of the sun. This couples the electron-hole gas radiatively to the cold sky. The open-circuit voltage is then no longer defined by the chemical equilibrium of the electron-hole gas with the sun. This reduces the open-circuit voltage and thus the power conversion efficiency to 45.9% for a 1 µm-thick device (see section 2.2.4 on p. 31).

Thermalization: Coupling the electron-hole gas to the Si lattice by assuming unity instead of optimum quantum yield introduces phonon generation that further reduces the cell efficiency to 27.1%. Unity quantum efficiency is a good approximation for terrestrial Si cells. Our experimental and theoretical investigations show a significant enhancement of the quantum yield due to impact ionization only at high photon energies that are not contained in the terrestrial solar spectrum (see Figure 2.17 on p. 32).

Auger recombination: Auger recombination is the dominating inevitable non-radiative recombination mechanism in crystalline Si [56, 19]. Adding this loss defines the ultimate power conversion efficiency limit for thin-film crystalline Si cells. For a 1 µm-thick cell exhibiting Lambertian light trapping we find η = 24.5% under one-sun illumination (see section 2.3.1 on p. 35). For optimum light trapping the ideal efficiency is even 26.7%. These ideal devices work at medium injection levels. The efficiency figures thus slightly depend on the as-yet unsatisfactory knowledge of the Auger recombination rates at intermediate injection levels (see section B.2.2 on p. 214).

Surface recombination: In practice, surface recombination is a particularly important loss mechanism. Assuming well passivated surfaces with S = 100 cm s^{-1} on both sides (acceptor concentration $N_A = 10^{16}$ cm^{-3}), the efficiency of the 1 μm-thick cell reduces to 18.2% (see section 2.3.2 on p. 40). This figure is not a fundamental limit, since surface recombination velocities smaller than 100 cm s^{-1} are possible. With today's technology, 100 cm s^{-1} is an outstandingly good value for a contacted Si surface.

Grain boundary recombination: Including grain boundary recombination with a recombination velocity of $S_{grb} = 10^4$ cm s^{-1} for columnar grains that are 10 times smaller than the film thickness, the efficiency of a 1 μm-thick film with Lambertian light trapping is estimated to be 12% (see section 2.3.3 on p. 45). A grain boundary recombination velocity of 10^4 cm s^{-1} is feasible for hydrogen-passivated high-temperature deposited polycrystalline thin-film cells.

No light trapping: For a planar polycrystalline thin-film with no light trapping our model determines an efficiency of only 4%.

Figure 6.1b shows the same gradual reduction of the device efficiency as Figure 6.1b for a cell ten times thicker with W_{eff} = 10 μm. The efficiencies are slightly higher, due to a larger optical absorption. The intrinsic efficiency limit of a 10 μm-thick cell with no defect recombination and Lambertian light trapping is 27.8% and thus only 13% relative higher than for a 1 μm-thick device. This shows that we should rather aim at devices in the 1 μm range, since, theoretically, we can gain an order of magnitude in material saving and reactor throughput without losing much efficiency.

Relative significance of the various loss mechanisms

Figure 6.2 illustrates the significance of the various loss mechanisms defined as the relative reduction of the efficiency when adding a loss mechanism to our model. This significance depends slightly on the order in which we introduce the mechanisms. We use the order from left to right in Figure 6.1. The losses that are not intrinsic, and that can thus, in principle, be avoided by introducing improved technologies, are marked by black bars.

Figure 6.2a shows the loss significance for devices of 1 μm thickness with the more significant losses sorted to the right. Light trapping is by far the most important feature to enable high cell efficiencies. The three major tasks to be solved for high efficiencies with thin films are therefore, in the order of their significance:

(i) Fabrication of an efficient light trapping scheme, e.g. by appropriate surface texturing.
(ii) Avoidance of grain boundary recombination, e.g. by growing monocrystalline material.
(iii) High-quality surface passivation, e.g. by applying silicon nitride dielectric layers.

Figure 6.2b shows that for 10 μm-thick cells light trapping is only half as important as for 1 μm-thick devices.

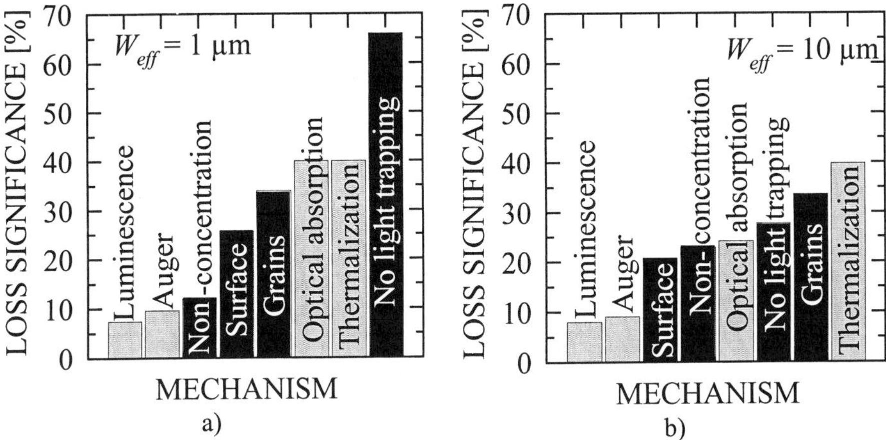

Figure 6.2. The significance of loss mechanisms measured as the relative decrease in cell efficiency when adding loss mechanisms in the order of Figure 6.1 from left to right. Black bars mark losses that may be reduced with an appropriate technology while gray bars mark intrinsic losses. a) Cell thickness W_{eff} = 1 µm; b) cell thickness W_{eff} = 10 µm.

6.2 Revealing the limitations of experimental cells

The identification of optical losses and recombination losses is an important prerequisite to optimizing thin-film crystalline Si cells in the laboratory. The quantum efficiency spectrum of a thin-film solar cell reflects both the optical properties and the transport properties, and is therefore a quantity that elucidates more of the device physics than the current-voltage curve.

Performing light beam-induced current mappings at various wavelengths provides lateral resolution and a depth resolution. For monocrystalline material the lateral resolution is not of interest and a quantum efficiency measurement (see section C.1 on p. 241) suffices.

6.2.1 Optics of thin-film cells

Ray-tracing analysis of faceted thin-film cells

We apply Monte Carlo ray-tracing (see section 3.5.1.1 on p. 74) to model the optical properties of thin-film crystalline Si cells. In order to test the ray-tracing approach, we fabricate crystalline Si films with well-defined geometries that are easily computer modeled. We consider float-zone Si wafers with periodic inverted pyramids (period of 13 µm) that are thinned down to about 45 µm (see Figure 3.8 on p. 75). The other test structure is a glass-encapsulated pyramidal film texture (see Figure 5.26 on p. 141) with a thickness W_{eff} = 10 µm and a texture period of 13 µm. We find good agreement between measured and calculated hemispherical reflectance spectra in both cases, thus demonstrating that Monte Carlo ray-tracing can model thin-film cells with texture periods and film thickness values as small as 10 µm.

For even thinner films and smaller texture periods we do not yet have a direct comparison of experiment and ray-tracing simulations. A comparison of incoherently and coherently calculated photogeneration current densities indicates, however, that ray-tracing permits us to calculate the photogeneration sufficiently accurately down to film thickness values of around 2 μm (see section A.2.2 on p. 191). Since photogeneration is an integral quantity, the interference effects average out when integrating over the wavelength. In other words, the coherence length of the broad-band solar spectrum is short. In addition, surface textures and isotropic illumination randomize the direction of light propagation and thus further weaken interference effects.

Detached back reflectors

We compare various types of back surface reflectors for crystalline thin-film cells (see section A.3 on p. 200): dielectric interlayers, reflectors from multi-layers with porous Si, and detached reflectors. The detached reflector is particularly attractive, due to its simple fabrication and high performance.

A detached back reflector consists of the free Si back surface mounted in front of a detached metal reflector. The gap between them is sufficiently large to avoid frustrated total internal reflection. The Si back surface may be textured and/or coated with a passivation layer. Most of the light is reflected by total internal reflection. The small fraction that passes the Si back surface is returned into the cell by the metal reflector. Using numerical ray-tracing, we calculate an effective back surface reflectance of 0.997 ± 0.001 for a pyramidal film texture with a detached Al reflector (see section A.3.3 on p. 206). This high reflectance value was calculated for light that is isotropically incident on the internal back surface. The reflection losses are proportional to the squared refractive index of the transparent gap material (see Eq. (A.43) on p. 207).

6.2.2 Quantum efficiency spectra

A powerful approach for the analysis of quantum efficiency spectra is the direct comparison of measured and simulated spectra.

Analytical modeling of the carrier generation rate: We give an analytical model for the carrier generation profile (see Eq. (3.70) on p. 77). This modcl contains many parameters, e.g. the angle of the first pass of light through the cell, the internal front surface reflectance for diffuse and specularly reflected light, the back surface reflectance, the Lambertian character of the back surface, and others. Most of these parameters need to be determined independently of the quantum efficiency measurement. One ore two of the more important parameters, e.g. the back surface reflectance, can then be determined by fitting the measured reflectance spectrum. We demonstrate good agreement of the carrier generation profiles determined from this analytical procedure with profiles calculated by ray-tracing (see Figure 3.10 on p. 81) for a 50 μm-thick crystalline Si cell with inverted pyramids on the front surface. This model is now also being applied by other groups [350].

Analytical modeling of the carrier transport: Our one-dimensional analytical model for the electronic transport considers minority carrier diffusion in the emitter, the base, and the back surface field layer (see section 3.5.2 on p. 82). The diffusion equations are solved for a carrier generation rate that we extract from the analysis of the reflectance spectrum. The strength of this approach is that the transport equations are solved analytically. It is thus possible to compare thousands of simulated spectra with the measured spectrum in just a few seconds of computation time.

Parameter confidence plots reveal fitting ambiguities: Iso-lines of the deviation of simulated and measured quantum efficiency are plotted as a function of the recombination parameters, e.g. as a function of the bulk diffusion length L and back surface recombination velocity S_b. We call such graphs parameter confidence plots (see Figure 3.14 and Figure 3.15 on p. 85); they identify graphically the set of parameter combinations that are commensurate with the measured data [25]. Such plots provide valuable information on the ambiguities of the fitting procedure and the confidence of the fit results.

Injection level-dependent surface recombination velocity: Crystalline thin-film cells with surfaces well passivated by SiO_2 show a quantum efficiency that strongly depends on the bias voltage applied to the cell (see Figure 3.16 on p. 87). This is a consequence of a surface recombination velocity at the SiO_2/Si interface that varies by orders of magnitude with the injection level (see Figure 3.18 on p. 89). The physical origin of this injection level dependence is the transition from electron-limited recombination with a large capture cross-section for electrons to hole-limited recombination with a small capture cross-section for holes. The reduction of the surface potential with increasing injection level also contributes to injection level dependence. The measured injection level dependence at the surface of our thin-film crystalline Si cells is explained quantitatively (see Figure 2.25 on p. 42) by recombination via midgap electronic dangling bond states with the extended Shockley-Read-Hall model (see B.3.1 on p. 220).

Surface passivation with silicon nitride is of greate interest to low-cost thin-film photovoltaics. At the silicon-silicon nitride interface, the capture cross-sections of electrons near the midgap are smaller than those of holes [187]. The injection level dependence is weaker than for surfaces passivated with silicon oxide (see Figure 2.26 on p. 43) and originates from the decrease in surface potential with increasing injection level (see Figure B.13 on p. 228).

Differential and actual recombination parameters: When the recombination rate at the cell surface depends non-linearly on the injection level, the quantum efficiency depends on the bias light intensity and the bias voltage. The surface recombination that we extract with a linear theory is not the actual recombination velocity $S = U_{sur} / \Delta n$, but the differential surface recombination $S_{diff} = dU_{sur} / d\Delta n$ (see Figure 3.17 on p. 88). Here, U_{sur} is the surface recombination rate and Δn the excess carrier concentration at the edge of the space charge region. In order to deduce the actual surface recombination velocity S, we integrate the differential values S_{diff} that we extract from measurements at various injection levels Δn (see Eq. (3.83) on p. 87). The necessity to distinguish actual and differential recombination parameters [30] is now generally accepted [225, 157].

6.2.3 Carrier recombination in polycrystalline cells

Grain boundary recombination at the interfaces of columnar grains makes the electronic transport three-dimensional.

Carrier recombination in the neutral base: We give an analytical solution of the minority carrier diffusion equation for polycrystalline thin-film cells (see Appendix B on p. 233) by applying a Fourier decomposition of the transport equation. With the help of the reciprocity theorem of charge carrier collection (see section 3.2 on p. 57) this analytical solution also provides the quantum efficiency of polycrystalline thin-film cells. Using our model, we extract grain boundary recombination velocities of $S_{grb} = 10^4$ to 10^5 cm s^{-1} for hydrogen-passivated polycrystalline thin-film Si cells (see section 4.1.3 on p. 96).

Carrier recombination in the space charge region: Carrier recombination at grain boundaries crossing the space charge region is also analytically modeled using the same approach as for the quasi-neutral base (see section B.4.3 on p. 237). Here we decompose Poisson's equation into its Fourier components. The assumption of an interface state density that does not change with energy leads to a linear relation of the Fermi level position and the grain boundary charges [35]. We also assume flat quasi-Fermi levels, which is equivalent to little recombination in the space charge region. Applying this model, we find that highly doped space charge regions exhibit less recombination (at a fixed splitting of the quasi-Fermi levels) than lowly doped space charge regions (see Figure 2.33 on p. 50). This finding is rather unexpected, since small-grained devices tend to use intrinsic material in the space charge region [253] in order to improve the device performance. We speculate that, similarly to amorphous Si cells, the dangling bond concentration at the grain boundary depends on the position of the Fermi level, thus leading to enhanced grain boundary recombination in doped material. Future investigations have to show whether tunneling-assisted Shockley-Read-Hall recombination, which is greater in highly doped material, is an important recombination mechanism in the experimental cells.

6.2.4 Effective diffusion lengths

Even with the improved multi-dimensional analytical models presented in this work, the routine analysis of internal quantum efficiency spectra requires much of apriori knowledge (e.g. doping concentration, mobility, emitter thickness) and, as revealed by parameter confidence plots, the fit results are seldom unique. This situation calls for an evaluation scheme that, at the expense of the amount of information deduced, is more robust and requires less apriori knowledge and less computational work.

Definitions of effective diffusion lengths: We define effective diffusion lengths L_Q and L_C, which are deduced from a plot of the inverse internal quantum efficiency against the optical absorption length (see section 3.1 on p. 55). The diode saturation current of the base also defines an effective diffusion length L_J. Table 6.1 gives an overview of the three effective diffusion lengths, their interrelationships and their interpretation.

Table 6.1. Various effective diffusion lengths as derived from the quantum efficiency data and current-voltage measurements.

	Quantum efficiency diffusion length L_Q	Current-voltage diffusion length L_J	Collection diffusion length L_C
Derived from	quantum efficiency	current-voltage	quantum efficiency
Definition	$\left.\frac{dIQE^{-1}}{d\alpha_s^{-1}}\right\vert_{\alpha_s^{-1}=0} = \frac{1}{L_Q}$	$j_o = \frac{qD_n n_o}{L_J}$	$\left.\frac{dIQE^{-1}}{d\alpha_s^{-1}}\right\vert_{\alpha_s^{-1}=\infty} = \frac{1}{L_C}$
Relationship	$L_Q = L_J$		$L_C \neq L_Q$
Interpretation	control recombination current		controls short-circuit current

Effective diffusion length from quantum efficiency equals effective diffusion length from dark saturation current: Applying the reciprocity theorem for charge carrier collection, we prove (see section 3.3.1 on p. 65) that the diffusion length L_Q extracted from the quantum efficiency data for strongly absorbed light is always identical to L_J. This relationship also holds for inhomogeneous semiconductors, e.g. for polycrystalline Si. All recombination rates have to be linear, however, in the excess carrier concentration Δn. The dark recombination current density, which is typically measured under forward injection conditions, may thus be deduced from the quantum efficiency data measured under short-circuit conditions.

Upper and/or lower limits to interface and volume recombination: Analytical expressions for the effective diffusion lengths L_Q and L_C in polycrystalline thin-film cells were deduced for the special case of columnar grains having a square cross-section (see Eqs. (C.19) and (C.25) on pp. 250 and 252). Using these expressions, we demonstrate that upper and/or lower limits to the bulk diffusion length L, the back surface recombination velocity S_b, and the grain boundary recombination velocity S_{grb} may be deduced from a measured value of the effective quantum efficiency diffusion length L_Q (see Figure 3.6 on p. 71).

6.2.5 Reciprocity theorem for charge carrier collection

The reciprocity theorem relates the local carrier collection efficiency to the excess carrier concentration under forward injection. The theorem was previously proven for free electrons and holes in their respective bands obeying Boltzmann statistics. We derive this reciprocity theorem from the principle of detailed balance (see section 3.2.1 on p. 58), following the work by Rau and Brendel [42]. Since the principle of detailed balance holds independently of the underlying statistics, we could generalize the reciprocity theorem to Fermi statistics (see section 3.2.2 on p. 64). Highly doped back surface field layers and highly doped emitters, as well as deep electronic states at grain boundaries or surfaces can now be analyzed with the reciprocity theorem.

Why quantum efficiency analysis is so powerful

With the reciprocity theorem we also understand why the quantum efficiency contains such a wealth of information (on linear cells): the inverse Laplace transform of the quantum efficiency spectrum $IQE(\alpha_s)$, when interpreted as a function of the absorption coefficient α_s, yields the local carrier collection efficiency η_c (see section C.4 on p. 254). The reciprocity theorem links the local $\eta_c(\boldsymbol{r})$ to the excess concentration $\Delta n(\boldsymbol{r})$ of the electrons (see section 3.2.2 on p. 64). This excess concentration $\Delta n(\boldsymbol{r})$ is proportional to the local recombination rate $\Delta n(\boldsymbol{r})/\tau(\boldsymbol{r})$. Quantum efficiency analysis thus provides, in principle, a spatially resolved insight on recombination rates deep in the device.

6.2.6 Discriminating surface and bulk recombination

An unambiguous determination of the bulk diffusion length L is not possible if the value of the diffusion length L exceeds the device thickness and if the surface recombination is not known. We introduced a non-destructive approach to drastically reduce surface recombination [45].

Applying corona charges to dielectrically coated Si surfaces electrostatically repels, depending on the polarity, the minority or majority carriers from the semiconductor surface and thus reduces surface recombination (see Figure 2.19 on p. 36). The surface concentration of deposited charges is limited by the dielectric strength of the insulating layer. Surface recombination velocities as small as 1 cm s^{-1} were achieved with this technique. This is a reduction by about three orders of magnitude relative to the case of a non-charged surface. The interface state density has to be smaller than the corona charge density to permit a shift of the quasi-Fermi level. At interface state densities that are too high, the corona charges find their counter charges in the interface states. In this case no band bending is induced and the electrostatic passivation fails.

Unfortunately, the corona charges are not stable for extended time periods. A technique to stabilize the corona charge would be of high practical importance for solar cell surface passivation.

6.3 Limitations of current thin-film approaches

In Chapter 4, we review the various experimental approaches to crystalline thin-film Si solar cells that are currently under investigation.

Classification of thin-film approaches

All thin-film cells require a substrate to enhance their mechanical strength. We broadly classify the various approaches according to the type of substrate used:

- High-temperature substrates (HTS) that withstand Si deposition temperatures $T_d > 800°C$,
- Low-temperature substrates (LTS) that require Si deposition at temperatures $T_d < 800°C$, and
- Layer transfer processes (LTP) that use a high-temperature resistant Si substrate for film growth and then transfer the film to a low-cost substrate such as glass or plastics.

Some characteristics of the three approaches are summarized in Table 6.2.

Table 6.2. Classification of thin-film approaches by the substrate, and typical values for the deposition temperature and grain size

	Low-T substrate (LTS)	High-T substrate (HTS)	Layer transfer process (LTP)
Preferred substrate	glass, plastics	ceramics, graphite	Si for growth, arbitrary for carrier
Deposition temperature	< 800°C	> 800 °C	> 800°C
Grain size	20 nm ... 1 μm smaller than thickness	1 μm ...1 mm larger than thickness	∞ larger than thickness

6.3.1 Physical requirements for high efficiency

On p. 162 we identified three physical requirements for thin-film cells with a high power conversion efficiency: (i) efficient light trapping, (ii) negligible grain boundary recombination, and (iii) efficient surface passivation. The capability to cope with these tasks is a better indicator for the potential of the three thin-film approaches than the mere value of the efficiency achieved today. Therefore we now compare the strengths and challenges for the LTS, the HTS, and the LTP approach with respect to issues (i) through (iii).

Light trapping: Figure 6.4 compiles the measured short-circuit current densities of crystalline thin-film cells as a function of film thickness. The lower line marks the current density of a device with thickness W_{eff} that collects all carriers generated by a single light pass through the cell (Model N from p. 16). The upper line is for the current density of a cell with Lambertian light trapping and unity collection efficiency (Model L from p. 16). All experimental results are closer to the no light trapping line than to the Lambertian line. The majority of the thin-film cells either have a poor optical design or a thickness that is not adapted to the material quality, e.g. the cells are too thick for the material quality achieved. These experimental results show that light trapping is *the* key task to improve today's thin-film cells.

Only the 2 μm-thick STAR cell [254] from an LTS approach and the PERL-type cells whose fabrication consumes a silicon wafer [351, 113, 48] show a significant amount of light trapping.

The record HTS cell [47, 236], for example, does not utilize any light trapping. Although suggestions were made to introduce light trapping in HTS cells with Bragg reflectors from sintered porous Si [29, 298], a substantial enhancement of the optical absorption by applying these measures was neither demonstrated nor simulated.

In this work we have achieved a short-circuit current density of 29 mA cm^{-2} with a waffle-shaped cell that is 17 μm thick using an LTP process. This value is currently only surpassed by the PERL-type thin-film Si cells that, unfortunately, consume a full Si wafer for their fabrication. Optical absorption measurements with waffle-shaped thin

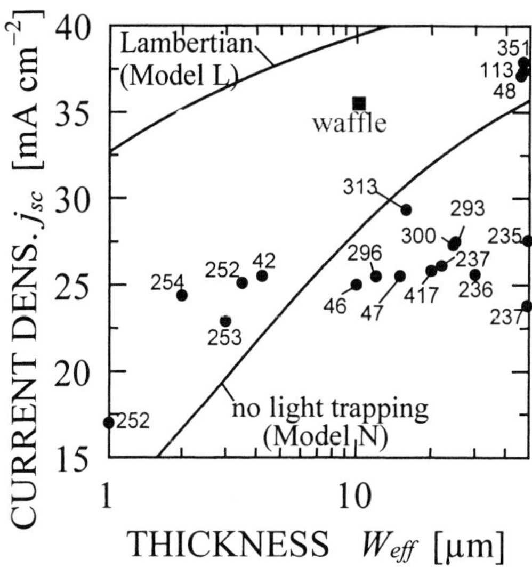

Figure 6.4. Measured short-circuit current densities of crystalline Si thin-film cells as a function of Si thickness. Data are labeled by the respective reference number. The square labeled "waffle" was simulated for a waffle-shaped film as shown in Figure 5.8.

films from the PSI process [53, 320] show a current density potential of 36 mA cm^{-2} for an effective film thickness of W_{eff} = 10 µm, even without using any antireflection coatings.

Grain boundary recombination: The importance of the grain boundary recombination depends on the grain size and the grain boundary recombination velocity. According to Ghosh's analysis conducted in 1980 [352], polycrystalline cells are generally limited by a high grain boundary recombination. The open circles in Figure 6.3 show the data collected by Ghosh et al. together with the open-circuit voltage V_{oc} (solid line) that we calculate with these author's model. For this calculation, we use the effective diffusion

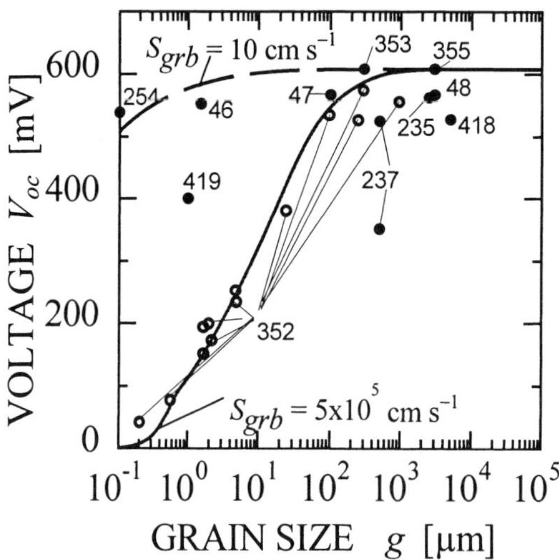

Figure 6.3. Experimental open-circuit voltage as a function of grain size. Data are labeled by the reference number. The solid line is calculated for a base doping of $N_A = 10^{17}$ cm^{-3}, base thickness W_{bas} = 10 µm, back surface recombination velocity S_b = 1000 cm s^{-1}, base diffusion length L_{bas} = 30 µm, and a grain boundary recombination velocity of $S_{grb} = 5\times10^5$ cm s^{-1}. To explain the LTS data [254, 46] a smaller grain boundary recombination velocity of S_{grb} = 10 cm s^{-1} has to be assumed (broken line).

length L_Q according to Eq. (C.24) on p. 252 in all formulas that Ghosh et al. used for the injection currents, and we use L_C according to Eq. (C.29) on p. 253 in the formulas that Gosh et al. used for the short-circuit current. Assuming grain boundary-limited recombination (S_{grb} = 5×10^5 cm s^{-1}), our simulation explains the data collected by Ghosh et al. The filled circles in Figure 6.3 are more recent measurements of open-circuit voltages in thin-film crystalline Si cells. We assumed L_b = 30 μm when calculating the solid line. Thus at at a large grain size a voltages below the solid line stem from intra-grain diffusion lengths L_b shorter than 30 μm. Experimental data with open-circuit voltages above the solid line are due to grain boundary recombination velocities smaller than 5×10^5 cm s^{-1}.

The HTS approach is capable of yielding grain sizes around 100 to 300 μm, values 10 times larger than the film thickness, whicht is typically 30 μm. The application of an analytical two-dimensional electronic transport model to the FORSOL cell [47] extracts grain boundary recombination velocities of 10^4 to 10^5 cm s^{-1} for hydrogen-passivated HTS cells. The HTS data fall close to the solid line. Grain boundary recombination can thus be significant in HTS cells [199]. Its importance can hopefully be decreased by further improving the hydrogen passivation. This assumption is supported by the report on a 16%-efficient cell fabricated by zone melt recrystallization on oxidized Si wafers [353, 289].

The reduction of the grain boundary recombination velocity in LTS cells to values that are orders of magnitude smaller than commonly observed in HTS cells is a major recent breakthrough in the development of thin-film Si solar cells. In order to explain voltages above 500 mV in grains that are only 0.1 to 1 μm in size [254, 46], we have to assume a grain boundary recombination velocity of 10 cm s^{-1} in Ghosh's model. Our two-dimensional analysis with a two-dimensional transport model (see section 4.2.3 on p. 104) results in grain boundary recombination velocities of a few hundered centimeters per second for the STAR cell [254].

Grain boundary recombination is absent in monocrystalline cells from the LTP class (see Figure 5.11 on p. 130). We therefore find no data points for LTP cells in Figure 6.3.

Figure 6.5 compiles the open-circuit voltages of thin-film crystalline Si solar cells as a function of the deposition temperature T_d. The figure shows a trend of increasing voltage with increasing deposition temperature. Beside the increase of grain size with deposition temperature T_d, another reason for this trend is the reduction of the intra-grain defect density with increasing T_d. At high deposition temperature T_d, the atoms are more likely to reach the regular lattice sites. In a low-temperature deposition, the Si atoms are frequently stuck to less ideal lattice positions, and crystal defects that reduce the intra-grain minority carrier lifetime are generated. Experimentally we find such a dependence of the density of structural defects on temperature, even in monocrystalline films (see Figure 5.13 on p. 131). Without analyzing the defect concentration *in* the grains, the enhanced open-circuit voltage should not be ascribed solely to an increasing grain size. The two highest reported voltages exceed 680 mV and are achieved with monocrystalline cells from thinned float-zone Si [113, 351]. This material grows at the melting point of Si.

Surface passivation: No systematic measurements of the surface recombination exist for crystalline thin-film cells.

A conceptual advantage of the layer transfer process is the free accessibility of both sides of the monocrystalline cell, which should permit high-level surface passivation with the same techniques as are applied to thick wafer cells. A surface recombination

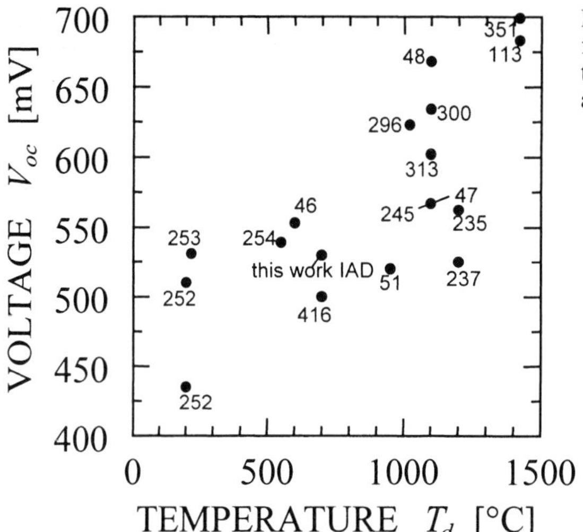

Figure 6.5. Open-circuit voltage as a function of deposition temperature of the active Si layer. Experimental data are labeled by the reference number.

velocity around 100 cm s^{-1} is feasible for monocrystalline Si and for LTP cells, with a doping concentration ranging from 10^{16} to 10^{17} cm^{-3} [190].

Our analysis shows that the bulk diffusion length L_b of the STAR cell (LTS approach) exceeds the device thickness, as does the effective diffusion length that results if we assume that only grain boundary recombination occurs. This means that the surface recombination velocity becomes an important factor which has so far not been addressed for LTS cells.

HTS cells generally rely on a back surface passivation with an electronically thick high-low junction (back surface field layer). Recombination velocities achieved with this concept are around 1000 cm s^{-1} [177]. With a Bragg reflector placed between the seed layer and the active layer [29], an efficient back surface passivation will be even harder to achieve, due to the large inner surface of the reflector.

6.3.2 Practical requirements for high-throughput fabrication

Besides the physical requirements for high efficiencies there are practical requirements for large-scale production such as high deposition rate, substrate availability, and simplicity of the cell process. We discuss these issues here.

Deposition rate: Figure 6.6 compiles the deposition rates R_d reported for crystalline thin-film cells as a function of deposition temperature T_d. The large deposition rate at high deposition temperature T_d is due to an enhanced mobility of the Si atoms that permits rapid growth. Plotting the rate R_d versus inverse temperature $1/T_d$ yields an activation energy of (1.1 ± 0.3) eV, which is a typical energy for atomic reordering processes.

The highest deposition rate of 5 μm min^{-1} was reported for high-temperature chemical vapor deposition (CVD) as applied in the HTS and LTP approach. Electric energy from thin-film cells fabricated by CVD has a cost advantage, since the high deposition temperature improves both the material quality (that is, cell efficiency) and the throughput of the epitaxy reactor (that is, production efficiency). High-temperature CVD is therefore the preferred deposition technique among the thin-film approaches that we listed in

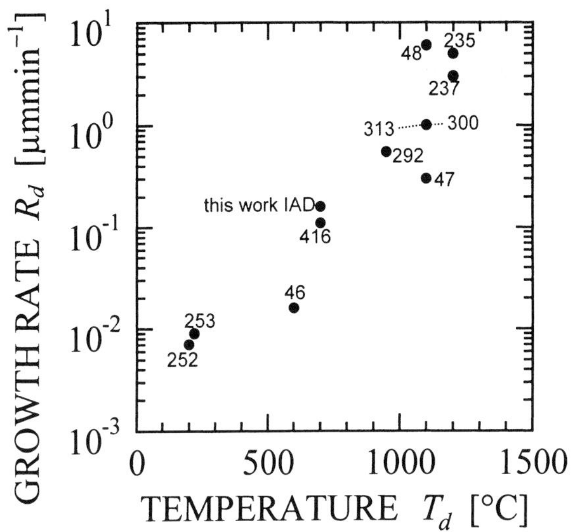

Figure 6.6. Growth rates of crystalline Si films for thin-film cells. Data are labeled by the reference number.

Table 6.3 on p. 175 and also among the layer transfer processes listed in Table 4.2 on p. 119. On the other hand, large-area reactors for high-temperature CVD with a throughput that is sufficient for photovoltaic applications are not yet developed.

The deposition rate for very high-frequency plasma-enhanced deposition at temperatures around 250°C [252] is three orders of magnitude lower than the deposition rate of the high-temperature CVD technique. Typical thickness values of LTS cells are only one order of magnitude below the thickness of HTS cells. Large-area plasma-enhanced deposition systems are available for the LTS approach.

Substrate availability: The substrate for thin-film deposition has to fulfill strict conditions (see section 4.1.1 on p. 92). These conditions are easier to satisfy if the processing temperature is low.

The major challenge for the HTS approach is the development of a low-cost, high-temperature resistant substrate that apparently does not yet exist. The reduction of the interaction of the substrate with the molten Si during the zone melt recrystallization (ZMR) process is still a challenging task.

Window glass that is available at low cost may only be used as the growth substrate for cells from the LTS approach.

The LTP approach uses monocrystalline Si wafers as a growth substrate that is readily available. In this work we demonstrated for the first time the re-use of the Si substrate wafer. However, a frequent (at least tenfold) re-use is required to meet the cost goals. None of the LTP processes that we discussed in section 4.3 on p. 108 has so far demonstrated such a frequent re-use of the growth substrate wafer by the fabrication of multiple working solar cells.

Cell process: Low-cost fabrication requires a simple cell process with only a few robust processing steps.

Zone melt recrystallization (ZMR) is apparently required for the HTS approach, since the power conversion efficiencies achieved without ZMR are still below 6%. The necessity for ZMR including capping layer deposition and removal adds to the complexity of the HTS approach.

The LTS approach apparently offers a process as simple as for amorphous Si solar cells. An advantage of the LTS approach is the availability of equipment for large-area deposition from amorphous Si technology. The integrated series connection of cells should also be feasible with a technology similar to the amorphous Si modules.

The process for monolithically integrated modules from the LTP approach that we developed at ZAE Bayern (see section 5.2.3 on p. 139) is quite simple compared to the HTS process. However, so far we use evaporated metal contacts which are probably not acceptable for mass production. Equipment for the high-throughput anodic etching of Si substrates, as well as equipment for an automated separation process, has already been developed for silicon-on-insulator products [292].

6.3.3 Device results

A collection of leading device results that were achieved with the three approaches are compiled in Table 6.3 on p. 175. All of this thin-film work is still in the laboratory phase with cell areas of only a few square centimeters.

The power conversion efficiencies achieved with the LTS approach [254] are 7 to 10%. Light trapping is superior to that of the other approaches. In combination with amorphous Si top cells deposited on the μc-Si bottom cells, efficiencies above 13% are possible in the near future [268, 354]. Cells from the LTS approach can replace amorphous solar cells due to their superior long-term stability. Due to the experience gained in fabricating amorphous Si solar modules, the LTS approach is more readily scalable to large production volumes than the HTS and the LTP approaches. A major challenge for this type of cells is the enhancement of the growth rate.

The power conversion efficiencies reached by the HTS approach [47, 236] are 8 to 11%. The power conversion of these cells is limited by the absence of any light trapping and by grain boundary recombination. Considering the encouraging cell efficiencies of 16% that were achieved in the thickness range of 50 to 100 μm [289, 355], the HTS approach is capable of reaching efficiencies similar to conventional wafer cells if the optical absorption is enhanced. The current practical limitation of the HTS approach is the non-availability of a low-cost substrate that withstands the high temperature during the ZMR process and permits the growth of high-quality material. There is no physical reason why such a substrate should not be developed. A substantial simplification of the cell fabrication process (three high-temperature CVD processes plus zone melt recrystallization) is certainly required to make this approach economically attractive.

The highest thin-film efficiencies of 12 to 14% were demonstrated with the LTP approach in this work and in Refs. [296, 300], using porous Si for layer transfer [52, 53]. Efficiencies of above 16% will soon be reached with this concept. The major challenge for this approach is the demonstration of a frequent re-use of the substrate. New processing equipment needs to be developed for large-scale fabrication. We expect the LTP approach to find its first applications as space solar cells (see section 5.4.5 on p. 154) and in consumer electronics (see section 5.2.3 on p. 139).

Table 6.3. List of leading device results achieved with thin-film crystalline Si solar cells until March 2001. See page 180 for updating comments.

Class	Substrate/Carrier	Temperature T [°C]	Rate R [µm min^{-1}]	Grain size G [µm]	Thickness W_{eff} [µm]	Quant. diffusion length L_Q/W_{eff}	Coll. diffusion length L_C/W_{eff}	Base diffusion length L_b [µm]	Current density j_{sc} [mA cm^{-2}]	Open-circuit voltage V_{oc} [mV]	Efficiency η [%]	Epitaxy	Reference Group
LTS	glass	<550		0.1[c]	2		16[c]	>3[c]	24.3	539	10.1	PE-CVD, STAR	[254] Kaneka
LTS	metal	600	0.016[a]	1.5	10	2.4[c]			25.0	553	9.2	SPC	[277] Sanyo
LTS	glass	220	0.009[b]	0.02	3[b]				22.9	531	8.5	VHF GD	[253] University of Neuchâtel
LTS	glass	200	0.007		3.5				25.1	453	7.5	VHF GD	[252] FZ-Jülich
HTS	silicon	1200	5	2500	50	1.0			27.6	562	11.5[d,e]	CVD	[235] ISE
HTS	graphite	1100	0.5	300	15	1.3[c]	0.6[c]	20	25.6	570	11.0[d,e]	CVD	[236]
HTS	ceramics (mullite)	1200	3	500	49			25[c]	23.8	525	8.2[d]	CVD	[237] CNRS
HTS	ceramics (SiSiC)	1150	6	2500	50				21.6	567	9.3[d,e]	CVD	[245] ISE
LTP	porous Si/ glass	1100	1	∞	15.5	1.4	7	>20	25.6	602	12.2	CVD, PSI	[234] ZAE
LTP	porous Si	950	0.55	∞	>10				27.5	520	9.3	LPE, SCLIPS	[306] Canon
LTP	porous Si/ plastics	1050		∞	12.5				25.5	623	12.5[O]	CVD, SPS	[296] Sony
LTP	porous Si/ glass	1100	1	∞	24.5	1.5	3	>26	27.3	634	14.0[O]	CVD, SPS	[300] University of Stuttgart

[a]The crystallization time of 10 h controls the rate of film production. [b]Estimates from other work of the same group. [c]Value does not hold for the cell for which the efficiency is quoted; instead the value is quoted or derived in Chapter 3 for a similar device from the same laboratory. [d]Process uses zone melt recrystallization. [e]Process uses photolithography.

6.4 Porous Si for layer transfer

Layer transfer processes (LTP) combine the LTS advantage of a readily available low-cost cell carrier with the HTS advantage of a high deposition temperature that permits a high material quality and a high deposition rate. The present major technological limitations of the non-availability of a low-cost substrate for the HTS approach and the limitations due to the currently small Si growth rate for LTS cells are thereby circumvented. In addition, grain boundary recombination is absent in monocrystalline thin-film Si cells fabricated by layer transfer.

In section 4.3 on p. 108 we discuss the various transfer concepts that are currently under investigation. One of these process is the so-called porous Si (PSI) process introduced by the author [53] and currently under development at ZAE Bayern.

Layer transfer with the PSI process

The concept of the PSI process (see Figure 5.1 on p. 121) is to grow a thin-epitaxial Si film on a textured monocrystalline Si wafer that has a porous surface layer. This layer functions as a seed for the epitaxial growth and as a perforation layer that permits the separation of the textured monocrystalline thin film from the growth substrate. The substrate wafer is re-used to form further cells.

The PSI process marked the first international presentation of the fabrication of a monocrystalline Si layer, using layer transfer with porous Si in a way that does not sacrifice the substrate wafer. Since that demonstration in 1997 the number of reports using porous Si for transfer has been rapidly increasing. Although the history of layer transfer processes is rather short in Si photovoltaics, record thin-film efficiencies have been reported for layer transfer using porous Si since 1998 [296, 300].

In Chapter 5 we report on the present status of the development of the PSI process. We use ion-assisted deposition (see section 5.1.2 on p. 126) and chemical vapor deposition (see section 5.1.3 on p. 132) for epitaxial growth on porous Si.

Ion-assisted deposition (IAD) for the PSI process

Ion-assisted Si deposition is a high-rate (0.3 μm min^{-1}), high-vacuum evaporation of Si. A fraction of 1 to 5% of the Si atoms is ionized and accelerated towards the substrate. This non-thermal kinetic energy is transferred into the surface of the growing film, thus permitting high-rate monocrystalline growth at a rather low sample temperature. We investigate ion-assisted deposition at 700°C because the re-construction of the porous Si is less severe at 700°C than at temperatures of 1100°C that are typical for CVD.

Monocrystalline IAD film in the micron thickness range: Monocrystalline Si waffles of 1 to 10 μm thickness with regular (see Figure 5.8 on p. 128) and random pyramids (see Figure 6.7) were fabricated up to sample diameters of 8.5 cm. It is an important feature of the PSI process that light trapping waffle-shaped films with an effective thickness in the micron range can be fabricated. Figure 6.7 shows such a 2 μm-thick waffle with residual porous Si on the bottom surface and metallization on the top surface. We achieve dislocation densities of 10^4 cm^{-2} in planar films detached from (100)-oriented porous Si substrates. A dislocation density of 10^7 cm^{-2} is typical for waffle-shaped films grown on (111)-oriented Si facets. The larger defect density is due to the smaller formation energy of stacking faults in films grown on (111)-oriented facets when compared to

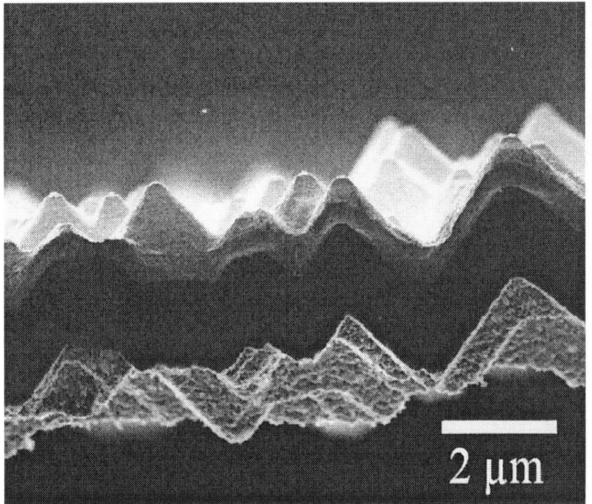

Figure 6.7. Monocrystalline randomly textured Si film fabricated by the PSI process. A metal layer is deposited onto the top surface. The bottom surface shows residual porous Si attached to the Si film

(100)-oriented facets. The dislocation density decreases with the deposition temperature (see Figure 5.13 on p. 131), thus indicating how to improve the material quality further.

Insufficient material quality of waffle-shaped IAD films: The largest minority carrier diffusion length we halve measured in devices from ion-assisted film deposition on textured substrates is (3±0.5) μm. This value is achieved without hydrogen passivation, P-gettering, or other high-temperature annealing. The diffusion length is at present limited by recombination at stacking faults and appears insufficient for high power conversion efficiencies.

Chemical vapor deposition (CVD) for the PSI process

Epitaxy by chemical vapor deposition at temperatures of 1100°C reconstructs the porous Si. The Sony group demonstrated that this sintering does not hinder a separation of epitaxial film [52, 296]. Annealing in hydrogen closes the surface of the porous Si [290], thus giving an ideal starting point for Si epitaxy (see Figure 5.5d on p. 124). Our Monte Carlo simulations prove that the reduction of the surface energy is the driving force for the surface closure and for the formation of a separation layer (see section 5.1.1.2 on p. 124).

Using CVD, we fabricate epitaxial films on randomly textured p^+-type Si wafers that have a porous surface. The film has randomly positioned inverted pyramids on one side and an almost planar surface on the opposite side (see Figure 4.23 on p. 113) [320].

Sufficient material quality of waffle-shaped CVD films: Transmission electron microscopy of these films shows no structural defects. Analyzing the same films by Secco etching gives a dislocation density below 10^3 cm^{-2}. Hence the CVD films on *textured* material have fewer structural defects than the IAD films on planar (100)-oriented substrates. The minority carrier diffusion length in waffle-shaped, randomly textured CVD films exceeds 20 μm in thin films of 15 μm thickness.

Light trapping in waffle-shaped thin films

We studied the light trapping in waffle-shaped thin crystalline Si films on the basis of optical reflectance measurements and ray-tracing.

Measurements: Experimental data are so far available only for the thickness range around W_{eff} = 10 μm. Measured and simulated reflectance spectra agree for periodic waffle shapes that are encapsulated under a planar cover-glass and use a detached Al reflector. The detached back surface reflector is an important optical design feature that, together with the waffle shape, permits photogenerated current densities as large as 36 mA cm^{-2}, even without applying antireflection coatings (see Figure 5.26 on p. 141). Absorption measurements of non-conformal waffles fabricated by CVD show a short-circuit current potential equal to that of conformal periodic IAD-grown waffles [320].

Simulations: For isotropic illumination we calculate a maximum current density of (30.5 ± 0.2) mA cm^{-2} by numerical ray-tracing for a W_{eff} = 1 μm waffle that has a detached Al reflector, a textured front glass and a single-layer silicon nitride coating on both Si surfaces. This is 85% of the theoretical maximum and 92% of the current from a Lambertian light trapping scheme. For higher film thickness values the gap to optimum light trapping is even smaller.

The photogenerated current density in triangular, pyramidal, and hexagonal conformal films of thickness W_f = 4 μm differs by less than 1 mA cm^{-2} (see Figure A.12 through Figure A.16 on p. 196-199). More important than the film shape is a large facet inclination (section 5.4.4 on p. 153) and a texture period that is similar to or smaller than the layer thickness (see Figure 5.32 on p. 148).

Solar cell results

At ZAE Bayern we developed a solar cell process for waffle-shaped CVD-grown films from the PSI process (see section 4.3.5 on p. 113). Cells from this process achieve an independently confirmed efficiency of 12.2% with an effective film thickness W_{eff} = 15.5 μm [234]. This is currently the highest power conversion efficiency reported for a crystalline thin-film cell that was fabricated without using photolithography. Neither photolithography nor high-temperature oxidation is compatible with low-cost thin-film photovoltaics. Our cells have an Al-covered textured back surface.

In contrast to CVD, the electronic quality of the IAD films is not yet sufficient to fabricate highly efficient thin-film cells. The best transferred cell we fabricated from an IAD-grown film is planar, 7 μm thick, and has a power conversion efficiency of $\eta = 4\%$. Quantum efficiency analysis reveals that the recombination in the bulk is the dominating loss.

Novel concepts for an integrated series connection

We developed novel processes for the series connection of thin-film solar cells from the PSI process.

Series connection of IAD films: The highly directed material flow in the IAD reactor permits us to define surface structures by shadow masks (shadow epitaxy). Using movable masks we realized for the first time a series connection in-situ during the epitaxy process (see Figure 5.18 on p. 134) [356]. The open-circuit voltage per cell is only 266 mV, due to the high stacking fault density in textured IAD films.

Series connection of CVD films: The shadow technique is not applicable to CVD, since the material transport is not as directed as for IAD. For the series connections of cells from CVD-grown films, we apply plasma etching in a novel processing sequence

[341] that is only applicable to thin-film cells from layer transfer (see Figure 5.25 on p. 139). Our 25 cm^2 mini-module consists of five series-connected cells and reaches a confirmed efficiency of 10.6%. The open-circuit voltage per cell is 607 mV. This is probably the first integrated thin-film module from monocrystalline and textured Si thin-film material.

Are we approaching the limits by layer transfer using the porous Si (PSI) process?
We determined a theoretical efficiency limit of 27.8% for a crystalline thin-film cell of 10 μm thickness that has Lambertian light trapping. The efficiency of our thin-film cell from the PSI process is currently only 12.2%. An efficiency of 14.0% was achieved with planar cells from a similar process [300]. There is currently still a factor of two between experiments and the theoretical optimum.

Quantum efficiency analysis: The quantum efficiency analysis of the 15.5 μm-thick waffle-shaped cell from the PSI process reveals a bulk diffusion length $L_b > 20$ μm and a back surface recombination velocity $S_b < 5000$ cm s^{-1}. These bounds on the recombination parameters are deduced using parameter confidence plots. The performance of the device is limited primarily by a low back surface reflectance that originates from a double reflection at the textured and Al-covered back surface.

Efficiency potential without improvements of the base material: A detached back reflector considerably improves the back reflectance (see section A.3.3 on p. 206). With a detached reflector, our simulation predicts an efficiency >15% (see Figure 5.33a on p. 152) with the parameters $L_b = 20$ μm and $S_b = 5000$ cm s^{-1}, the worst case that is in agreement with our measurement. The simulated open-circuit voltage is 630 mV. This value is quite close to the experimental value of 607 mV. The difference of 23 mV is caused mainly by an as yet unoptimized emitter doping profile. The simulated short-circuit current is 32 mA cm^{-2} and exceeds the experimental value of 25.6 mA cm^{-2} primarily due to the model assumption of an improved back reflectance. Hence an efficiency of >15% is technically feasible with a simple process, even without improving the current material quality. This expected efficiency exceeds that of standard industrial wafer cells that are 20 times thicker than our thin-film cells.

Efficiency potential with improved material and high-efficiency processing: With Lambertian light trapping the photogeneration in a cell with $W_{eff} = 10$ μm is 90% of that in a cell with $W_{eff} = 500$ μm. Similarly to the PERL cell, the CVD-grown waffle cells are monocrystalline, and have one surface bounded by (111) facets while the opposite surface is almost planar. We thus expect approximately 90% of the efficiency of the PERL cells, if PERL-type processing is applied to thin films. An efficiency of 22% is therefore about the best that we can expect for 10 μm-thick waffle-shaped Si cells (see Figure 5.33a on p. 152).

6.5 Update

To illustrate for the reader the rapid advance in the field of thin-film solar cells we mention two recent results concerning layer transfer devices.

The efficiency record for thin-film crystalline Si cells was 14.0% when the work on this book was finished in March 2001 (see Table 6.3 on p. 175). Only a couple of months later an efficiency of 16.6% was announced for a 45 µm-thick thin-film cell that was fabricated using a process with photolithographic steps [357].

The record for photolithography-free processed thin-film cells was 12.5 % in March 2001 (see Table 6.3 on p. 175). Recently our team at ZAE Bayern fabricated a 25 µm-thick device with a power conversion efficiency of 15.4% (independently confirmed at FhG-ISE in Freiburg, Germany) [358]. This device was fabricated by layer transfer using porous Si and without any photolithography. In addition we demonstrated for the first time the fourfold use of a silicon growth substrate [359].

While these are the leading device results for thin-film crystalline Si solar cells in June 2002, I am shure that these records will not hold for long.

APPENDIX A

A Light trapping

A.1 Lambertian light trapping

In Chapter 2 on p. 16 we defined Lambertian light trapping. An analytic formula for the optical absorption of a Lambertian light trapping scheme (Eq. (2.5) on p. 16) was given without a derivation. Here we follow our work in Ref [342] and derive this formula for a Lambertian light trapping scheme with a detached back surface reflector and a non-ideal front surface reflectance.

Figure A.1 gives a schematic representation of a thin silicon film that is deposited on a transparent textured substrate (denoted by texture). The transparent substrate (texture) has a reflector of reflectance R_m on the back. The silicon film has an index of refraction n_s and the optical absorption coefficient α_s. The optical properties of a coverglass (if any) and of the front surface of the Si film are merged into an effective, angle-averaged transmittance T_f that applies to light entering from the front (air) into the Si film. The model also uses an effective transmittance T_t for fully randomized light to pass from the texture (the material of refractive index $n_t < n_s$ between the metal reflector and the back surface of the cell) into the Si film. The metal reflector has a reflectance $R_m = 1 - T_m$. The symbol T_m describes the transmittance into the metal film and is identical to the optical absorption in the metal reflector. We consider the monochromatic light power fluxes I_{id} downwards and I_{iu} upwards. The indices $i = f, s, t$ denote fluxes in the front f, the silicon film s, and the texture t, respectively. The front material has a unity refraction index. In the silicon film, we further distinguish the power fluxes I_{sdt} and I_{sut} at the top of

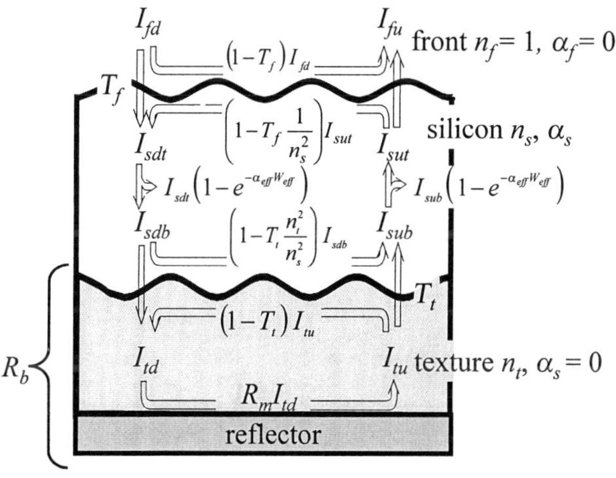

Figure A.1. Schematic representation of a silicon film deposited onto a transparent and textured substrate (denoted texture). The refractive indices n of the front f, the Si film s, and the texture t are $n_f = 1$, n_s, and n_t, respectively. The silicon/texture/metal reflector system has an effective back surface reflectance R_b. Redrawn from Ref. [342].

the film, and the fluxes I_{sdb} and I_{sub} at the bottom of the film. We introduce the effective back surface reflectance

$$R_b = I_{sub} / I_{sdb} \tag{A.1}$$

as a cumulative description of the Si/texture/reflector system. The power flow diagram of Figure A.1 with the eight intensities I defines eight linear equations. We explain only one of these equations. The upward energy flux at the bottom of the silicon film

$$I_{sub} = I_{tu}\, T_t + I_{sdb}\left(1 - T_t\,(n_t/n_s)^2\right) \tag{A.2}$$

has two components: the upward energy flux I_{tu} in the texture that is transmitted into the film with transmittance T_t, and the energy flux reflected at the silicon/texture interface. The energy flux transmitted from the silicon into the texture is $I_{sdb}\ T_t\ n_t^2/n_s^2$, because for fully randomized light the fraction n_t^2/n_s^2 of all rays falls into the loss cone that has a transmittance T_t. Consequently, the energy reflected is $(I_{sdb} - I_{sdb}\ T_t\ n_t^2/n_s^2)$ and hence the second term of Eq. (A.2) is also explained. The other seven equations

$$I_{sdt} = I_{sdt}\left(1 - e^{-\alpha_{eff} W_{eff}}\right) + I_{sdb} \tag{A.3}$$

$$I_{td} = I_{sdb}\, T_t\,(n_t/n_s)^2 + (1 - T_t)I_{tu} \tag{A.4}$$

$$I_{tu} = R_m I_{td} \tag{A.5}$$

$$I_{sub} = I_{tu}\, T_t + I_{sdb}\left(1 - T_t\,(n_t/n_s)^2\right) \tag{A.6}$$

$$I_{sub} = I_{sub}\left(1 - e^{-\alpha_{eff} W_{eff}}\right) + I_{sut} \tag{A.7}$$

$$I_{fu} = I_{sut}\, T_f\,(1/n_s)^2 + I_{fd}\left(1 - T_f\right) \tag{A.8}$$

and

$$I_{fd} = 1 \tag{A.9}$$

are derived in a similar manner. Some of the equations contain an effective absorption coefficient α_{eff} that we define in the next section, as to make the energy fluxes absorbed in the Si film $I_{sub}\left(1 - \exp\left(-\alpha_{eff} W_{eff}\right)\right)$ and $I_{sdt}\left(1 - \exp\left(-\alpha_{eff} W_{eff}\right)\right)$. The system of linear Eqs. (A.2)-(A.9) has a unique solution [342].

A.1.1 Active absorption

We define the active absorption as being caused by the generation of electron-hole pairs. It does not include the absorption in the metal reflector. The active absorption

$$A = \left(1 - \exp(-\alpha_{eff} W_{eff})\right)\left(I_{sub} + I_{sdt}\right) / I_{fd} \tag{A.10}$$

in the Si film of effective thickness W_{eff} depends on the two fluxes I_{sub} and I_{sdt} in the Si film and the incoming energy flux I_{fd}. Here, the angle-averaged effective absorption coefficient α_{eff} is defined by the relation [360]

$$T_r = \exp(-\alpha_{eff} W_{eff}) = 2\int_0^{\pi/2} \sin(\vartheta)\cos(\vartheta)\exp(-\alpha_s W_{eff}/\cos(\vartheta))d\vartheta \tag{A.11}$$

where T_r denotes the transmittance of fully randomized light through a planar film of thickness W_{eff}. Two partial integrations of Eq. (2.6) yield [111]

$$T_r(\alpha_s W_{eff}) = \exp(-\alpha_s W_{eff})(1-\alpha_s W_{eff}) + (\alpha_s W_{eff})^2 E_1(\alpha_s W_{eff}) \tag{A.12}$$

where $E_1(z) = \int_z^\infty t^{-1}\exp(-t)\,dt$ is the exponential integral function that we evaluate to a precision of 10^{-5} with a simple polynomial approximation given in Ref. [361]. The isotropic light propagation enhances the path length and therefore $T_r = \exp(-\alpha_{eff} W_{eff}) < \exp(-\alpha_s W_{eff})$. The absorption enhancement factor due to isotropic propagation is shown in Figure A.2. The enhancement factor is 2 for weakly absorbed light (see Eq. (2.8)) corresponding to an "average" angle of 60° since 1/cos(60°) = 2. The absorption is not enhanced for strongly absorbed light; hence the enhancement factor is unity for $\alpha_s W_{eff} >> 1$. The active absorption

$$A = \frac{(1-T_r)(1+R_b T_r) n_s^2 T_f}{n_s^2 - R_b(n_s^2 - T_f)\,T_r^2} \tag{A.13}$$

follows from Eq. (A.10) by inserting the intensity variables I that are a solution of Eqs. (A.2) through (A.9). So far, the treatment is exact and Eq. (2.5) represents an easy-

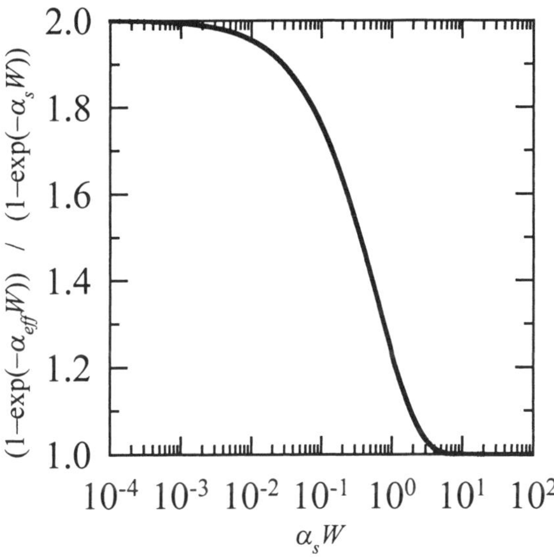

Figure A.2. Absorption enhancement due to isotropic light propagation relative to the absorption caused by a path length W.

to-use analytical expression for the absorption of a Lambertian light trapping scheme. This analytic treatment of Lambertian light trapping was first published in 1999 [242]. For very thin films, or, equivalently, for very weakly absorbed light we have $\alpha_s W_{eff} << 1$. Then Eq. (2.5) is further simplified by $T_r = 1 - \alpha_{eff} W_{eff} = 1 - 2\,\alpha_s W_{eff}$ and Eq. (A.13) becomes

$$A_{W_{eff}\to 0} = \frac{2\alpha_s n_s^2 (1 + R_b) T_f W_{eff}}{n_s^2 - n_s^2 R_b + R_b T_f + 4\alpha_s n_s^2 W_{eff}} \quad \text{(A.14)}$$

We name the limit $W_{eff} \rightarrow 0$ the thin-film limit. Equation (2.5) is a generalization of the formula

$$A_{W_{eff}\to 0} = \frac{4n_s^{?} W_{eff} \alpha_s T_f}{4n_s^2 W_{eff} \alpha_s + T_f} \quad \text{(A.15)}$$

derived by Luque [415]. The validity of the latter is restricted to the special case of unity back surface reflectance $R_b = 1$. For the general case of arbitrary $\alpha_s W_{eff}$ and non-unity back reflectance our Eq. (A.13) applies.

A.1.2 Path length distribution in the thin-film limit

The path length that a light ray propagates in the silicon film between entering and leaving the structure depends on the position, the angle of incidence, and the state of polarization. For the ensemble of all rays we find a path length distribution $f(l)$ that determines the optical absorption

$$A(\lambda) = 1 - \int_0^\infty f(l) \exp(-\alpha_s(\lambda) l) dl \quad \text{(A.16)}$$

Here α_s is the absorption coefficient of the Si film at wavelength λ. In the near-infrared part of the solar spectrum, where light trapping is particularly important, the wavelength dependence of the index of refraction is weak. The directions of light propagation are governed by Snell's law and are therefore similar for different wavelengths λ. In consequence, the path length distribution $f(l)$ is almost wavelength-independent. All information on the light trapping performance of a texture is contained in the path length distribution $f(l)$ that may be determined at a single wavelength [342]. Hence, the path length distribution $f(l)$ deserves some more consideration. The first moment of the distribution $f(l)$ is the average path length

$$\bar{l} = \int_0^\infty l f(l) dl \quad \text{(A.17)}$$

and the second moment is the standard deviation

$$\sigma = \sqrt{\int_0^{\infty} \left(l-\bar{l}\right)^2 f(l) dl} \tag{A.18}$$

From Eqs. (A.16), (A.18), and (A.19) we obtain

$$A \approx 1-\left(1+\frac{\sigma^2 \alpha_s^2}{2}\right)\exp\left(-\alpha_s \bar{l}\right) \tag{A.19}$$

by expanding for low absorption coefficients α_s. Textures with a large average path length $\bar{l}$ and a small standard deviation σ are best for high optical absorption A. The active absorption A is related to the path length distribution $f(l)$ via a Laplace transformation. Hence, the path length distribution

$$f(l)_{W_{eff}\to 0} = \frac{1}{2\pi i}\int_{c-i\infty}^{c+i\infty}\left(1 - A(\alpha_s)\right)\exp[-\alpha_s l]\, d\alpha \tag{A.20}$$

is the inverse Laplace transformation of the absorption A for any real $c > 0$. Equation (A.20) gives a new possibility to determine the path length distribution from an experimental absorption spectrum. However, we find numerical stability problems similar to the determination of the local carrier collection efficiency via inverse Laplace transformation of experimental internal quantum efficiency spectra [209, 362, 210]. From Eqs. (A.20) and (A.14) we calculate an exponential path length distribution

$$\begin{aligned} f(l)_{W_{eff}\to 0} = {} & \frac{\left(1+R_b\right)T_f\left(n_s^2 - n_s^2 R_b + R_b T_f\right)}{8\, n_s^2\, W_{eff}}\exp\left[\frac{-l\left(n_s^2 - n_s^2 R_b + R_b T_f\right)}{4 n_s^2 W_{eff}}\right] \\ & + \left(1 - \frac{T_f}{2} - \frac{R_b T_f}{2}\right)\delta(l) \end{aligned} \tag{A.21}$$

that becomes

$$f(l)_{W_{eff}\to 0} = \frac{1}{4\, n_s^2\, W_{eff}}\exp\left[\frac{-l}{4 n_s^2 W_{eff}}\right] \tag{A.22}$$

for the special case $R_b = T_f = 1$. In general, however, as described by Eq. (A.21) $f(l)$ depends on the effective back reflectance R_b and the effective front surface transmittance T_f. The Dirac delta function $\delta(l)$ appears for non-zero front surface reflectance $R_f = 1 - T_f$ or non-unity back surface reflectance R_b. The meaning is that a fraction R_f of rays has zero path length at unity back surface reflectance $R_b = 1$. We have not yet found a satisfactory interpretation of the appearance of R_b in the argument of the Dirac delta function. We should, however, keep in mind that Eq. (A.21) is valid in the thin-film limit only. In this limit, the light has to travel a negligible distance from the front surface to the back surface, which could explain the appearance of R_b in the argument of the Dirac delta function.

If the path length distribution $f(l)$ is exponential, then the fraction of rays

$$F(l) = \int_l^\infty f(m)dm \tag{A.23}$$

that is still inside the cell after having propagated a path length l is also an exponential function. This is exactly what we found for the geometrical light trapping schemes by numerical ray-tracing [342].

The slope and the ordinate intersection of the path length distribution (when plotted as $\ln(f)$ vs. l) can, in principle, be used to extract the effective back surface reflectance R_b and the effective front surface transmittance T_f. A direct determination of R_b and T_f from a fit of Eq. (A.14) to the measured active absorption is, however, more reliable.

From Eqs. (A.17) and (A.21) we calculate the average path length

$$\bar{l}_{W_{eff}\to 0} = \frac{2\,n_s^2\,(1+R_b)\,T_f\,W_{eff}}{n_s^2 - n_s^2 R_b + R_b\,T_f} \tag{A.24}$$

in the thin-film limit. Please note that the average path length is $\bar{l} = 4\,n_s^2 W_{eff}$ for unity back surface reflectance $R_b = 1$. This value is independent of the front surface transmittance T_f, because the low front surface transmittance T_f is counterbalanced by an increased internal front reflectance $R_f = 1 - T_f$ for all rays within the loss cone. By inspection of Eq. (A.24), we verify that $\bar{l} \leq 4\,n_s^2 W_{eff}$ for all values of T_f and R_b. Hence, our analytical treatment reproduces the fact that the theoretical maximum of the average path length is $4\,n_s^2\,W_{eff}$ [16].

The standard deviation

$$\sigma_{W_{eff}\to 0} = \sqrt{\frac{4\,n_s^4(1+R_b)\,T_f\,(8-4\,T_f-4\,R_bT_f+T_f^2+2\,R_bT_f^2+R_b^2\,T_f^2)\,W_{eff}^2}{\left(n_s^2-n_s^2R_b+R_bT_f\right)^2}} \tag{A.25}$$

follows from Eqs. (A.18) and (A.21). For unity back surface reflectance $R_b = 1$, the standard deviation increases with decreasing transmittance T_f. Hence, the smallest standard deviation achievable for a fully randomizing Lambertian texture with maximum average path length $\bar{l} = 4\,n_s^2\,W_{eff}$ is $\sigma = 4\,n_s^2\,W_{eff} = \bar{l}$. This minimum standard deviation corresponds to the maximum optical absorption according to Eq. (A.14) and is achieved for unity front surface transmittance $T_f = 1$ only. Low front surface reflectance $R_f = 1 - T_f$ is therefore a precondition for a low σ and, trivially, for a high light trapping efficiency.

A.1.3 Carrier generation profile

The approximate formula describing the carrier generation profile

$$g(Z) = \frac{T_f}{W_{eff}\,\ln(T_r)}\,\frac{n_s^2\left(T_r^{\,Z/W_{eff}} + R_bT_r^{\,2-Z/W_{eff}}\right)}{R_b\left(n_s^2 - T_f\right)T_r^2 - n_s^2} \tag{A.26}$$

for Lambertian light trapping is derived on p. 82. We check this analytical generation profile against numerical ray-tracing of a Lambertian thin film. Figure A.3 shows good agreement between the carrier generation rate of a 4 µm-thick cell as calculated by ray-tracing and as calculated with the analytical formula of Eq. (A.26).

The reflectance of a Lambertian light trapping scheme is

$$R = 1 - T_f \frac{n_s^2 \left(1 - R_b T_r^2\right)}{n_s^2 - R_b \left(n_s^2 - T_f\right) T_r^2} \tag{A.27}$$

This generation profile and this reflectance spectrum hold for weak and for strong optical absorption.

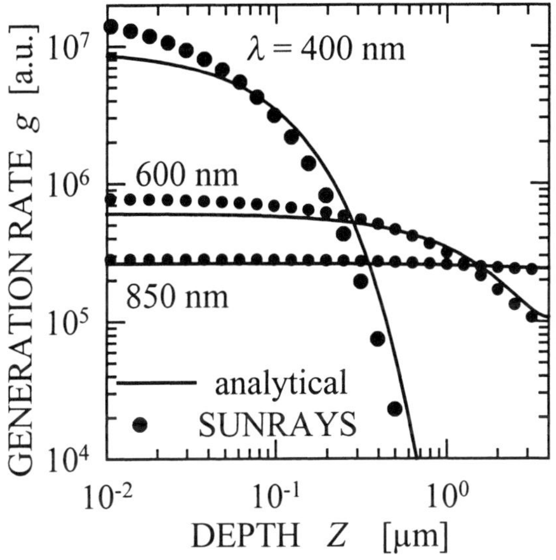

Figure A.3. The shape of the analytical generation profile and the generation profile calculated by ray-tracing agree.

A.2 Geometrical light trapping

A.2.1 Maximum average path length

As shown above, in a Lambertian light trapping scheme weakly absorbed light has an average path length $\bar{l} = 4\, n_s^2 W_{eff}$. Miñano and Luque showed that this value is the *maximum* average path length of *any* geometrical light trapping scheme [110, 360, 16]. Their proof assumes a Si film with an ideal antireflection coating that has zero reflectance at all angles of incidence. This is not the situation encountered in today's thin-film solar cells, which have a surface reflectance of a few percent. The question arises whether a finite surface reflection could possibly change the theoretical maximum of the average path length in the cell.

In this section, we generalize the previous theorem to thin films with non-ideal antireflective properties. We also give a generalization for films with spatially inhomogeneous refractive index. The knowledge of the upper limit of the path length enhancement permits us to estimate the scope for further improvements in geometrical light trapping. In order to conduct the proof, we give a short introduction to the terminology of geometrical optics [363, 364].

Light rays and étendue

Figure A.4 illustrates a ray of solid angle $d\Omega$ that propagates through the area element $dx\, dy$. A ray traversing a two-dimensional surface is identified by the point of intersection (x,y) of the ray and that surface and by its direction of propagation (p,q,r). The direction cosine r is abundant because $p^2 + q^2 + r^2 = 1$. Hence the ray is defined by four numbers (x,y,p,q). The power transmitted through the area element $dx\, dy$ is

$$dP = R\, n^2\, dx\, dy\, dp\, dq \tag{A.28}$$

where $dp\, dq = r\, d\Omega$ denotes the solid angle element $d\Omega$ projected onto the area element $dx\, dy$, n is the refractive index in the center of that area element, and R is the radiance. The product

$$d\mathcal{E} = n^2\, dx\, dy\, dp\, dq \tag{A.29}$$

is defined as the étendue [16] and hence $dP = R\, d\mathcal{E}$. For isotropic radiation the radiance R is independent of position and direction.

A basic theorem of geometrical optics is the conservation of phase space volume $d\mathcal{E}$ [364, 16]. For a ray traversing an area $dA_1 = dx_1\, dy_1$ with étendue $d\mathcal{E}_1$ first and an area $dA_2 = dx_2\, dy_2$ later with étendue $d\mathcal{E}_2$, it holds that $d\mathcal{E}_1 = d\mathcal{E}_2$. The proof of this theorem is analogous to the proof of Liouville's theorem in classical mechanics where the phase space corresponds to the manifold spanned by the coordinates $(x,\, y,\, n\, p,\, n\, q)$ [364]. For loss-free systems, it follows from $dP_1 = dP_2$ that $R_1 = R_2$. The radiance R is a constant of the optical system.

Generalized derivation of the maximum average path length

Figure A.5 shows a piece of Si of arbitrary shape that is illuminated through a planar

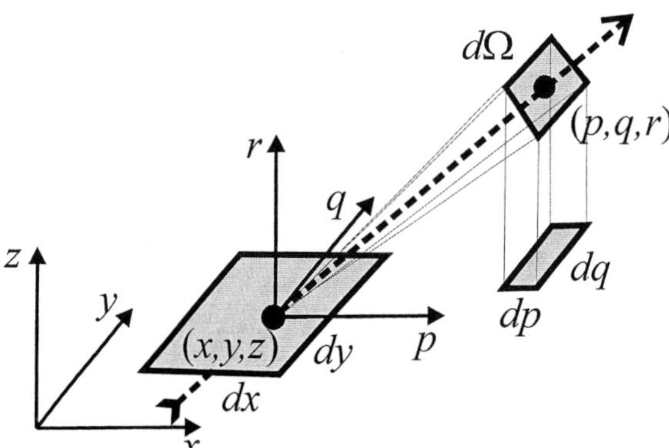

Figure A.4. The broken line symbolizes a light ray that illuminates an area $dA = dx\, dy$ in the direction (p,q,r) under a solid angle $d\Omega$ that has the projection $dp\, dq$ in the direction of the normal of surface dA. The étendue $\mathcal{E}$ is defined as the squared index of refraction n_s^2 times area dA and projected solid angle $dp\, dq = r\, d\Omega$.

face of area A_c. This planar area is not necessarily identical to the film surface. For a film textured with inverted pyramids, the area A_c would, for example, be like a sheet of paper placed on the textured cell surface. Each point of area A_c is illuminated from the directions covered by the solid angle Ω. The illuminating rays are not show in Figure A.5. The illuminating radiance R_o is equal for all illuminating rays. Hence the étendue illuminating the cell is

$$\mathcal{E}_{il} = \int_{\Omega} dp\, dq \int_{A} dx\, dy \tag{A.30}$$

The refractive index is omitted in this equation since we assume illumination in air that has a unity refractive index.

Consider an arbitrary point (x_1,y_1,z_1) inside the cell volume and an arbitrary direction (p_1,q_1). This particular ray (x_1,y_1p_1,q_1) is indicated as a dashed arrow in Figure A.5 and has the étendue $d\mathcal{E}_1 = n_s^2\, dx_1\, dy_1\, dp_1\, dq_1$ and the radiance R_1. The ray exists since otherwise the path length could be enhanced by changing the geometry to illuminate this ray also. Now we trace this ray backwards. In the example shown in Figure A.5, the ray hits the side wall at (x_2,y_2,z_2). Since the side wall is partially transmitting, the transmitted ray has the radiance T_2R_1. The étendue of this transmitted ray is $d\mathcal{E}_2 = d\mathcal{E}_1$. Even if the direction of the transmitted ray falls into Ω, this ray does not belong to the set of rays that illuminate the cell, because the point (x_2,y_2,z_2) does not belong to A_c. Similarly, other rays may escape through A_c but not through Ω. Hence, we remain with the reflected ray that has the étendue $d\mathcal{E}_2 = d\mathcal{E}_1$ and radiance $R_2 =(1 - T_2)\, R_1$. This ray hits the area A_c at point (x_3,y_3,z_3), where it is partially transmitted into the solid angle Ω with étendue $d\mathcal{E}_3 = d\mathcal{E}_2 = d\mathcal{E}_1$ and radiance $R_3 = (1 - T_3)\, R_2 = (1 - T_3)\,(1 - T_2)\, R_1$. The generalization of our example to an arbitrarily number of branching points is clear: all rays are traced backwards until they leave the cell having the étendue $d\mathcal{E}_1$. All backward traced rays fall in two groups: Those that leave the structure through A_c and into Ω have in sum the radiance R_{in}. The second group of rays does not fall into A_c and/or not into Ω and have, in sum, the radiance R_{out}.

We now trace those rays that emerge through A_c and Ω with the illuminating radiance R_o forward to point (x_1,y_1,z_1). As the transmission probability and the reflection prob-

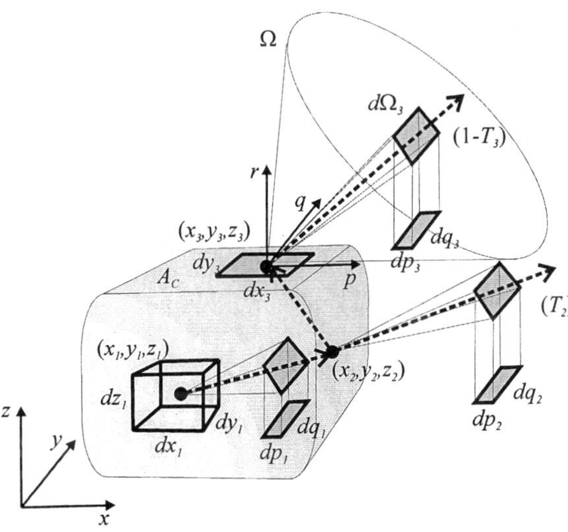

Figure A.5. The Si film (gray block) is illuminated at constant radiance through an area A_c and a solid angle Ω. We trace a ray at point (x_1,y_1,z_1) backwards. The direction is (p_1,q_1) and the étendue is $\mathcal{E}_1 = n_l^2\, dx_1\, dy_1\, dp_1\, dq_1$. At point (x_2,y_2,z_2), a fraction T_2 of the radiance leaves the film, while the fraction $(1 - T_2)$ is traced back to (x_3,y_3,z_3). From there, the ray falls into Ω.

ability are equal for rays propagating forwards or backwards along a given direction we conclude that

$$R_1 = \frac{R_{in}}{R_{in} + R_{out}} R_o \tag{A.31}$$

Hence the maximum possible value for R_1 is R_o. This maximum is achieved only if all branching rays leave the cell through A and Ω, and thus $R_{out} = 0$. So there is no need to have a zero front surface reflection in cells with maximum average path length. Losses due to absorption, e.g. in antireflection layers, are of course prohibited in a cell with maximum average path length.

Now consider a small area $dx\ dy$ around point (x_1, y_1, z_1) that is the top surface of a box of volume $dx\ dy\ dz$. Assuming a maximum path length, we know that the radiance is R_o for all rays (x_1, y_1, p, q) (with the possible exception of a sub-manifold with zero phase space volume). Since the top face of the small box is illuminated by isotropic radiation of radiance R_o, we neither add nor subtract any rays if we consider the side walls and the bottom of the box to have unity reflectance from inside and outside. The average path length of the rays inside the box is

$$dl = 4\ dz \tag{A.32}$$

following Eq. (2.8). The extra factor 4 accounts for the reflector at the bottom of the volume dV and for the oblique traversal. The fraction

$$dN = d\mathcal{E} / \mathcal{E}_{il} \tag{A.33}$$

of rays that enter the volume dV_c through the front surface is the ratio of phase space

$$d\mathcal{E} = \pi\ n_s^2\ dx\ dy \tag{A.34}$$

to the illuminating phase space $\mathcal{E}_{il}$ of rays that enter through A_c and Ω. Hence the maximum path length

$$\bar{l}_{max} = \int_V dl\ dN = \int_{V_c} 4\,dz \frac{\pi\, n_s^2 dx\, dy}{\mathcal{E}_{il}} = \frac{4\pi < n_s^2 > V_c}{\mathcal{E}_{il}} \tag{A.35}$$

where $< n_s^2 >$ denotes the spatial average of the refractive index in the thin film. For isotropic illumination the illumination étendue is $\mathcal{E}_{il} = \pi\ A_c$. Hence the maximum average path length is

$$\bar{l}_{max} = 4 < n_s^2 > \frac{V_c}{A_c} = 4 < n_s^2 > W_{eff} \tag{A.36}$$

where the effective film thickness W_{eff} is defined as the ratio of film volume V_c to cell area A_c. For non-homogeneously illuminated cells the average path length can be larger. For normal illumination $\mathcal{E}_{il} = 0$ and no upper limit for the average path length can be derived from Eq. (A.35). However, as discussed in section A.2.3 on p. 193, the case of isotropic illumination is more relevant than the case of normal illumination.

The enhancement of the average path length by restricting $\mathcal{E}_{il}$ in angular space or in real space with concentrators and other means is discussed in Ref. [415].

Limiting optical absorption

Knowing the average path length $\bar{l}$, we derive an upper limit for the optical absorption. From Eq. (A.16) we find

$$\begin{aligned} A &= 1-\exp\left(-\alpha_s \bar{l}\right)\int_0^\infty f(l)\exp\left(-\alpha_s\left(l-\bar{l}\right)\right)dl \\ &\leq 1-\exp\left(-\alpha_s \bar{l}\right)\int_0^\infty f(l)\left(1-\alpha_s\left(l-\bar{l}\right)\right)dl \\ &= 1-\exp\left(-\alpha_s \bar{l}\right) \\ &\leq 1-\exp\left(-\alpha_s \bar{l}_{\max}\right) \\ &= 1-\exp\left(-\alpha_s 4 n_s^2 W_{eff}\right) \end{aligned} \tag{A.37}$$

where the relation $\exp(-x) \geq 1 - x$ is used from the first to the second line [110]. The maximum absorption A is reached if *all* rays that are isotropically incident on the cell have the same path length $l = 4\, n_s^2\, W_{eff}$. No light trapping texture with this property has been invented yet. It was, however, shown that this maximum is *not* achievable with a cell that is macroscopically planar [111]: rays traversing such a planar slab sufficiently obliquely will always have path lengths $l > 4\, n_s^2\, W_{eff}$, and consequently not all of the rays have the length $l = 4\, n_s^2\, W_{eff}$, which is necessary to achieve $A = 1 - \exp(-\alpha_s\, 4\, n_s^2\, W_{eff})$. We consider the as-yet unknown light trapping texture with all path lengths $l = 4\, n_s^2\, W_{eff}$ to be the optimum geometrical light trapping scheme. The maximum short-circuit current densities to be expected from such optimum geometrical light trapping are displayed in Figure 2.8 on p. 20.

A.2.2 Coherent versus incoherent simulations

A plane Si layer of thickness $W = 4$ µm shows interference fringes in the transmittance, the reflectance, and the absorption spectra. A numerical simulation will reproduce these interference fringes only if the phase of electromagnetic radiation is accounted for. We call such a simulation coherent, as opposed to incoherent simulations [218].

In order to estimate the influence of phase effects we compare coherent and incoherent calculations for a special case of an Encapsulated-V texture [27] with a large period $p = 500$ µm and a facet inclination angle $\alpha = 60°$. This V-shaped thin-film texture is depicted in Figure A.11 on p. 195. The program SUNRAYS is capable of simulating coherently if we consider the Si film to be part of the antireflection coating system.

Figure A.6 shows that the absorption of a $W_f = 4$ µm-thick cell is almost identical for coherent and incoherent simulations. Interference fringes are hardly present due to the wide range of angles of incidence caused by the randomization process in the glass superstrate and the motion of the sun. The maximum current densities j_{sc}^* of coherent and

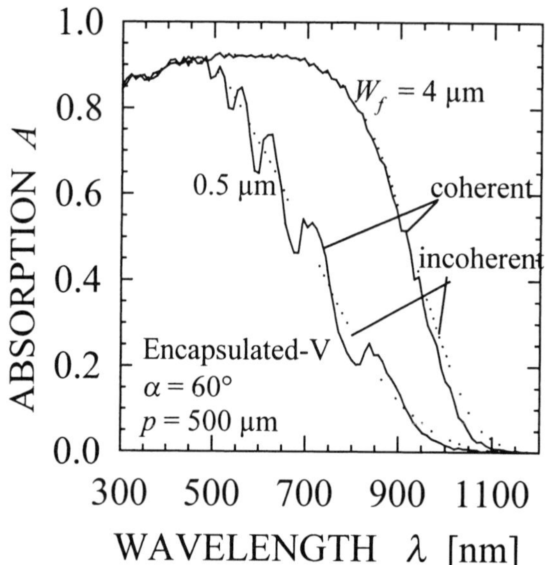

Figure A.6. Coherent and incoherent calculated absorption of the Encapsulated-V texture for Si films with thickness W_f. The 4 μm-thick film hardly shows interference effects in the coherent calculation since many angles of light incidence are involved. Thinner films with W_f = 0.5 μm do show interference effects in the coherent calculation.

incoherent simulations differ by 1% only. The slightly larger current density for the incoherent calculated structure is caused by the effect of light randomization within the silicon substrate. This effect is not included in our coherent simulation. For an even smaller film thickness W_{eff} = 0.5 μm, interference fringes are clearly visible in the coherent calculation. However, the absorption oscillates around the incoherent calculated values. The integrated current density j_{sc}^* differs by 0.5% only. The coherence length of the broad-band solar spectrum is sufficiently small to have all interference effects averaging out when calculating the photogeneration current density j_{sc}^*. To summarize, the error in the determination of current densities j_{sc}^* when replacing coherent simulations by incoherent simulations is around 1% down to a film thickness of W_f = 0.5 μm under the year's average illumination.

Geometrical ray-tracing is also not applicable if the texture period p becomes smaller or similar in size to the wavelength λ. A recent numerical study (Figure 5 in Ref. [365]) shows the reflectance as calculated by the coupled-wave theory for a grooved Si surface with facet angle α = 54.7° and texture periods ranging from p = 0.5 to 4 μm under normal incidence of perpendicularly polarized light with wavelength λ = 820 nm. The reflectance values calculated by the coherent coupled wave theory and by incoherent ray-tracing differ by less than 2% absolute. Again, the differences in j_{sc}^* will be even smaller under broad-band solar illumination with a wide range of angles of incidence (due to the motion of the sun). Randomization of light propagation in the cover-glass and in the Si film will also tend to average out phase effects. We speculate that in terms of j_{sc}^*, incoherent simulations are a valid approximation down to $p \approx 2$ μm. However, a quantitative comparison of the photogeneration j_{sc}^* from the exact coupled wave theory and from an approximate incoherent ray-tracing remains a task for future work. Recent numerical tools that solve Maxwell's equations in three-dimensional space could be used for this purpose [366, 367].

A.2.3 Illumination geometry

Many authors model the light trapping performance of textured solar cells under normal illumination (e.g. Ref. [2]), because this condition is commonly used to measure the efficiency of a cell. Low-cost thin-film solar cells are likely to be used without tracking systems. Normal incidence occurs only once a year. Hence we should not neglect the movement of the sun during the day and the year. Figure A.7 compares the projected solid angle $\Omega = 0$ of normal illumination (white cross) with the projected solid angle of an extended source that is apparently generated by the daily and yearly movement of the sun (dark gray area). The module is assumed here to lie flat at the equator and the p-axis is east-west oriented. The vertical (light gray) band and the horizontal (dark gray) band are equivalent for textures with a fourfold symmetry, such as the frequently used pyramidal textures. Hence during a year the solar module sees sunlight impinging from all directions except from the four white fields in the circle depicted in Figure A.7. These white fields account for only 22% of the circle's area and even these fields are partially filled by diffuse light impinging on the module. Isotropic illumination is therefore the more relevant illumination condition when compared to normal incidence. In this work, we model the movement of the sun during the year as described in Ref. [368]. We use an extended light source with the projected solid angle $\Omega_{ya}' = 2.4$ (dark gray band in Figure A.7) and constant radiance at all wavelengths. The spectral power distribution of our "year's average illumination" is that for global air mass 1.5, as compiled in Ref. [13].

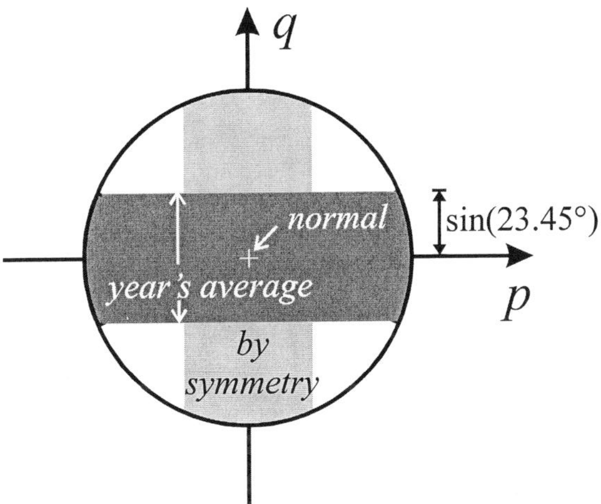

Figure A.7. Comparison of the projected solid angle for three illumination conditions. For normal illumination the angle $\Omega' = 0$. All directions of sunlight during a year subtend Ω' denoted by the dark gray band. For textures with a fourfold symmetry, the light gray and the dark gray bands are equivalent. The full circle $\Omega' = \pi$ is subtended for isotropic illumination.

Impact of illumination geometry and encapsulation on photogeneration

A light trapping scheme that permits large current densities for normally incident light may perform poorly, if the cells are under illumination from an extended source and under encapsulation. To demonstrate this, we compare the previously discussed inverted pyramid texture with the simple prism groove (SPG) texture [369] that has one facet inclined by $\gamma = 29°$. Opposing facets include an angle of 90° as shown on the right-hand side of Figure A.8. The textures have a period of 10 μm, no ridge tops, a 109 nm-thick SiO_2 film on the front surface, and an Al reflector separated by 200 nm of SiO_2 from the silicon back surface.

Figure A.9 shows that under the normal incidence, and without encapsulation (conditions that are typical for efficiency measurements in the laboratory) the inverted pyramids perform better than the simple prism grooves for all thickness values. For encapsulated modules under an extended light source that models the year's average illumination (conditions that are more realistic for outdoor applications of thin-film cells) the current density of the inverted pyramid texture is inferior to the SPG texture.

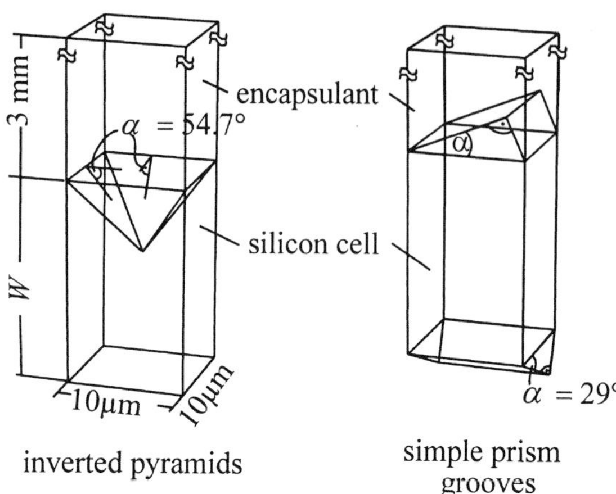

Figure A.8. Unit cells of light trapping textures.

Figure A.10 reveals the reason for the better performance of the SPG texture: this texture keeps all rays inside the cell for at least four passes [369]. More than 80% of the rays stay inside the cell for at least eight passes. In contrast,

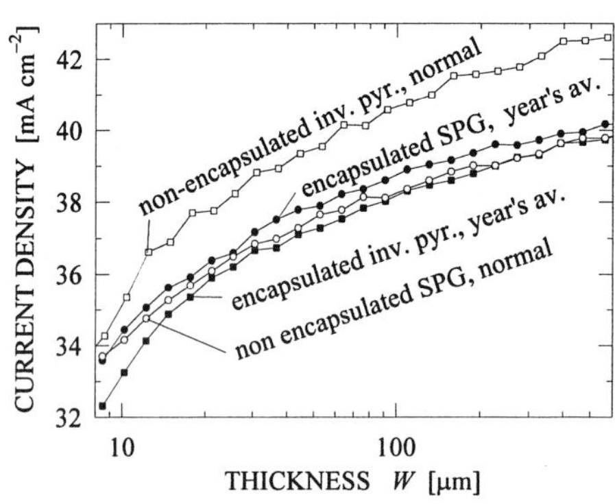

Figure A.9. Calculated photogeneration current densities for silicon cells of thickness W. The statistical error is 0.07 mA cm^{-2}.

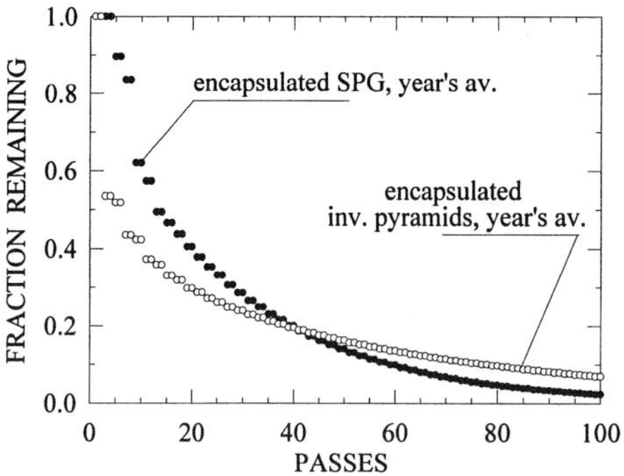

Figure A.10. Fraction of light rays remaining inside the cell as a function of the number of passes through the cell.

the inverted pyramid texture loses more than 40% of the rays after the first two passes through the cell, which is similar to the case without encapsulation [2].

Recently, a surface texture that keeps all normally incident light rays inside the Si film for at least four passes through the cell has been discovered [370]. This texture is particularly interesting for achieving highest-efficiency laboratory cells, since it does not use an encapsulating glass cover and is optimized for normal incidence of light.

A.2.4 Conformal film textures

We call a surface texture conformal if both surfaces of the film are parallel. Such films are fabricated by depositing Si onto a textured substrate. Figure A.11 shows the four light trapping textures we investigate here. We choose textures with 2-, 3,- 4-, and

6-fold rotational symmetry. This choice is motivated by the fact that these unit cells can be arranged periodically to fill a plane completely. Each of the textures has three parameters that define the geometry: texture period p, facet angle α, and film thickness W_f. Here, we measure W_f perpendicularly to the facets.

Textures investigated

The Encapsulated-V texture utilizes a structured glass superstrate and a detached back surface reflector [27]. The inclination of the facet angle α increases the cell surface and cell volume by a factor of $1/\cos(\alpha)$. The space between the Si film and the back surface reflector is filled with air. The Encapsulated-V texture has translational symmetry and is a two-dimensional (2D) texture. It transforms into a three-dimensional (3D) texture if the surface of the glass is rough [27]. The V-shaped Si films fabricated by etching of Si wafers were first analyzed by Uematsu [221]. Naturally, the grooves of the Encapsulated-V textures are best aligned east-west for optimum photogeneration.

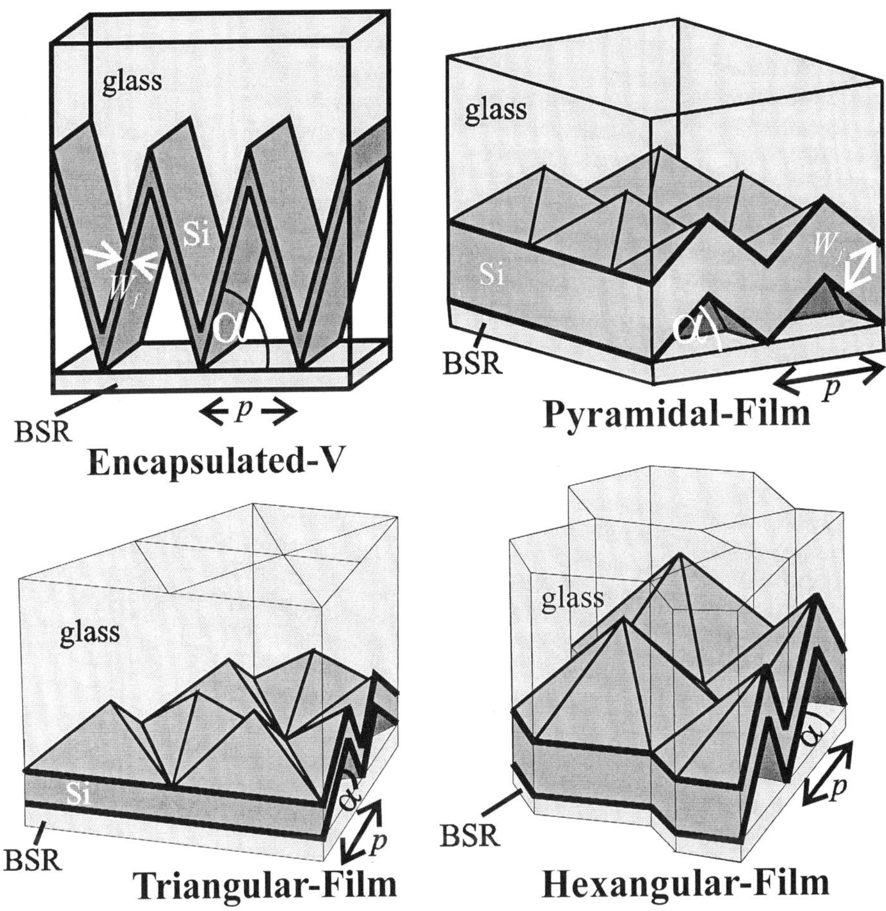

Figure A.11. Four types of conformal film textures: a thin Si films covers a textured glass substrate. The texture period p is larger than the film thickness W_f. A detached back surface reflector reduces metal reflection losses. Facets are inclined by an angle α relative to the macroscopic cell surface. The tips of the structures are in the center of the unit cells.

The Triangular-Film texture shown in Figure A.11 consists of a glass structured with three-sided inverted pyramids with facets inclined to an angle α. The unit cell has a triangular base with equally sized sides. The tip of the triangle is centered relative to the base. Textures with a triangular base were first investigated in 1960 by Dale and Rudenberg [371] for reflection control in thick Si wafers. Recently, films with triangular-based pyramids were shown to allow good light trapping in conjunction with double-layer antireflection coatings on the front and an Ag reflector on the back of the film [107]. Here, we do not use antireflection coatings and apply a detached Al reflector instead of an Ag reflector adjacent to the Si.

The Pyramidal-Film texture shown in Figure A.11 consists of a glass structured with four-sided inverted pyramids. The facets of the pyramids are inclined by the angle α against the plane front surface of the glass superstrate. The bases of the individual pyramids are squares. The Pyramidal-Film texture was introduced, independently of each other, by Thorp [107] and the author [109].

The Hexangular-Film texture shown in Figure A.11 consists of six-sided inverted pyramids with facets of inclination α. The base of each pyramid is a hexagon.

Simulation conditions

All four textures are simulated with no antireflection coatings at the air/glass and the glass/Si interfaces. We assume a 109 nm-thick SiO_2 layer for surface passivation at the Si/air interface. The Al back surface reflector (BSR) is detached from the Si layer. Detached reflectors are analyzed separately in section A.3.3 on p. 206. In view of the short diffusion lengths L that are currently achieved with thin Si films deposited on glass substrates [109], we consider thin layers with a thickness $W_f = 4$ µm. All results are qualitatively the same for other thickness values. The glass superstrates have a thickness of 2 mm. We assume a year's average illumination [214] in this study. One side wall of the unit cells shown in Figure A.11 is oriented in the east-west direction.

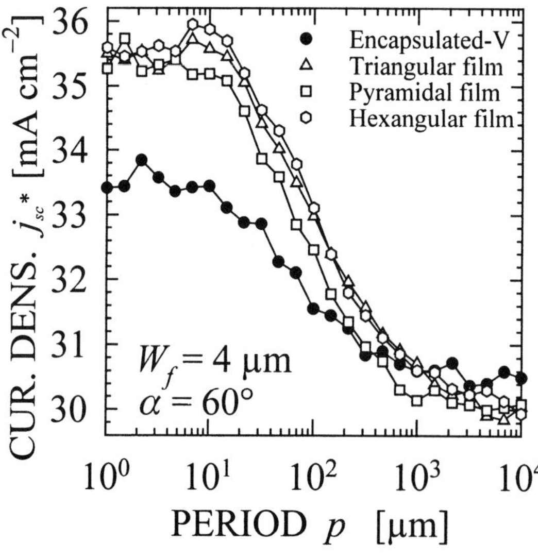

Figure A.12. All four textures show an increase in maximum short-circuit current density j_{sc}^* with decreasing period p. The two-dimensional Encapsulated-V texture shows smaller current densities than the three-dimensional textures.

Impact of texture period p on photogeneration j_{sc}^*

Figure A.12 shows the dependence of the maximum short-circuit current density j_{sc}^* on the texture period p. All the textures displayed in Figure A.11 show larger photogenerated current densities j_{sc}^* at smaller texture periods p [108, 109]. We explained the reason for this along with Figure 5.32 on p. 148. The photogeneration current density j_{sc}^* saturates at a high current value for $p < 10$ µm ($p/W_f < 0.4$) and saturates at a low value for $p > 1000$ µm ($p/W_f > 250$). At small texture periods the textures fall into two classes: the 3D textures, which show large current densities j_{sc}^*, and the 2D Encapsulated-V texture, which shows a smaller current density. At large texture periods the 2D Encapsulated-V texture performs slightly better than the 3D textures.

While small texture periods give the best light trapping, coarse textures with periods of 0.5 to 5 mm are much easier to fabricate, e.g. by low-cost techniques such as the embossing of hot and soft glass by patterned steel rollers. Figure A.12 might help to identify a good compromise between optimum performance and technical feasibility. First experiments on small-scale embossed sol-gel glasses [345] and on small-scale embossed Borofloat glasses [372] verified the expected absorption enhancement.

Impact of texture period p on average path length $\bar{l}$

Figure A.13 reveals the origin of the smaller current densities j_{sc}^* generated by the two-dimensional Encapsulated-V texture. The average path length (49 ± 2.6) W_{eff} of the 3D textures is larger than the average path length (12 ± 1.5) W_{eff} of the Encapsulated-V texture by a factor of 4. This factor originates from the dimensionality of the textures [16]. As expressed by Eq. (A.36) on p. 190, the theoretical limit to the average path length $\bar{l}$ under isotropic illumination is $l_{max} = 4\, n_s^2\, W_{eff} \approx 50\, W_{eff}$ for three-dimensional textures, while it is $\bar{l}_{2D,max} = 4\, n_s \approx 14\, W_{eff}$ in the two-dimensional case. All of the four textures almost reach the theoretical limit that is set by the dimensionality for all texture periods p. The period p dependence of the current density j_{sc}^* shown in Figure A.12 is

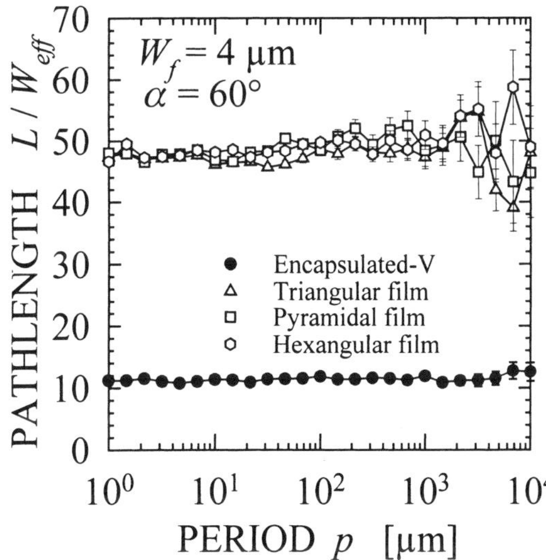

Figure A.13. Average path length under A year's average illumination for the textures shown in Figure A.11.

consequently *not* caused by a p-dependent average path length $\bar{l}$. The similarity of the average path length despite a vastly different photogeneration demonstrates that a large value of $\bar{l}$ is a necessary but not sufficient condition for good light trapping.

Figure A.14 shows the standard deviation σ of the path length distribution $f(l)$ for the textures shown in Figure A.11. All textures exhibit a standard deviation σ that is independent of period p for $p < 10$ µm ($p/W_f < 2.5$) and increases with $p^{0.5}$ for $p > 100$ µm ($p/W_f < 25$). The reason for this increase is as follows: the larger the period p, the fewer rays are trapped at the intersections of adjacent facets. However, those few rays that are trapped have to propagate a long way in one facet until they are able to escape from the Si layer. Therefore, the few rays that are trapped in large-period textures have an extraordinarily long path length. This discrepancy between many rays with a short path length and a few with a long path length increases the standard deviation σ. The Encapsulated-V texture shows a smaller standard deviation σ than the other textures. Small standard deviations σ are beneficial for good absorption (Eq. (A.19)), and hence the p dependence of the standard deviation σ causes the p dependence of the maximum current density j_{sc}^* shown in Figure A.12. At very large periods p the advantage of small σ apparently counterbalances the disadvantage of having a smaller average path length $\bar{l}$, because at large periods the Encapsulated-V texture shows larger current densities than the other textures.

Impact of facet angle α on photogeneration j_{sc}^*

Due to improved light trapping (see Figure A.12), the current densities of fine textures ($p < 15$ µm) are approximately 5 mA cm^{-2} larger than those of coarse textures ($p > 100$ µm) . We therefore concentrate on small periods ($p = 15$ µm, $W_f = 4$ µm) for the investigation of the dependence of the photogeneration on the facet angle α. At large angles (which means sharp textures), the current densities j_{sc}^* are largest, as shown in

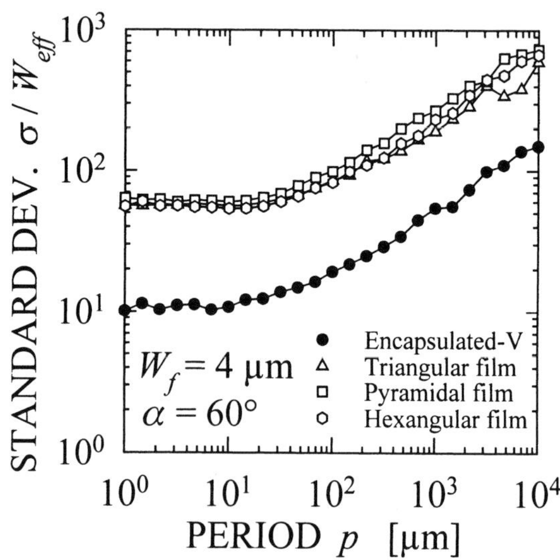

Figure A.14. At periods $p > 10$ µm the standard deviation σ of the path length distribution $p(l)$ increases with increasing texture period and is proportional to $p^{0.5}$ for $p > 300$ µm.

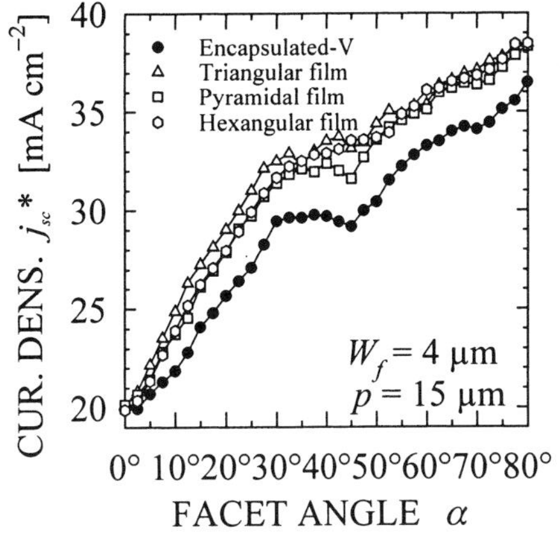

Figure A.15. Current densities j_{sc}^* of up to 38 mA cm^{-2} are feasible with steep facet angles $\alpha = 75°$ and small periods $p = 15$ µm.

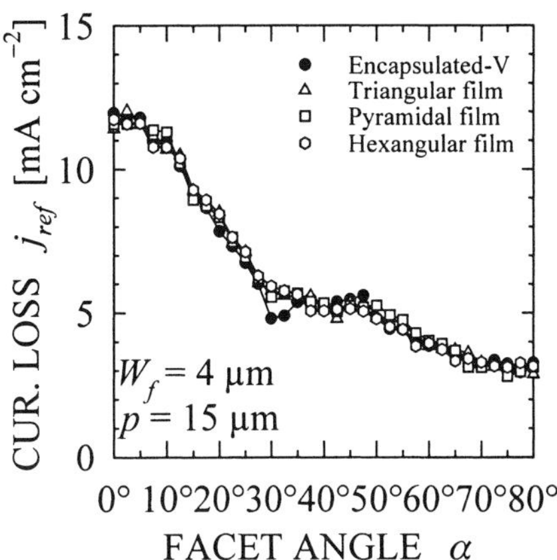

Figure A.16. Current density loss j_{ref} due to surface reflectance for textures with period p = 15 µm.

Figure A.15. A photogeneration of 38 mA cm^{-2} is possible from $W_f = 4$ µm-thick c-Si films without the use of antireflection coatings at $\alpha = 80°$. The Pyramidal-Film texture and the Encapsulated-V texture show a dip of the current density j_{sc}^* at a facet angle α = 45°, a behavior that was also calculated in Refs. [111, 108]. At an angle $\alpha = 45°$ a double reflection reverses the direction of light propagation and does not contribute to a randomization of light propagation in the film. Hence, light trapping is poor and the current density is small. This dip at $\alpha = 45°$ is absent if the base of the Pyramidal-Film texture is chosen to be quadrangular, but not square. This measure avoids an exact inversion of the direction of propagation in reflection at two opposing facets.

Impact of facet angle α on reflection loss j_{ref}^*

All light rays that leave the cell structure without entering the Si at least once contribute to the front surface reflectance R_s. The reflectance spectrum $R_s(\lambda)$ corresponds to a current loss

$$j_{ref}^* = \frac{q}{hc} \int_0^{1250nm} \lambda' \, R_S(\lambda') I_{AM1.5G}(\lambda') d\lambda' \tag{A.38}$$

Figure A.16 shows the current loss j_{ref} that decreases with increasing facet angle α, since the light has more attempts to enter the Si film. However, even for very sharp angles α the current loss j_{ref} is still 3 mA cm^{-2}, due to the reflectance at the air/glass interface. Cover-glasses with textured front surfaces or antireflection coatings have been considered to reduce these losses [88, 373, 27, 107].

A.3 Back surface reflector

Efficient light trapping not only requires a long path length but also needs a back surface reflector with large reflectance.

Current loss due to non-ideal back reflectance

Figures A.17a and A.17b show the current loss due to a less-than-unity back surface reflectance in a Lambertian light trapping scheme in absolute and relative units. We use Eq. (A.13) on p. 183 for a front surface transmittance $T_f = 1$ to calculate the data. The loss is given relative to the case of unity back reflectance R_b. The current loss increases with decreasing thickness because thinner films rely more heavily on light trapping. For thick films the back reflector is not necessary and the current loss is almost zero. These results underline the necessity of highly reflective back surface reflectors in thin-film Si solar cells. The formation of high-efficiency reflectors is a standard problem of thin-film optics [374]. The standard solution is to use dielectric multi-layers. These multi-layers exhibit a maximum reflectance at their design wavelength λ_0.

Estimate of design wavelength λ_0 in thin-film cells of thickness W_{eff}

Let us consider a cell of thickness $W_{eff} = 4$ µm with unity front surface transmission T_f. The absorption caused by the first pass of the light through the cell is denoted by a dotted line in Figure A.18. This is the total absorption if the back reflectance is zero. The gain in absorption due to a reflector with unity back reflectance is shown as the solid line and is caused by light absorption during all subsequent passes. The total absorption of an ideal Lambertian cell is the sum of both contributions (broken line). If the reflector cannot have a high reflectance at all wavelengths, it should be designed to have its maximum reflectance at approximately 900 nm (see arrow).

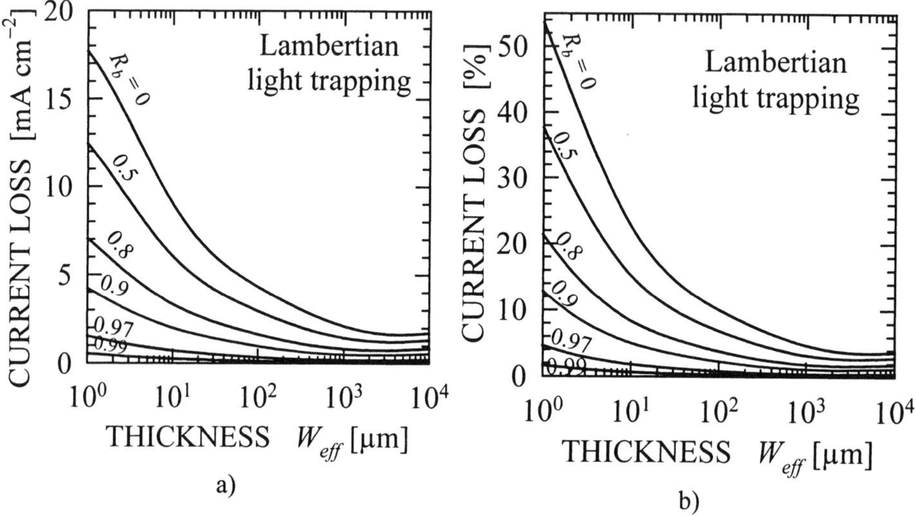

Figure A.17. Photogeneration loss of a solar cell with Lambertian light trapping and back surface reflectance R_b relative to the ideal case of unity back reflectance: a) absolute loss expressed as a current density; b) relative loss.

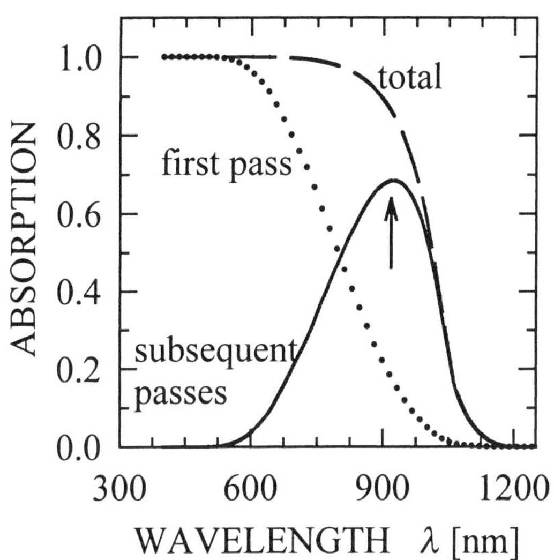

Figure A.18. Absorption of a W_{eff} = 4 μm-thick Lambertian cell with unity back reflectance (broken line) and with zero back reflectance (dotted line). Gain due to reflector (solid line) peaks at the wavelength marked by the arrow.

Figure A.19 shows the design wavelength as a function of the thickness W_{eff} for back reflectors of various reflectance values R_b. Lambertian and planar cells are considered. For the determination of λ_0 we weight the gain in absorption with the photon flux of a black body at 6000 K. The maximum of absorption enhancement times photon flux is defined as the design wavelength. The design wavelength λ_0 turns out to be shorter for thinner cells, because here the reflector is already necessary to absorb effectively the short-wavelength light, which has more photons than long-wavelength light. Figure A.19 may serve as an estimate for the design wavelength λ_0. A precise determination of λ_0 requires account to be taken of the wavelength dependence of the reflectance R_b.

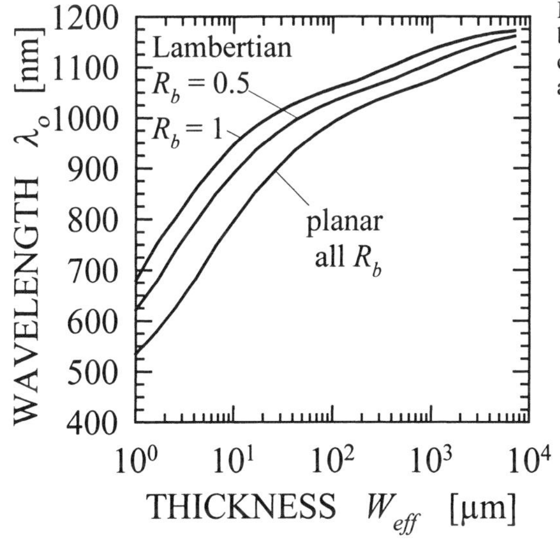

Figure A.19. Design wavelength λ_0 of back surface reflectors of thin-film cells with Lambertian light trapping and without light trapping (planar).

Dielectric multi-layers, multi-layers from porous Si, and detached reflectors are three approaches to achieve a high back reflectance. We will discuss each of these approaches in the subsequent sections.

A.3.1 Dielectric interlayers

The highly efficient thin-film cells from thinned FZ wafers with energy conversion efficiencies >20% use an Al metal reflector with a SiO_2 layer between the metal and the Si [113, 351]. The reflectance of the Si/SiO_2/Al system for non-polarized light of wavelength $\lambda = 850$ nm is shown in Figure A.20 as a function of the thickness t of the oxide layer and for various angles of incidence. The periodic interference fringes at angles smaller than the critical angle of total internal reflection arcsin(1.46/3.65) = 23.6° are due to the standing wave in the oxide. For angles larger than 23.6°, the evanescent wave in the oxide interacts with the Al layer to cause frustrated total internal reflection. At large thickness values $t > 400$ nm, the decay length of the evanescent wave is smaller than the oxide thickness and hence the reflectance is unity.

Let us now consider the high-efficiency cell with inverted pyramids on the front and a planar back as shown in Figure 3.8 on p. 75. Light that impinges normally on the cell reaches the back reflector at an angle of 39.6°, as calculated with Eq. (3.66) on p. 76. If the cell is encapsulated by a material of index of refraction of 1.46, we find an angle of incidence that is 32.3°. In both cases loss-free total internal reflection occurs, provided an oxide thickness $t > 300$ nm avoids frustrated total internal reflection.

For isotropic internal incidence of light on the back reflector the reflectance

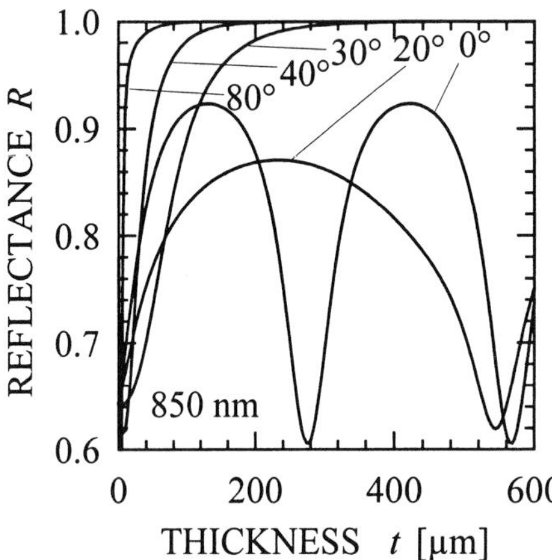

Figure A.20. Internal reflectance R for non-polarized light approaching the Si/SiO_2/Al system from within Si as a function of the oxide thickness t for various angles of incidence. The wavelength is 850 nm.

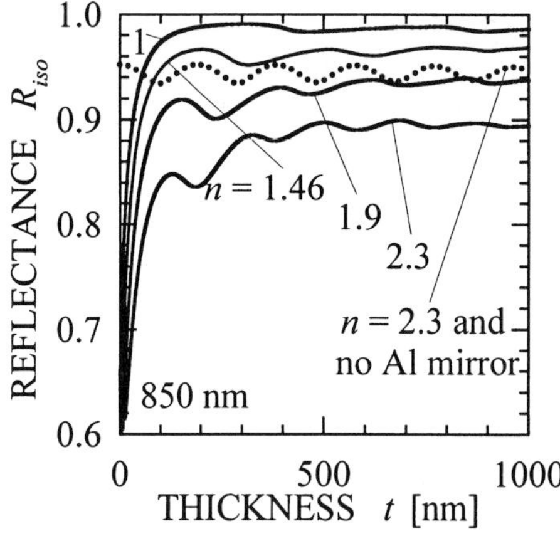

Figure A.21. Internal reflectance R_{iso} at λ = 850 nm for isotropic incidence on a Si/dielectric film/Al system. The index of refraction of the dielectric layer varies from n = 1 to 2.3. The dotted line is for a dielectric layer with n = 2.3 and no Al layer.

$$R_{iso} = \int_0^{\pi/2} R_b(\vartheta)\sin\vartheta\cos\vartheta\, d\vartheta \Big/ \int_0^{\pi/2} \sin\vartheta\cos\vartheta\, d\vartheta \tag{A.39}$$

is the étendue-weighted average over all angles. The line denoted by n = 1.46 in Figure A.21 shows this isotropic reflectance R_{iso} for a SiO_2 dielectric layer (n = 1.46) of thickness t. The reflectance has a first local maximum at t = 200 nm. Further thickness enhancement does not improve the reflectance value.

Silicon nitride is more likely to be used as a dielectric layer in thin-film Si photovoltaics than SiO_2, as discussed in the section on surface passivation on p. 42. Figure A.21 therefore also shows the isotropic reflectance for dielectric layers with the index of refraction n = 1.9 (stoichiometric silicon nitride Si_3N_4:H) and n = 2.3 (silicon-rich silicn nitride Si_xN_4:H, x >3). The higher the index of refraction, the poorer the reflectance. This is because the reflection at both the Si/SiN_x and the SiN_x/Al interfaces decrease with increasing n, because the index of the dielectric layer is smaller than the indices of Si and Al. A dielectric layer with n = 2.3 with an Al metal reflector is not a good choice. Simply omitting the Al layer enhances the reflectance (dotted line in Figure A.21) and simplifies the process. The reflectance at t = 0 is the isotropic internal reflectance of a planar Si/air interface. The reflectance value is R_{iso} = 0.952.

Let us now study the wavelength dependence of the back reflectance. Figure A.22 shows the internal reflectance of a Si/Al, a Si/Au, and a Si/Ag interface for light impinging normally from the Si side. No intermediate dielectric layers are assumed. A reflectance better than 92% is achieved for Au and Ag. Both materials are too expensive for use in low-cost thin-film devices. The minimum in the reflectance of Al at 820 nm is caused by an inter-band transition in the metal [375]. At 850 nm the reflectance is only 64%. A Lambertian cell with such a poor back reflectance and a thickness W_{eff} = 4 µm loses about 20% of its photogeneration. An Al reflector is therefore certainly not a good choice for thin-film cells. Annealing of a Si/Al interface further reduces the reflectance. The insertion of a dielectric layer is one possibility to enhance the reflectance (see Figure A.21) and to reduce the surface recombination.

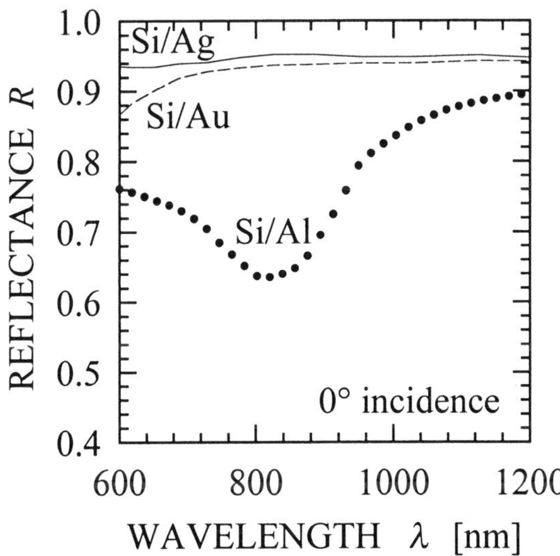

Figure A.22. Spectral dependence of the internal reflectance under normal incidence of the Si/Al (dotted line), Si/Au (broken line),and Si/Ag (solid line) interface without a dielectric layer between the metal and the silicon.

A.3.2 Porous reflector

One approach to thin-film Si photovoltaics that we discuss in Chapter 4 on p. 91, is the deposition of an epitaxial thin Si layer onto a low-cost Si substrate [376, 177, 377, 378, 379, 380]. The homo-epitaxial interface between the substrate and the thin film does not reflect the visible light even if the substrate is highly doped, because the change in optical constants caused by the free carriers is negligible for the reflection of visible light. Thus a significant fraction of the solar photons is absorbed in the substrate. Due to the low electronic quality and/or Auger recombination, the photogenerated electron-hole pairs will not make it to the contacts. We estimated the substrate losses and substrate contributions to the photocurrent for textured and planar cells grown on highly doped Si substrates in Ref. [381].

A reflector for thin-film cells on highly doped carrier wafers and for thin-film cells on thick low-quality Si seed layers was realized by Mauk et al. [382]. A tungsten film with stripe openings is evaporated onto the seed layer. The Si film nucleates in the openings and laterally overgrows the metal. Thus a buried metal reflector is fabricated. However, most Si epitaxy techniques apply high-temperature processes at 1000 to 1150 °C that would drive the metal into the active Si layer. To minimize the photogeneration in the Si substrate [381], the reflector should be positioned close to the interface of the substrate and epitaxial layer. Recently, we suggested the use of porous Si layers between the substrate and the epitaxial layer as a reflector [29].

Figure A.23 illustrates the principle of the porous Si reflector. A single layer or a multi-layer stack of porous Si films is introduced between the Si solar cell and the Si substrate. Each porous layer is characterized by its porosity p, its thickness d, and its wavelength-dependent optical constants n and k. See section 5.1.1 starting on p. 122 for more information on porous Si. The refractive index of porous Si decreases with increasing porosity. The typical diameter of the voids as well as of the Si filaments in the porous film is much smaller than the optical wavelengths. The effective medium theory therefore applies. Following Theiss [383] we apply the Looyenga formula [384], for the

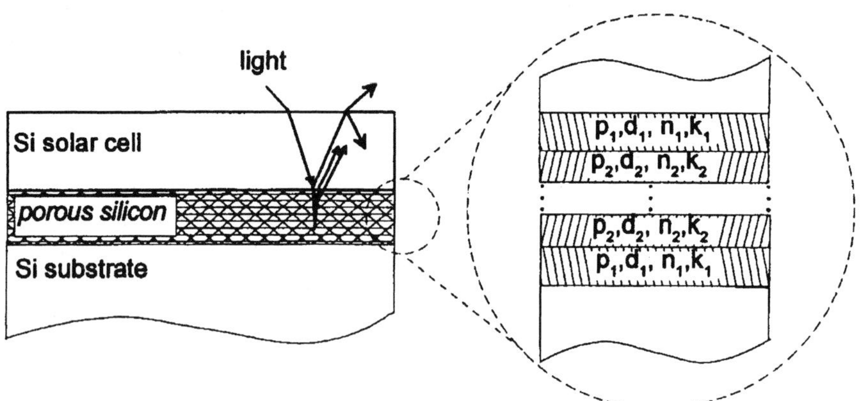

Figure A.23. Si cell on a Si substrate having a porous Si light reflector at the back. The reflector consists of several double layers with porosity p_1, p_2 and thickness d_1, d_2, respectively, as shown in the magnified portion. The reflector is located between the p-type Si epitaxial solar cell and the p^+-type Si substrate. A sintered porous reflector scatters light and introduces light trapping. Figure from Ref. [29].

effective dielectric function

$$\varepsilon_{eff} = \left(p\varepsilon_{air}^{1/3} + (1-p)\varepsilon_s^{1/3}\right)^3 \tag{A.40}$$

of a film with porosity p. The optical constants n and k of the porous Si are related to the effective dielectric constant

$$\varepsilon_{eff} = (n + \mathrm{i}\,k)^2 \tag{A.41}$$

where i denotes the imaginary unit. The dielectric functions of air and bulk Si are denoted by $\varepsilon_{air} = 1$ and ε_s respectively. Hence, a single porous layer or a multi-layer system can be designed to function as an interference filter [385]. If the optical thickness $n \times d$ of the porous Si layer equals half the design wavelength $\lambda_0/2$, we obtain a maximum reflectance at wavelength λ_0. The greater the porosity of the layer, the smaller the refractive index, and the higher is the reflectance. The range of porosity values that is experimentally accessible has an upper bound, because highly porous layers are mechanically unstable [53]. In order to circumvent the stability problem we suggest the use of a large number of layers having small porosity instead of a single layer having a large porosity. A multi-layer system consisting of layers that alternate in porosity p_1 and p_2 is shown in Figure A.23. Each layer has the optical thickness $\lambda_0/4$, where λ_0 denotes the design wavelength. Such porous Bragg reflectors are known from the literature [385].

We analyze a Bragg reflector that consists of 20 porous Si layers with alternating values of porosity $p_1 = 0.4$ and $p_2 = 0.6$. The substrate is a 500 μm-thick, bifacially polished, $10^{19}\,\mathrm{cm}^{-3}$ boron-doped Si wafer. The symbols in Figure A.24 show the reflectance measured after the formation of the Bragg reflector and *without* additional epitaxy on top of the reflector. The reflectance reaches almost unity at the design wavelength $\lambda_0 =$ 940 nm. The Bragg reflector causes interference fringes in the 400 to 900 nm range. The measurement agrees well with a multi-layer simulation [29].

On top of the Bragg reflector we fabricate a 3 μm-thick Si film by chemical vapor epitaxy at 1000°C. The epitaxy process is described in Ref. [378]. The reflectance spectrum measured after epitaxy is depicted by the symbols in Figure A.25. As expected, the reflectance is enhanced in the 900 to 1100 nm range, where the absorption in the thin Si layer is weak. We find a decrease in infrared reflectivity relative to the case before epitaxy. The reflectance is less than 0.90 at all wavelengths. In addition to the interference fringes caused by the Bragg reflector, we find fringes caused by the 3 μm-thick epitaxial layer. Both types of fringes are superimposed. The reflectance maximum shifts from 940 nm before epitaxy to about 1000 nm after epitaxy. The agreement with the simulation is not as good as without epitaxial layer. The high-temperature processing leads to a sintering of the porous Si, which makes the modeling with abrupt interfaces from high to low porosity inappropriate. Figures 5.6c and 5.6d on p. 125 show SEM micrographs of a Bragg reflector from porous Si prior to epitaxy and after epitaxy. The restructuring of the porous Si causes voids that scatter light in Si. Such light scattering introduces some degree of light trapping, as depicted for the ray propagating to the left in Figure A.23.

At ZAE Bayern we fabricated thin-film cells with a base thickness of 15 μm on a highly doped p^+-type planar Si substrate with and without a porous reflector. The cell with the reflector exhibited a short-circuit current that was 5% higher than the reference cell without a reflector. While this is a proof of principle, the current enhancement that is achievable with such light-scattering back reflectors requires further investigation.

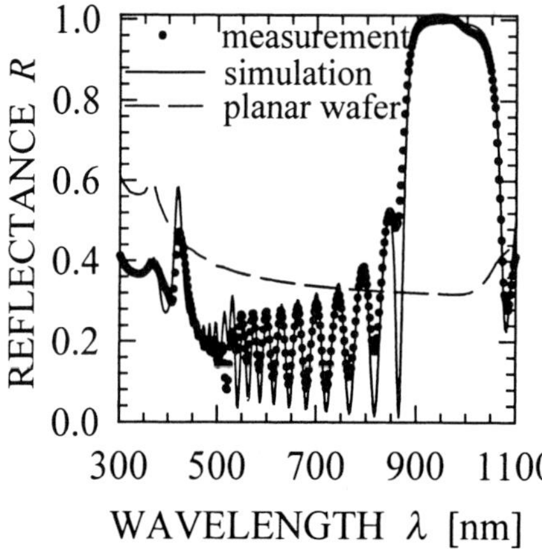

Figure A.24. The symbols show the reflectance measured for a system of 20 porous Si layers. The continuous line is a simulation using alternating porosity values of p_1 = 0.6 and p_2 = 0.4, and thickness values of d_1 = 125 nm and d_2 = 100 nm. The dashed curve indicates the reflectance of a bare silicon wafer without a reflector. Data from Ref. [29].

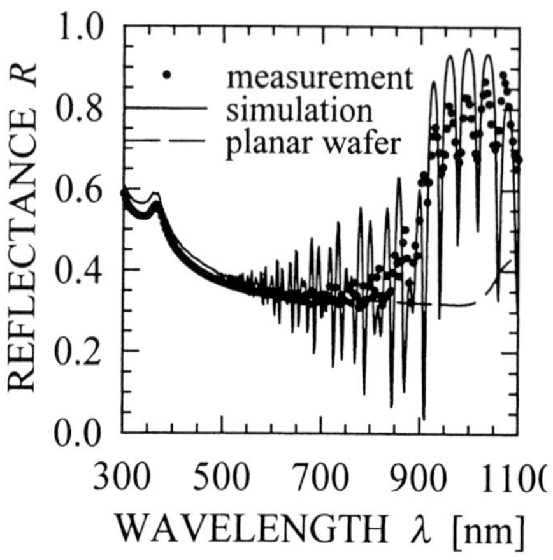

Figure A.25. Reflectance spectrum of a porous Si multi-layer interference filter after epitaxial growth of a 3 µm-thick Si layer. The symbols are experimental data, while the continuous line is a simulation for porosity p_1 = 0.54 and p_2 = 0.36, and thickness values of d_1 = 120 nm and d_2 = 95 nm. The dashed line shows the reflectance of a bare silicon wafer without a reflector. Data from Ref. [29].

A.3.3 Detached reflector

A particularly simple approach is to place *nothing* behind the cell. Provided the direction of light propagation is sufficiently randomized by the front surface, only a fraction $1/n_s^2$ of the isotropically impinging light escapes through the back of the cell, where n_s is the silicon index of refraction. The effective back reflectance is thus $R_b = 1 - 1/n_s^2 = 0.92$. For diffused light the deposition of a reflecting layer makes sense from the optical point of view only if its reflectance is greater than 0.92. A detached reflector that is placed behind the cell may serve to reflect the escaping fraction $1/n_S^2$ back into the cell. The distance of the detached metal reflector from the back surface should measure at least a few wavelengths to avoid attenuated total internal reflection.

The metal reflectors with a thick intermediate dielectric layer, which we discuss in the framework of Fresnel equations in section A.3.1 starting on p. 202, uses the same principle. Here, however, we analyze these so-called detached back surface reflectors in the framework of Lambertian light trapping, which we discuss in more detail in section A.1 on p. 181. With the definition of the effective back reflectance by Eq. (A.1) on p. 182 and with the solution of the linear system of Eqs. (A.2) through (A.9) we find the effective back surface reflectance

$$R_b = 1 - \left(\frac{n_t}{n_s}\right)^2 \left(\frac{1}{1-R_m} + \frac{1}{T_t} - 1\right)^{-1} \tag{A.42}$$

For the special case of a back surface with an ideal antireflection coating (T_t = 1 in Figure A.1), this simplifies to

$$R_b = 1 - \left(\frac{n_t}{n_s}\right)^2 (1 - R_m) \tag{A.43}$$

The interpretation is as follows: for fully randomized light the intensity, or equivalently the path length, is proportional to the square of the refraction index n_s^2 [95], as discussed before in section A.1. An index of refraction n_t in front of the metal reflector that is smaller than the index n_s for Si (see Figure A.1 on p. 181) reduces the optical loss in the metal by a factor $(n_t/n_s)^2$, due to a reduced density of propagation modes (see Eq. (5.1) on p. 144). Hence, the smaller the value for the index of refraction n_t, the larger the effective back reflectance R_b.

We verify this analytical treatment by numerical ray-tracing. The insert in Figure A.27 shows a Si back surface textured with square-based inverted pyramids of facet angle 54.7°. This surface is positioned in front of a metal with reflectance R_m. Metal and textured surface are separated by a distance greater than the wavelength. The gap is filled with a medium of refractive index n_t that has no optical absorption.

The simulated reflectance R_b shown in Figure A.26 is higher than the reflectance R_m of the metal reflector. The reflectance R_b increases with a decreasing index of refraction n_t of the transparent medium. For an air/Al interface that has a reflectance of approximately $R_m = 0.95$, the effective back surface reflectance is $R_b = 0.997 \pm 0.001$, with the gap filled by air. With an encapsulant or SiO_2 in the gap (n_t = 1.46) we calculate R_b =

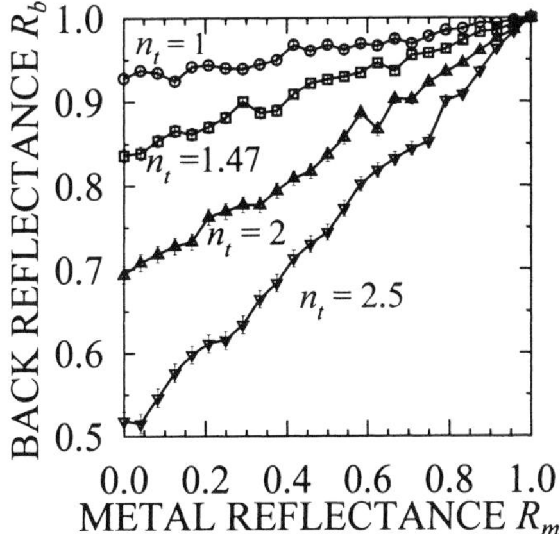

Figure A.26. The reflectance R_b of the detached back surface reflector is higher than the reflectance R_m of the metal reflector. R_b increases with decreasing index of refraction n_t of the transparent medium. See insert of Figure A.27.

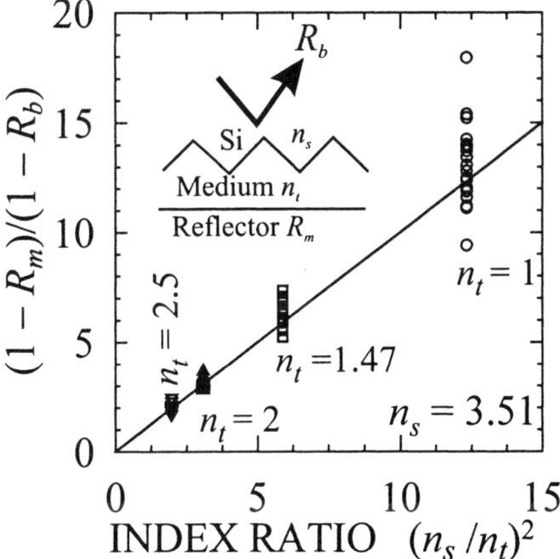

Figure A.27. Ratio of the optical losses with a plane metal reflector of reflectance R_m *adjacent* to the Si back surface and with a *detached* back surface reflector of reflectance R_m. This ratio is determined by the ratio of refractive indices n_s for Si and n_t for the transparent medium.

0.992 ± 0.001. Consequently, an effective back surface reflectance exceeding 99% is feasible with the concept of a detached back surface reflector. This high reflectance is a broad-band reflectance, since no "resonant" optically thin layers are involved and since the indices of refraction n_s and n_t have a weak dependence on wavelength in the near-infrared spectral range.

To verify Eq. (5.1) we re-plot the data from Figure A.26 in Figure A.27. All data points fall approximately on the line $(1 - R_m)/(1 - R_b) = (n_s/n_t)^2$. The large scatter for small n_t values results from the statistical error of $(1 - R_m)$, which is particularly large if R_m is close to unity. Hence, the reduction of reflection losses by the ratio of indices squared is verified by numerical ray-tracing.

APPENDIX B

B Recombination

Recombination in semiconductors is the transition of an electron from the conduction band to the valence band, or equivalently, the annihilation of an electron from the conduction band with a hole from the valence band. Recombination also occurs in semiconductors at thermal equilibrium to counterbalance the photogeneration at thermal equilibrium that is caused by the black body radiation at the equilibrium temperature. Under non-equilibrium conditions, e.g. under illumination or under carrier injection, the carrier concentrations are higher than in equilibrium. Recombination then tends to restore the equilibrium carrier concentrations.

Figure B.1 shows various energy levels that are frequently used. By definition the energy of electrons increases towards the top. The figure assumes a long carrier lifetime, resulting in flat quasi-Fermi levels E_{Fn} and E_{Fp}. The splitting of the quasi-Fermi levels $E_{Fn} - E_{Fp}$ equals the free energy of an electron-hole pair at the respective position. A position-dependent carrier concentration is caused by a non-constant electrostatic potential Φ. The intrinsic Fermi level E_i equals the Fermi level of an intrinsic semiconductor at thermal equilibrium and is close to the midgap in Si.

B.1 Carrier concentrations

The recombination rate depends on the concentrations of electrons and holes. In this section we give standard formulas that we frequently use in this work for simulating the recombination rates in thin-film crystalline Si cells.

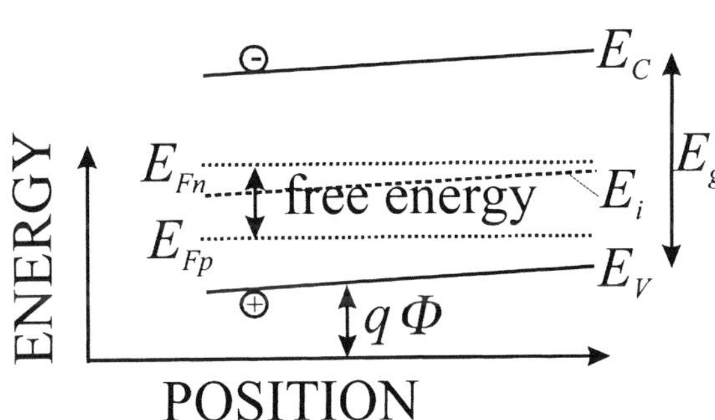

Figure B.1. Position-energy diagram illustrating the conduction band edge E_C, the valence band edge E_V, the intrinsic Fermi level E_i, the bandgap energy E_g, the quasi-Fermi energy E_{Fn} for electrons and E_{Fp} for holes, and the electrostatic potential $q\Phi$. The splitting of the quasi-Fermi levels is the free energy of an electron-hole pair.

B.1.1 Electron and hole concentration

The electron concentration

$$n = n_i \exp\left(\frac{E_{Fn} - E_i}{kT}\right) \tag{B.1}$$

and the concentration of holes

$$p = n_i \exp\left(\frac{E_i - E_{Fp}}{kT}\right) \tag{B.2}$$

increase exponentially with a decreasing distance from their quasi-Fermi levels to their respective band edges. In the above equations n_i denotes the intrinsic carrier concentration that we discuss in the next section.

From the above two equations and the charge neutrality condition we derive the carrier concentrations

$$n = \frac{N_D - N_A}{2} + \sqrt{\left(\frac{N_D - N_A}{2}\right)^2 + n_i^2 \exp\left(\frac{E_{Fn} - E_{Fp}}{kT}\right)} \tag{B.3}$$

and

$$p = \frac{N_A - N_D}{2} + \sqrt{\left(\frac{N_A - N_D}{2}\right)^2 + n_i^2 \exp\left(\frac{E_{Fn} - E_{Fp}}{kT}\right)} \tag{B.4}$$

as a function of the acceptor concentration N_A, the donor concentration N_D, and the splitting of the quasi-Fermi levels $E_{Fn} - E_{Fp}$. The output voltage U of a cell is related to the splitting of the quasi-Fermi levels. For infinite mobility we find $q\,U = E_{Fn} - E_{Fp}$. The above expressions are therefore helpful to calculate the recombination rates in cells of infinite mobility, known doping, and known applied voltage.

B.1.2 Intrinsic carrier concentration

The intrinsic carrier concentration

$$n_i = \sqrt{N_C N_V} \exp\left(-\frac{E_g}{2kT}\right) \tag{B.5}$$

is a fundamental material parameter that depends on the Si band structure via the effective density of states N_C and N_V and the Si bandgap energy E_g. Equation (B.5) holds for a non-degenerate semiconductor. Note the exponential temperature dependence of n_i

which requires accurate temperature control during all measurements of recombination rates.

The value $n_i = 1.45\times10^{10}$ cm^{-3} has for many years been generally accepted [79] as being independent of the doping concentration and of the carrier injection level.

Measurements on highly efficient Si solar cells led to a major revision of the accepted n_i value [386, 70]. For these measurements, special solar cells were constructed that permit accurate measurement of the base recombination. The base recombination is proportional to n_i^2, since the recombination rate is proportional to the product of electron and hole concentrations. With all other parameters of the cell determined by independent measurements, the n_i value was calculated from the measured saturation current. Application of this procedure to cells of various base doping concentrations N_A resulted in the intrinsic carrier concentrations shown in Figure B.2. The average of the experimental values is $n_i = 1\times10^{10}$ cm^{-3}, a value that is now often used to model recombination in Si.

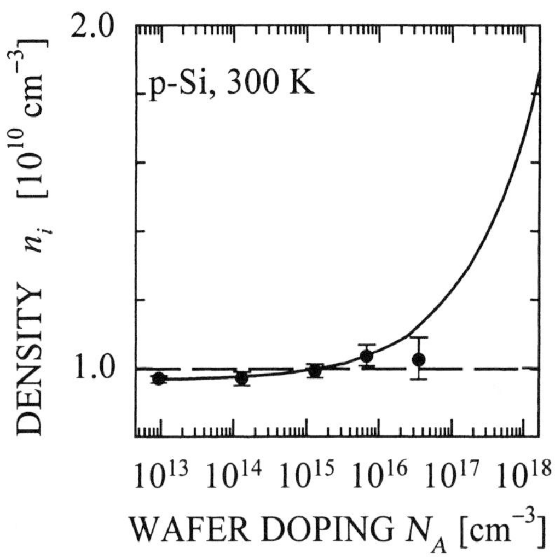

Figure B.2. Experimental intrinsic carrier concentration from Refs. [386, 70, 387] (dots), average experimental value (broken line), and Schenk's theory [388] (solid line). Figure redrawn from Ref. [389].

Injection-dependent intrinsic carrier concentration

With increasing free carrier concentration the Coulomb interaction of electrons and ions is shielded. In consequence, the energy gap reduces and, according to Eq. (B.5), the value of n_i increases. Schenk and co-workers recently calculated the bandgap narrowing ΔE_g on purely quantum mechanical grounds as a function of doping concentrations N_A and N_D, electron concentration n, and hole concentration p [388]. Schenk also gave a parameterization of ΔE_g. Following Altermat et al. [389], we use this parameterization to calculate the solid line in Figure B.2 that shows the intrinsic carrier concentration

$$n_i = n_{io} \exp\left(\frac{\Delta E_g}{2kT}\right) \qquad \text{(B.6)}$$

with $n_{io} = 0.965 \times 10^{10}$ cm^{-3} [389] as a function of the acceptor concentration N_A for the low-injection case. The slight tendency of the experimental n_i values to increase with doping N_A is correctly described by Schenk's theory. Unfortunately, no experimental n_i data are yet available for grater doping.

Figure B.3 shows the n_i values calculated for p-type and n-type Si as a function of the doping and of the quasi-Fermi level splitting $(E_{Fn} - E_{Fp})/q$. Silicon solar cells are gener-

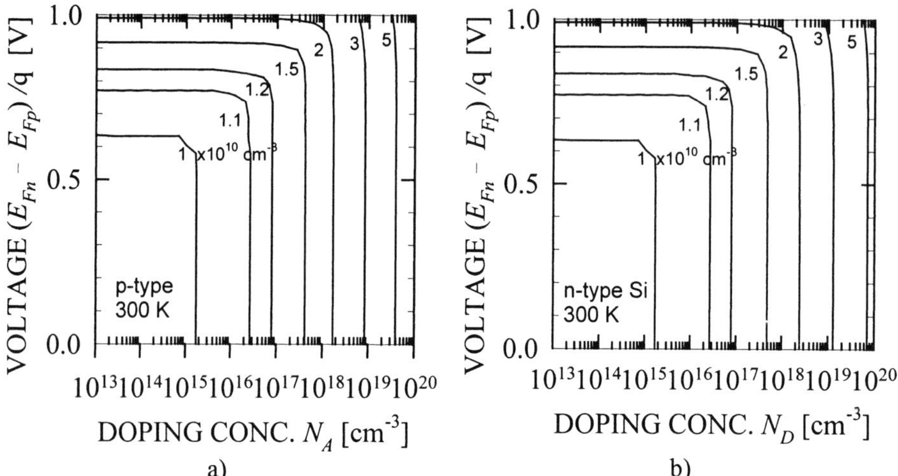

Figure B.3. Intrinsic carrier concentration as a function of doping density and quasi-Fermi level splitting U: a) p-type Si and b) n-type Si.

ally operated at voltages below 0.7 V. For p-type Si with an acceptor concentration $N_A > 10^{16}$ cm^{-3} the n_i value is almost independent of the voltage $(E_{Fn} - E_{Fp})/q$, which means independent of the carrier injection level. For intrinsic Si the n_i value varies by less than 5% up to 0.7 V, and thus the recombination rate calculated ignoring the weak injection level dependence is less than 10% wrong. This variation in n_i is equivalent to an error in a simulated open-circuit voltage of less than 3 mV. We therefore neglect the injection dependence of n_i in this work.

In contrast, the dependence of n_i on the doping concentration that we depict in Figure B.4 is fully accounted for in all our modeling. However, instead of using Schenk's rather complex parameterization, we use

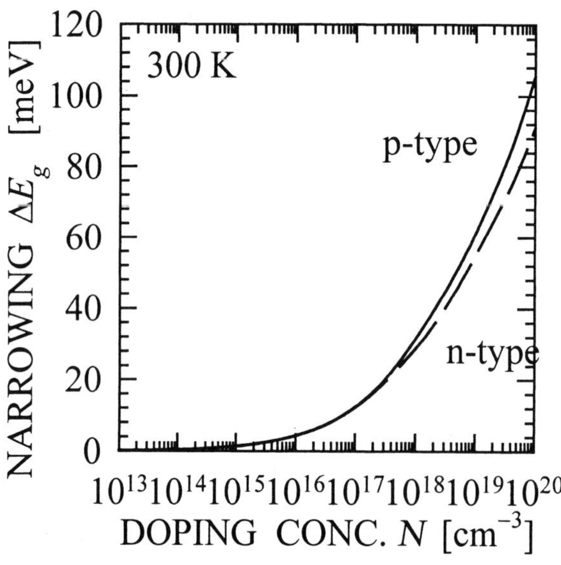

Figure B.4. Bandgap narrowing for p-type and n-type Si at 300 K calculated with Schenk's quantum mechanical model for low-injection conditions.

$$\Delta E_g\,/\,\text{meV} = -1416.5 + 131.08\ \ln\left(N_A/\text{cm}^{-3}\right) - 4.0411\ \ln^2\left(N_A/\text{cm}^{-3}\right) + 0.041526\ \ln^3\left(N_A/\text{cm}^{-3}\right) \tag{B.7}$$

for p-type, and

$$\Delta E_g\,/\,\text{meV} = -993.94 + 94.070\ \ln\left(N_D/\text{cm}^{-3}\right) - 2.9654\ \ln^2\left(N_D/\text{cm}^{-3}\right) + 0.031147\ln^3\left(N_D/\text{cm}^{-3}\right) \tag{B.8}$$

for n-type Si. This parameterization agrees with that of Schenk to within 1 meV for the doping range shown in Figure B.4. The bandgap narrowing ΔE_g calculated with Schenk's model for the low-injection case was shown to agree well with the experimental data collected by Klaassen et al. [388, 331].

B.2 Mechanisms

B.2.1 Radiative recombination

The spontaneous transition of an electron from the conduction band into the valence band is called radiative recombination. This process is shown schematically in Figure 2.12a on p. 24. Assuming independent particles, the net recombination rate

$$U_{Rad} = B\left(n\,p - n_o p_o\right) \tag{B.9}$$

is the difference between the *recombination* rate $B\,n\,p$, which is proportional to the electron and hole concentrations n and p, respectively, and the thermal *generation* rate $B\,n_o\,p_o = B\,n_i^2$ [115]. The symbols n_o and p_o denote the equilibrium concentrations. The constant B depends on the semiconductor material and is also temperature-dependent. In direct semiconductors, radiative electron-hole recombination is a two-particle process; hence B is large, e.g. $B = 3\times10^{-10}$ cm^3 s^{-1} for GaAs at room temperature [390]. In contrast, for Si that is an indirect semiconductor, a phonon is required for the radiative recombination process to conserve the crystal momentum. A three-particle process is less likely, and hence the value B is expected to be smaller than in GaAs.

To calculate n_i, we consider a piece of Si in the dark that is in thermal equilibrium with its environment at 300 K. The thermal generation rate $B\,n_i^2$ is then caused by black body radiation at 300 K. Let $n_{\gamma c}(E, 0)$ denote the photon flux per étendue at photon energy E for zero quasi-Fermi level splitting (compare Eq. (2.18) on p. 25). Then the constant B fulfills the relation

$$B n_i^2 = q\,2\,\pi \int_{E=0}^{\infty} n_{\gamma c}\left(E,0\right)\alpha_S\left(E\right) n_S^2\left(E\right) dE \tag{B.10}$$

where α_S and n_S denote the absorption coefficient and the refractive index of Si. The factor $2\,\pi$ accounts for the fact that at thermal equilibrium the propagation of radiation is fully isotropic (the projection of the half-sphere has an area π; the factor is thus 2π for the full sphere representing isotropic light). For the optical constants of Si [5] and the intrinsic carrier concentration $n_i = 0.965\times10^{10}$ cm^{-3} of intrinsic Si we calculate $B = 2.4\times10^{-15}$ cm^{-3}s^{-1}. We use this value in our simulations because it is consistent with the optical properties of Si used in this work.

Ruf et al. measured $B = 5\times10^{-15}$ cm^{-3} s^{-1} [391]. Other experimentalists measured values close to $B = 10^{-14}$ cm^{-3} s^{-1} [392, 393, 394]. The difference is possibly due to the different wafer doping used in the various experiments. Lower doping causes less screening and the charge carriers are no longer independent as is assumed by the ansatz in Eq. (B.9). Thus electrons gather around the holes and vice versa, thereby enhancing radiative recombination.

B.2.2 Auger recombination

The Auger recombination process is depicted schematically in Figure 2.12b on p. 24. Auger recombination involves three particles: two holes and one electron, or two electrons and one hole. An electron-hole pair recombines and the third particle receives the energy and the momentum of the recombining pair. Assuming non-interacting free carriers, the Auger recombination rate is proportional to each of the three carrier concentrations. The coefficients of proportionality C_p and C_n are defined by Eq. (2.32) on p. 36. Literature values of C_p and C_n for Si are listed in Table B.1.

Table B.2 lists ambipolar Auger coefficients C_A, as measured for highly injected Si wafers of low resistivity. From Eq. (2.32) we would expect an Auger coefficient $C_n + C_p$ for high injection. However, the measured values are larger due to an incompletely shielded Coulomb interaction. In contrast to Refs. [166] and [167], Johnsson et al. [395] used infrared absorption measurements in switched pin-diodes from electron-irradiated intrinsic Si. We suspect the latter results of being less accurate, because Si material with a small Shockley-Read-Hall lifetime was used that makes it more difficult to extract the Auger lifetime.

Table B.1. Published low-level injection Auger coefficients C_n and C_p of crystalline Si at 300 K [21].

Reference	Coefficient C_n [10^{-31} cm^6 s^{-1}]	Coefficient C_p [10^{-31} cm^6 s^{-1}]	Doping range [cm^{-3}]
[158]	2.8	0.99	5×10^{18}-1×10^{20}
[161]	1.7	1.2	2×10^{18}-2×10^{19}
[162]	1.8	-	6×10^{18}-6×10^{19}
[163]	2.2	-	2.5×10^{18}-8×10^{19}
[164]	1.6	-	5×10^{18}-2×10^{20}

Table B.2. High-level injection Auger coefficients C_n and C_p of crystalline Si at 300 K.

Reference	Coefficient C_A [10^{-31} cm^6 s^{-1}]	Injection range [cm^{-3}]
[166]	16.6	10^{15}-2×10^{17}
[167]	20	5×10^{15}-10^{17}
[395]	11	10^{17}

We use the parameterization suggested by Schmidt et al. [71] to account for the Coulomb enhancement of Auger recombination [165, 159]. The formulas used in Chapter 2 are

$$R_{Aug} = g_{ehh} C_p \left(p^2 n - p_o^2 n_o\right) + g_{eeh} C_n \left(n^2 p - n_o^2 p_o\right) \tag{B.11}$$

with

$$g_{eeh} = 1 + 44 \left[1 - \tanh\left(\left\{\frac{n+p}{5\times10^{16}\,\text{cm}^{-3}}\right\}^{0.34}\right)\right] \tag{B.12}$$

and

$$g_{ehh} = 1 + 44 \left[1 - \tanh\left(\left\{\frac{p+n}{5\times10^{16}\,\text{cm}^{-3}}\right\}^{0.29}\right)\right] \tag{B.13}$$

Under very high injection or doping (n, $p > 5\times10^{18}$ cm^{-3}), these formulas yield unity Coulomb enhancement factors g_{eeh} and g_{hee} due to full screening. For a high-injection condition at $n = p = 10^{17}\,\text{cm}^{-3}$, the ambipolar Auger coefficient is $C_a = C_n + C_p = 17.6\times10^{-31}$ cm^6 s^{-1}, in fair agreement with the data measured by Sinton and Swanson [166], and Yablonovitch and Gmitter [167]. The parameterizations (B.12) and (B.13) are, however, not particularly accurate for intermediate injection conditions [71].

B.2.3 Defect recombination

Dislocations, surfaces, interfaces, and grain boundaries, as well as point defects such as vacancies or impurities, cause deviations from the ideal periodicity of the semiconductor lattice. Electronic states that would not exist in an ideal and infinitely extended Si crystal are introduced. Electronic defect states in the bandgap facilitate carrier recombination. The energy of the recombining electron-hole pair is transferred to the lattice. We call the defect recombination extrinsic since it is, at least in principle, avoidable. Grain boundaries, for example, may be avoided by growing single crystals. A strictly zero defect concentration is only achievable in small samples because all defects have a non-zero equilibrium concentration.

B.2.3.1 Bulk defects

The model to describe the recombination via defect states at energy E_t in the bandgap was developed by Shockley, Read [169], and Hall [170]. The defect states are assumed to be non-interacting gap states. The gap state captures holes and electrons with the capture cross-section σ_p and σ_n alternately. The process is shown schematically in Figure 2.12a on p. 24. The thermally activated release of a captured electron into the conduction band or a captured hole into the valence band is also possible. In the stationary state the occupation of the defect states as well as the carrier concentration in the bands is constant (in time average). This condition leads to the Shockley-Read-Hall (SRH) recombination rate

$$R_{SRH} = N_t \frac{\left(n\,p - n_i^2\right)\sigma_n \sigma_p v_{th}}{\left(n + n_1\right)\sigma_n + \left(p + p_1\right)\sigma_p} \tag{B.14}$$

where v_{th} denotes the thermal velocity of electrons and holes (the difference in thermal velocity of electrons and holes due to different effective masses is ignored here), n_i is the intrinsic carrier concentration, and n and p are the electron and the hole concentrations at the location of the defect, respectively. The values

$$n_1 = N_c \exp\left(\frac{E_t - E_C}{kT}\right) \tag{B.15}$$

and

$$p_1 = N_v \exp\left(\frac{E_V - E_t}{kT}\right) \tag{B.16}$$

depend on the effective density of states N_c and N_v, and the energy E_C and E_V of the conduction and the valence band, respectively. The capture rates, when expressed in terms of capture times, are

$$\tau_{no} = 1/(N_t v_{th} \sigma_n) \tag{B.17}$$

and

$$\tau_{po} = 1/(N_t v_{th} \sigma_p) \tag{B.18}$$

This capture time is the time for which a single carrier exists before it is captured by one of the defect states, if all defect states were empty. The SRH carrier lifetime is

$$\tau_{SRH} = \frac{n - n_0}{R_{SRH}} = \left(\frac{p + p_1}{p + n_o}\right)\tau_{no} + \left(\frac{n + n_1}{n + p_o}\right)\tau_{po} \tag{B.19}$$

where n_o and p_o denote the equilibrium concentrations.

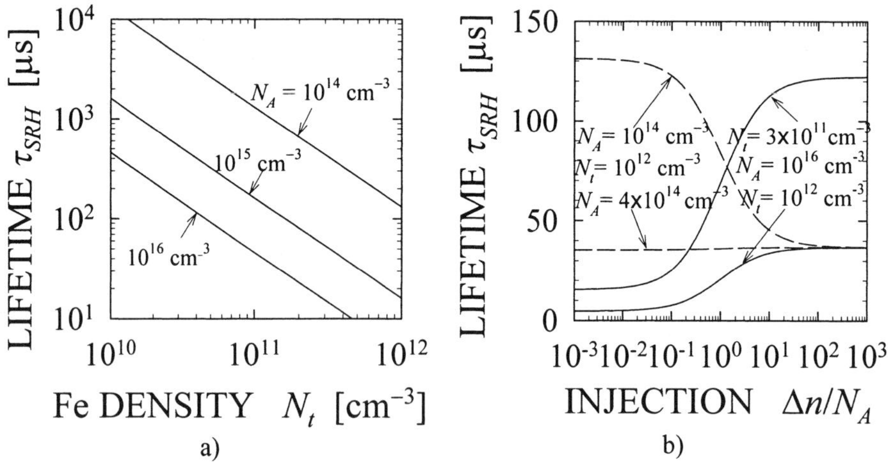

Figure B.5. Calculated SRH lifetime in p-type Si at 300 K: a) dependence on defect density N_t at low-level injection for various doping concentrations N_A; b) dependence on excess minority carrier injection $\Delta n = (n - n_o)$ at various defect concentrations N_t. The parameters are: $v_{th}\times\sigma_n = 3\times10^{-7}$ cm^3 s^{-1}, $v_{th}\times\sigma_n = 3\times10^{-8}$ cm^3 s^{-1} and $E_C - E_t = -0.29$ eV. Figure redrawn from Ref. [151].

Example

Interstitial Fe has a trap energy $E_t - E_v = 0.39$ eV, and the Fe-B pair has levels at $E_t - E_v = 0.1$eV and $E_C - E_t = -0.29$ eV [396]. It is the latter defect that is made responsible for carrier recombination in Fe-contaminated p-type Si at 300 K [397]. Figure B.5a shows the SRH lifetime as calculated by Eq. (B.19). A large defect concentration N_t and a small majority carrier concentration N_A cause a small lifetime. Figure B.5b shows the dependence of the lifetime on the injection level $\Delta n/N_A = (n - n_o)/N_A$. Depending on the acceptor concentration, the lifetime increases or decreases with increasing injection. This complex interdependency of the SRH recombination statistics makes a numerical analysis of the recombination rate using Eq. (B.14) highly recommendable.

A list of frequently observed trap levels and their capture cross-sections can be found in Ref. [398].

B.2.3.2 Surface defects at flat band conditions

Thin-film cells have a large surface-to-volume ratio. Accurate modeling of surface recombination is therefore important. The definition of the surface recombination S is given by Eq. (2.37) on p. 40.

Figure B.6 shows the position-energy diagram of a dielectric layer, e.g. silicon dioxide (SiO_2) or silicon nitride (SiN_x), on silicon. Two aspects are different for surface recombination when compared to the previously discussed recombination at bulk defects: at surfaces, we always find an energetically broad distribution of defects due to various bonding configurations and due to statistical local bond distortions. The second complication is that the carriers have to pass a space charge region to reach the defect at the interface. Let us first consider the flat-band case without such a space charge region. We include bending later, on p. 220.

The density of electronic states D_{it} per energy and area varies with energy and is indicated by the density of lines in Figure B.6. The line lengths symbolize the energy-

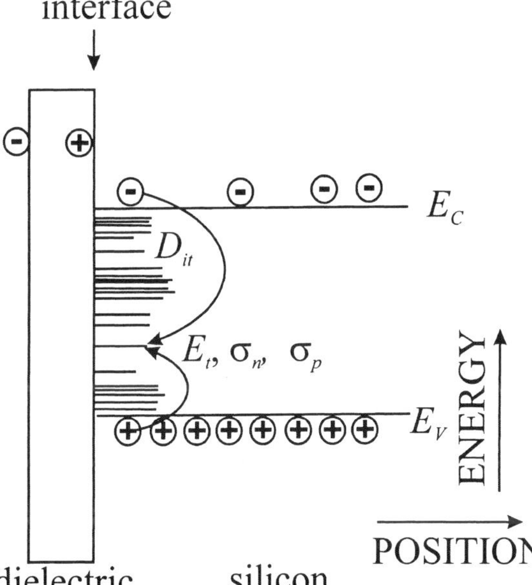

Figure B.6. Position-energy diagram of the interface between silicon and a dielectric film. Band bending due to interface states and surface charges is omitted.

dependent capture cross-sections σ_p and σ_n. The situation is complex, because structurally different defects may cause interface states of different capture cross-sections at a single energy. To simplify the notation we assume that σ_p and σ_n are unique functions of energy E. Using the SRH recombination statistics, the surface recombination rate

$$U_{sur} = \left(n_{sur} p_{sur} - n_i^2\right) \int_{E_V}^{E_C} \frac{\sigma_n(E)\sigma_p(E) v_{th} D_{it}(E)}{\left(n_{sur} + n_1(E)\right)\sigma_n(E) + \left(p_{sur} + p_1(E)\right)\sigma_p(E)} dE \tag{B.20}$$

is now the integral over the recombination rates via all recombination centers. Here, n_{sur} and p_{sur} are the surface (or interface) concentrations of the electrons and the holes, respectively. The excess concentrations

$$\Delta n_{sur} = n_{sur} - n_o = p_{sur} - p_o = \Delta p_{sur} \tag{B.21}$$

are equal for electrons and holes since charge neutrality also holds at the interface due to our flat-band assumption. From Eq. (B.20) we find for the surface recombination velocity

$$S = \left(n_o + p_o + \Delta n_s\right) \int_{E_V}^{E_C} \frac{\sigma_n(E)\sigma_p(E) v_{th} D_{it}(E)}{\left(n_o + \Delta n_{sur} + n_1(E)\right)\sigma_n(E) + \left(p_o + \Delta n_{sur} + p_1(E)\right)\sigma_p(E)} dE \tag{B.22}$$

The neutrality relation of Eq. (B.21) makes $S(\Delta n_{sur})$ a function of a single parameter Δn_s. Without neutrality at the interface (e.g. including space charge regions), we had a two-parameter dependence $S(\Delta n_{sur}, \Delta p_{sur})$. Under steady-state conditions, the occupation of monovalent acceptor states with electrons is

$$f_A(E) = \frac{\sigma_n n_{sur} + \sigma_p p_1}{\sigma_n\left(n_{sur} + n_1\right) + \sigma_p\left(p_{sur} + p_1\right)} \tag{B.23}$$

while the occupation of donor states with holes is

$$f_D(E) = \frac{\sigma_n n_1 + \sigma_p p_{sur}}{\sigma_n\left(n_{sur} + n_1\right) + \sigma_p\left(p_{sur} + p_1\right)} \tag{B.24}$$

The occupation only depends on the ratio σ_n/σ_p and hence $f_A(E_t)$ and $f_D(E_t)$ are functions of the trap energy E_t and the ratio $R_\sigma = \sigma_n/\sigma_p$ [399].

Figure B.7 shows the occupation probability f_A versus trap energy E_t for energy-independent interface state density D_{it} and for an energy-independent ratio $R_\sigma = \sigma_n/\sigma_p = 3$. In Figure B.7a the occupation (which is proportional to the charge density at constant D_{it}) is given for intrinsic Si at thermal equilibrium (broken lines) and for an excess carrier concentration $\Delta n_{sur} = 10^{15}$ cm^{-3} (solid line). In equilibrium the Fermi level E_F is close to the midgap. Under illumination the quasi-Fermi levels for electrons E_{Fn} and holes E_{Fp} split. Above the electron quasi-Fermi level E_{Fn} the interface states are empty. Below the quasi-Fermi level of the holes E_{Fp} the states are occupied by electrons. Between the two quasi-Fermi levels the occupation has an almost constant value

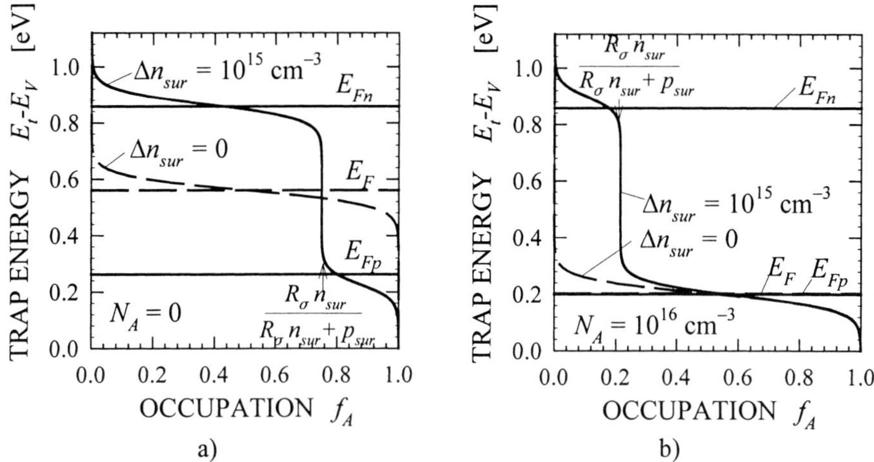

Figure B.7. Occupation probability f_A versus trap energy $E_t - E_v$ for an interface state distribution with energy-independent ratio $R_\sigma = \sigma_n/\sigma_p = 3$: a) intrinsic Si with an excess carrier concentration $\Delta n_{sur} = 0$ (broken line) and $\Delta n_{sur} = 10^{15}$ cm^{-3} (solid line); b) p-type Si with $N_A = 10^{16}$ cm^{-3} and excess electron concentration $\Delta n_{sur} = 0$ (broken line) and $\Delta n_{sur} = 10^{15}$ cm^{-3} (solid line).

$$f_A(E) \approx \frac{\sigma_n n_{sur}}{\sigma_n n_{sur} + \sigma_p p_{sur}} = \frac{R_\sigma n_{sur}}{R_\sigma n_{sur} + p_{sur}} \tag{B.25}$$

that results from (B.23) by neglecting the small terms with subscript 1.

In Figure B.7b we consider the same situation as in Figure B.7a, but now for a p-type Si with an acceptor concentration $N_A = 10^{16}$ cm^{-3}. The hole concentration is now larger than the electron concentration and the equilibrium Fermi level shifts towards the valence band. According to (B.25) the almost constant occupation f_A between the two quasi-Fermi levels shifts to smaller values as compared to Figure B.7a.

The transition regions from the occupation value zero to $R_\sigma n_{sur}/(R_\sigma n_{sur} + p)$, and from the occupation value $R_\sigma n_{sur}/(R_\sigma n_{sur} + p)$ to unity, are slightly closer to the band edges than the respective quasi-Fermi levels [399]. An exact agreement between these transition energies and the quasi-Fermi levels is not to be expected, because the quasi-Fermi levels were defined to give the correct occupation in the bands and not in the trap states. It is, however, possible to define additional Fermi levels for the trap states [399].

The approximate total charge Q in an interface of constant and energy-independent interface state density D_{it} is then, for intrinsic Si,

$$Q = Q_o + qD_{it} \frac{R_\sigma n_{sur} - p_{sur}}{R_\sigma n_{sur} + p_{sur}} \frac{q(E_{Fn} - E_{Fp})}{2} = Q_{dark} + qD_{it} \frac{R_\sigma - 1}{R_\sigma + 1} \frac{q(E_{Fn} - E_{Fp})}{2} \tag{B.26}$$

since $n_{sur} = p_{sur}$, and for p-type Si it is

$$Q = Q_o + qD_{it} \frac{R_\sigma n_{sur}}{R_\sigma n_{sur} + p_{sur}} q(E_{Fn} - E_{Fp}) \tag{B.27}$$

The symbol Q_o denotes the interface charge at equilibrium.

B.3 Extended Shockley-Read-Hall recombination

B.3.1 Surface recombination

The discussion of SRH recombination at Si surfaces or interfaces on p. 217 neglected the presence of space charge regions. In Figure B.8 we consider the more general case of a silicon surface with band bending due to charges of a passivating dielectric film of thickness t. This film may or may not be metallized. The system has four charged layers as shown in the upper part of Figure B.8.

(i) The surface charge density Q_G (in C cm^{-2}) on the outer surface of the dielectric layer. If the dielectric layer is metallized, this charge could be induced by the work function difference of the metal and the Si and/or by applying a gate voltage U_G [184]. For a non-metallized dielectric the charge Q_G could be deposited with a corona discharge chamber (see Figure 2.19 on p. 36).

(ii) The fixed charge density Q_f that resides in the dielectric layer and typically extends from the interface to a depth d into the insulator. This charge stems from charged defects in the dielectric layer. If the energy of these states is outside the bandgap the state of charge is fixed and cannot be changed by varying the carrier injection level in the Si. States inside the gap can change their occupation, if they are positioned within one tunneling distance from the interface.

(iii) The charge density Q_{it} in the interface states, which depends on the position of the quasi-Fermi levels in the semiconductor. And finally,

(iv) the charge Q_{sc} in the space charge region of the semiconductor.

Because of the presence of the charges Q_G, Q_f, and Q_{it}, the Si surface layer is in general charged: band bending that is shown in the lower part of Figure B.8 occurs. The assumption $\Delta n_{sur} = \Delta p_{sur}$ that we used to derive Eq. (B.22) becomes inappropriate. We therefore have to calculate the surface carrier concentrations n_{sur} and p_{sur} from the concentrations Δn_W and Δp_W at the edge of the space charge region that is positioned at W_{scr}. Knowing n_{sur} and p_{sur}, we determine the surface recombination rate U_{sur} using Eq. (B.20).

Calculation of quasi-Fermi levels E_{Fn} and E_{Fp}

At W_{scr} the quasi-Fermi levels are

$$E_{Fn} = -\frac{kT}{q}\ln\left(\frac{n_W}{n_i}\right) \quad \text{and} \quad E_{Fp} = +\frac{kT}{q}\ln\left(\frac{p_W}{n_i}\right) \tag{B.28}$$

for electrons and holes, respectively, if we consider the intrinsic level $E_i = 0$ in the neutral region as the reference level of the energy scale (compare also Figure B.1 on p. 209). At the edge of the space charge region, quasi-neutrality $\Delta n_W = \Delta n_W$ holds, and hence E_{Fn} and E_{Fp} are unique functions of the excess minority carrier concentration Δn_W for a given doping concentration N_A (see Eq. (B.3) on p. 210).

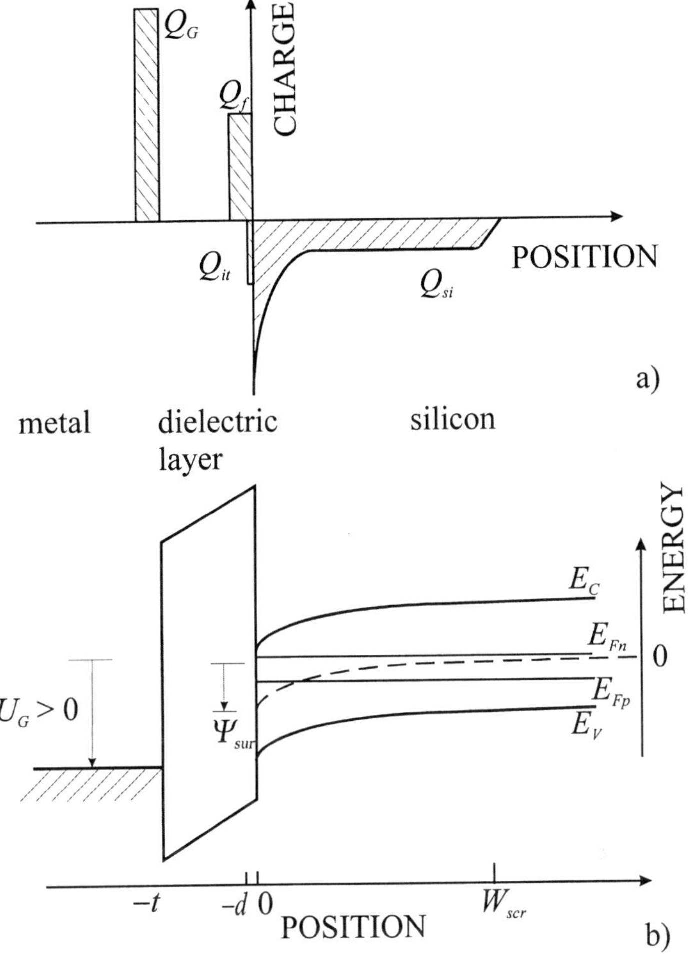

Figure B.8. a) Charges and b) energy band diagram of a metal-oxide-semiconductor structure under voltage bias U_G. After Ref. [184].

Calculation of surface potential Ψ_{sur}

The surface potential Ψ_{sur} has a value that makes the total charge

$$Q_G + Q_f + Q_{it}(E_{Fn}, E_{Fp}, \Psi_{sur}) + Q_{sc}(E_{Fn}, E_{Fp}, \Psi_{sur}) = 0 \tag{B.29}$$

vanish. A non-zero total charge would create a spatially homogeneous electric field extending to infinity, a configuration that is energetically impossible.

Calculation of semiconductor charge Q_{sc}

The semiconductor charge

$$Q_{sc} = -\frac{\Psi_{sur}}{|\Psi_{sur}|}\sqrt{\frac{2kTn_i\varepsilon_0\varepsilon_S}{q^2}F(\Psi_{sur}, E_{Fn}, E_{Fp})} \tag{B.30}$$

in the space charge region of the semiconductor depends on the function [184]

$$F(\Psi_{sur}, E_{Fn}, E_{Fp}) = e^{\beta(E_{Fp}-\Psi_{sur})} - e^{\beta E_{Fp}} + e^{\beta(\Psi_{sur}-E_{Fn})} - e^{-\beta E_{Fn}} + \beta\Psi_{sur}\frac{N_A}{n_i}, \tag{B.31}$$

where β denotes the thermal voltage, N_A is the acceptor concentration (negative for n-type material), and ε_o and ε_s are the permittivity in vacuum and the relative permittivity of Si, respectively. We assume spatially constant quasi-Fermi levels E_{Fn} and E_{Fp} across the space charge region. This assumption is only valid for sufficiently small recombination rates in the space charge region and at the surface. Large recombination rates will reduce the splitting $E_{Fn} - E_{Fp}$ towards the surface. Hence we slightly overestimate the actual recombination rate and the effective surface recombination velocity, when assuming flat quasi-Fermi levels.

Calculation of the interface charge Q_{it}

We now consider the charge Q_{it} in the interface states. The interface state density

$$D_{it} = D_A + D_D \tag{B.32}$$

has a contribution D_A from acceptor states and a contribution D_D from donor states. The interface charge

$$Q_{it} = -\int_{E_V}^{E_C} D_A(E)f_a(E)dE + \int_{E_V}^{E_C} D_D(E)f_d(E)dE \tag{B.33}$$

depends on the occupation function defined by Eqs. (B.23) and (B.24) on page 218.

Determination of charges Q_G and Q_f

An expression for the functional dependence of the gate charge Q_g of a metallized surface is given in Ref. [184]. If Q_g is a corona charge on a non-metallized surface, it does not depend on Ψ_{sur} and is determined experimentally from an integration of the charging current over the charging time. The fixed charge Q_f in the dielectric layer is also independent of Ψ_{sur} and is experimentally accessible by capacitance-voltage measurements.

Calculation of surface carrier concentrations n_{sur} and p_{sur}

Knowing the functional dependence of all the charges on the surface potential, we determine Ψ_{sur} numerically as the root of Eq. (B.29). Then the surface carrier concentrations

$$n_{sur} = n_W(E_{Fn})\ \exp(+q\Psi_{sur}/kT) \quad \text{and} \quad p_{sur} = p_W(E_{Fp})\ \exp(-q\Psi_{sur}/kT) \tag{B.34}$$

that depend exponentially on the surface potential are also known, and the surface recombination rate U_{sur} follows from Eq. (B.20) exactly as for the flat-band case. Finally, the effective surface recombination velocity is $S_{eff} = U_{sur}/\Delta n_W$.

Defects at the Si/SiO$_2$ interface

Our 20.6%-efficient thin-film Si cell has an Si/SiO$_2$/Al back surface reflector [113]. In order to elucidate the physical origin of the back surface recombination, we start with a short review of the nature of the defects at the Si/SiO$_2$ interface.

The energy levels of chemically bonded atoms split into a bonding and an anti-bonding state. The bonding state is lower in energy and neutral when occupied. The anti-bonding state is neutral when empty. The regular tetrahedral bonding in crystalline Si causes the energy gap, which is free of electronic states. Stretching weakens the bonds and thus reduces the energy difference between bonding and anti-bonding states.

Figure B.9 shows the interface state density de-convoluted according to the model of Füssel et al. [400]. The interface states U_t that are observed at the band edges, and that decay in state density exponentially towards the midgap, originate from stretched Si bonds.

The U_t defects are amphoteric because they behave donor-like for states close to the valence band (neutral when occupied) and acceptor-like for states near the conduction band. The U_t defect is an intrinsic defect since it contains only Si atoms. The defect density and energetic position is only slightly sensitive to the preparation of the oxide. The interface state density peaks at the band edges and hence the recombination activity

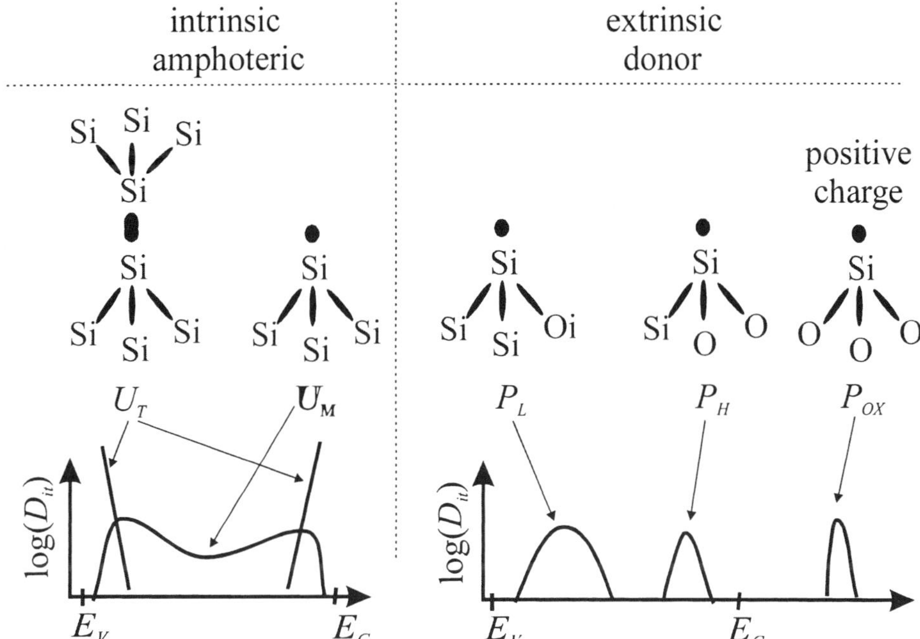

Figure B.9. Microscopic origin (top) of interface states density (bottom) according to the model of Füssel et al. [400]: intrinsic interface defects contain Si atoms only. Their nature is amphoteric. Extrinsic interface defects contain O back bonds. They are donors. The P_{OX} state is energetically located in the conduction band and causes the positive oxide charge.

is lower than for the other defects that peak in density close to the midgap. Stretched bonds do not cause energy states in the middle of the gap, because such a strong stretching would cause a rupture of the bonds, thus generating Si dangling bonds.

The Si dangling bond defect is also an intrinsic defect that Füssel et al. called the U_m defect (the name P_{b0} center is also in use). Being a midgap state, this defect often dominates the surface recombination for alnealed Si/SiO_2 interfaces. The U_m defect is amphoteric in nature with donor-like behavior in the lower and acceptor-like behavior in the upper part of the bandgap. The interface state density of the U_m center is a U-shaped, doubly peaked distribution.

Oxygen atoms at the Si/SiO_2 interface cause extrinsic interface defects. Silicon dangling bonds with one oxygen back bond have a Gaussian shaped interface state density distribution that peaks in the lower half of the bandgap at about $E_t - E_V = 0.4$ eV. The standard deviation of 0.12 eV is due to variations of the local bonding environment [401]. This so-called P_L center (or P_{b1} center) is a donor-like bonding state. The corresponding anti-bonding state lies within the conduction band and is inactive for recombination processes.

Dangling Si bonds with two oxygen back bonds have a Gaussian interface state density distribution that peaks in the upper half of the bandgap at $E_t - E_V = 0.7$ eV and at a width of 0.08 eV. This so-called P_H center is also donor-like. The corresponding anti-bonding state lies in the conduction band.

Silicon dangling bonds with three oxygen back bonds have the bonding and the anti-bonding state within the conduction band. The bonding states (P_{ox} -states) are always unoccupied, thus causing a fixed positive charge that is in general found at the Si/SiO_2 interface. This positive charge is independent of the position of the Fermi level.

Flietner and Füssel performed interface state measurements on a variety of differently prepared interfaces. In all cases they were able to model the interface state density as a superposition of the interface state densities of the U_T, U_M, P_H, and P_L center [401, 400].

Experimental data on the interface state density

Figure B.10 shows experimental interface state densities measured on alnealed Si/SiO_2 interfaces. The data represented by diamonds were measured by Füssel on an oxide fabricated in our lab and processed similarly to the 20.6%-efficient thin-film cell [4]. In this case the state distribution peaks at $E_t - E_V = 0.75$ eV, indicating a superposition of U_T and P_H centers. The data for the triangles were measured for an oxide prepared at 1050°C with 2 vol% trichlorethylene (TCA). These data are from Ref. [176]. After cooling the samples in an Ar atmosphere, Al was evaporated and the sample was annealed at 450°C in forming gas. This so-called alneal procedure often generates an interface symmetrical to the midgap. In this case, the U_M defects have a midgap interface state density of 2×10^{10} cm^{-2} eV^{-1} and dominate the recombination.

Experimental data on capture cross-sections

Capture cross-sections are the second important ingredient for modeling SRH recombination. Unfortunately, an accurate measurement is difficult. Cross-sections published in the literature vary by many orders of magnitude. One reason for the scatter of data published in the literature is the high sensitivity of the interface properties on sample preparation. Different types of defects prevail, depending on the sample treatment, e.g. alneal or not. Different capture cross-sections are expected from different defects. Hence it is advisable to measure the cross-sections and the interface state density with exactly the same sample as is used for the determination of the surface recombination. The ex-

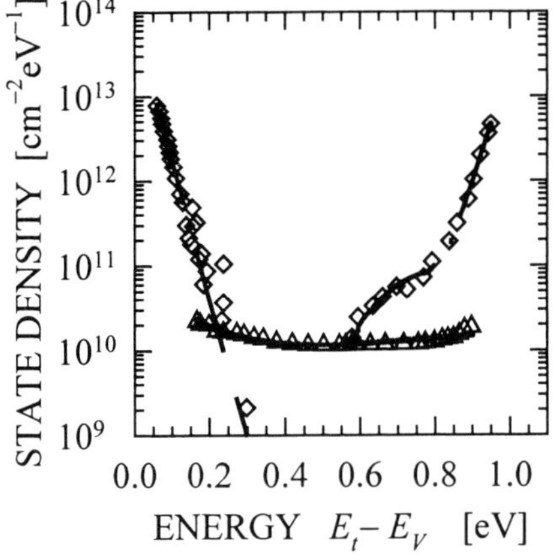

Figure B.10. Interface state density of alnealed $Si/SiO_2/Al$ samples as determined from capacitance-voltage measurements; diamonds: as reported in Ref. [4]; triangles: as reported in Ref. [176].

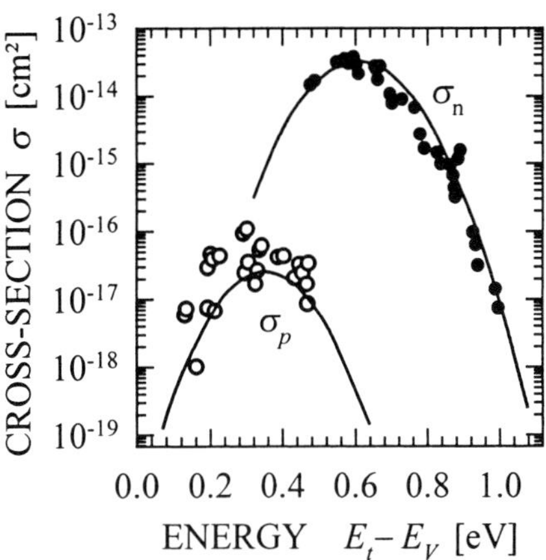

Figure B.11. Measured capture cross-section for electrons σ_n and holes σ_p. Data and parabolic fit from Ref. [176].

perimental data in Figure B.11 show the capture cross-sections σ_n and σ_p for electrons and holes, respectively, for the sample that has the interface state density marked by triangles in Figure B.10 [176]. Unfortunately, it is not possible to measure the cross-section for minority carriers with deep level transient spectroscopy (DLTS). Using n-type and p-type samples it is possible to determine σ_n in the upper half of the band and σ_p in the lower part. Extrapolations (solid lines) with little experimental or theoretical justification are necessary to estimate σ_n in the lower half of the band and σ_p in the upper half of the band. Here σ_n is three orders of magnitude larger than σ_p at the midgap. An average value for the σ_n / σ_p reported in the literature is 100 [188].

Fitting experimental data

Experimental data for the injection level dependence of the surface recombination velocity at a Si/SiO_2 interface are shown in Figure 2.25 on p. 42. In this and the subsequent paragraphs we apply the extended SRH recombination model to these experimental data.

We assume energy-independent values for D_{it}, σ_n, and σ_p, since the interface parameters of exactly the cell under test were not measured. We fix the ratio σ_n / σ_p at 100 [188]. The thin-film cell has p-type doping with an acceptor concentration $N_A = 4\times10^{16}$ cm^{-3}. The fitted injection level dependence is also shown in Figure 2.25 on p. 42. The theoretical curve describes the measurement to within the measurement uncertainty. The three fitting parameters are $D_{it} = 8 \times 10^9$ cm^{-2}eV^{-1}, $\sigma_n = 10^{-14}$ cm^{-2}, and the fixed oxide charge density $Q_f = q\ 9.9\times10^{10}$ cm^{-2}.

We can also model the same experimental data with either of the two interface state densities of Figure B.10 instead of using a constant D_{it}. Other values for the oxide charge, the interface state density, and the capture cross-section ratios result. Thus, an apriori knowledge of σ_n and σ_p is required to model the interface recombination meaningfully. If this knowledge is not available, we suggest the use of energy-independent interface parameters as the least prejudiced choice.

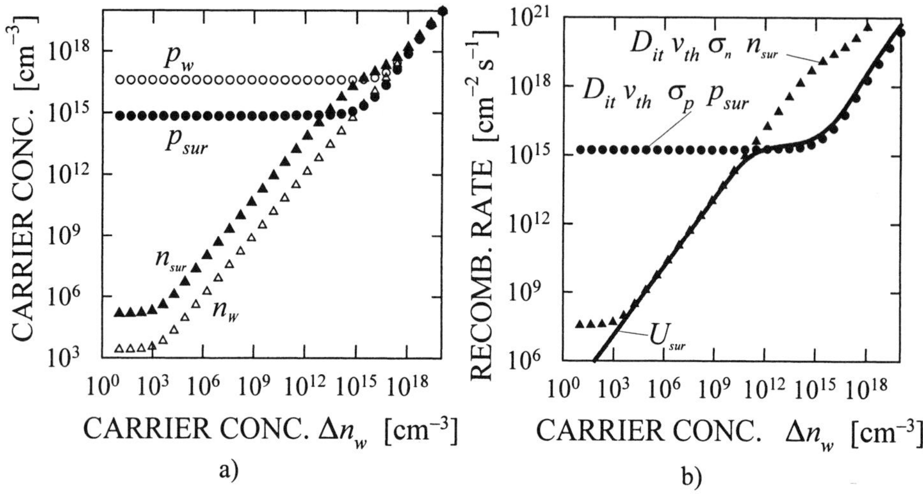

Figure B.12. a) Electron and hole concentrations as a function of excess carrier concentration n for the simulation shown in Figure 2.25 on p. 42. The open symbols denote the concentrations p_w and n_w at the edge of the space charge region while the concentrations p_{sur} and n_{sur} are taken at the surface. b) Capture rates for holes, capture rates for electrons, and total recombination rate U_s.

Interpretation of the injection level dependence

As a result of the best fit, the carrier concentrations n_{sur} and p_{sur} at the Si/SiO_2 interface and n_W and p_W at the edge of the space charge region are known. Figure B.12a shows these concentrations. The hole concentration is equal to the acceptor concentration N_A at low injection levels. We use the term injection level synonymously with the term excess carrier concentration Δn_W. The electron concentration is orders of magnitude smaller and increases linearly with the injection level Δn_W. At high injections $\Delta n_W >> N_A$ the electron and hole concentrations become equal.

The concentrations n_{sur} and p_{sur} at the surface are different from n_W and p_W at the edge of the space charge region. The positive oxide charge, as well as the work function difference of the Al reflector and the p-type Si, causes a surface potential of approximately $\Psi_{sur} = 0.1$ V. This potential (band bending) enhances the electron concentration, reduces the hole concentration at the surface, and introduces a region of injection levels with $p_{sur} < n_{sur}$. We call this the cross-over region. In our example this region is two orders of magnitude wide in Δn_W.

Figure B.12b illustrates the capture rates of electrons (triangles) and holes (circles) that are calculated from the surface concentrations n_{sur} and p_{sur} by multiplication with the respective capture cross-sections, thermal velocity, and interface state density. Since the capture cross-section σ_n of electrons is a factor of 100 larger than σ_p for holes, the cross-over region widens by two further orders of magnitude and thus becomes four orders of magnitude wide. Recombination requires hole and electron capture. The smaller of the two capture rates is the bottleneck for the total recombination rate U_s (solid line).

Electron-limited recombination at low injection

Up to about $\Delta n_W = 10^{11}$ cm^{-3}, the electron capture is lower and controls U_s. The surface recombination velocity

$$S = E_g\ D_{it} v_{th} \sigma_n \exp\left(\frac{q\Psi_{sur}}{kT}\right) \quad \text{(B.35)}$$

is high due to the large cross-section for electrons and the enhancement of minority carrier concentration due to the band bending. For our example, we calculate S_{eff} = 1.3×10^4 cm s^{-1} from Eq. (B.35), in good agreement with the simulation shown in Figure 2.25 on p. 42.

Hole-limited recombination an intermediate injection in the cross-over region

In the cross-over region, hole capture controls the surface recombination. Due to the cross-over, hole capture is independent of Δn_W in this region. Robinson et al. term this effect recombination rate saturation mechanism [188]. As the effective recombination velocity is defined with reference to the excess carrier concentration Δn_W at the edge of the space charge region, we find a surface recombination

$$S_{eff} = E_g D_{it} v_{th} \sigma_p p_W \exp\left(\frac{q\Psi_{urs}}{kT}\right) / \Delta n_W \quad \text{(B.36)}$$

that decreases linearly with Δn_W. As the cross-over is four orders of magnitude wide we expect a decrease in S by four orders of magnitude in agreement with the simulated S_{eff} values shown in Figure 2.25 on p. 42.

The band bending vanishes in the high-injection region. We find $n_{sur} = p_{sur}$. However, it is still the holes that control the total recombination rate U_{sur}, due to the smaller capture cross-section. In this region the surface recombination velocity is

$$S_{eff} = E_g\ D_{it} v_{th} \sigma_p \quad \text{(B.37)}$$

which yields $S = 3$ cm s^{-1} in our example, a value that is also in approximate agreement with the simulation. Since we measured no data in the high-injection region, it is not clear whether our sample really has such a low surface recombination. From the experiment and the simulation we learn that the injection dependence of S_{eff} has two origins:

(i) the difference in capture cross-sections and
(ii) the band bending.

Both of these effects work in the same direction and thus enhance the injection dependence for the $Si/SiO_2/Al$ interface on p-type Si that we investigated here. Both effects may, however, also work into opposite directions, e.g. in n-type material.

Capture rates at the Si/SiN_x interface

Figure B.13 shows the capture rates for electrons and holes at a silicon/silicon nitride interface. The simulation parameters are the same as those used for the simulation of the injection level-dependent surface recombination velocity of the 1.5 Ω cm sample in Figure 2.26 on p. 43. At low injection levels the band bending is $\Psi_{sur} \cong 0.1$ V, similar to the oxidized sample. The cross-over region does not exist, however, because the holes have a larger capture cross-section than the electrons [187]. The surface recombination rate U_{sur} is limited by the supply of electrons at all injection levels. The decrease in surface recombination velocity is solely due to the reduction of the surface potential when switching from low- to high-injection conditions. Consequently the reduction of the

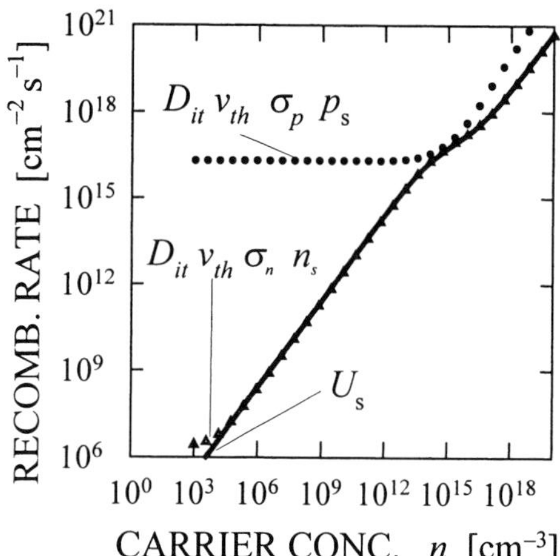

Figure B.13. Capture rates for holes and electrons and total recombination rate U_s of a Si/SiN$_x$ interface. The simulation parameters are the same as for the simulation of the 1.5 Ω cm sample in Figure 2.26 on p. 43.

surface recombination velocity is only by two orders of magnitude, rather than four orders of magnitude for the oxidized sample.

B.3.2 Grain boundary recombination

We model grain boundary recombination with the extended Shockley-Read-Hall model. The position of the various energy levels at a grain boundary in p-type Si is shown in Figure 2.32 on p. 49. Due to our assumption of flat quasi-Fermi levels, this model is limited to low interface recombination rates. An extension to high recombination rates is outside the scope of this work. Numeric simulators, such as the programs PC1D [402] or DESIS [344], may be used to study high recombination rates. Here we have to keep in mind that our analytical simulations are only qualitatively correct for the case of large interface state densities.

Doping dependence

The doping concentration of the cell is easy to control during the growth of the thin-film absorber. Therefore its impact on grain boundary recombination is an important matter. Figure B.14a shows the surface potential Ψ_{sur} as a function of the acceptor concentration N_A. The surface potential increases with the interface state density, since more interface states accommodate more charges, which induce a larger band bending. All curves show a maximum: for low doping concentrations the barrier height Ψ_{sur} increases logarithmically with the doping concentration N_A, and for large doping concentrations it decreases again. In order to understand this behavior, we consider the charges at the grain boundary Q_{it} and in the semiconductor Q_{sc}. Band bending is of the order of $q\Psi_{sur} = 0.1\, E_g$. Hence a typical charge at the interface is

$$Q_{it} = q\, 0.1\, D_{it}\, E_g \tag{B.38}$$

The charge in the two space charge regions on either side of the grain boundary is

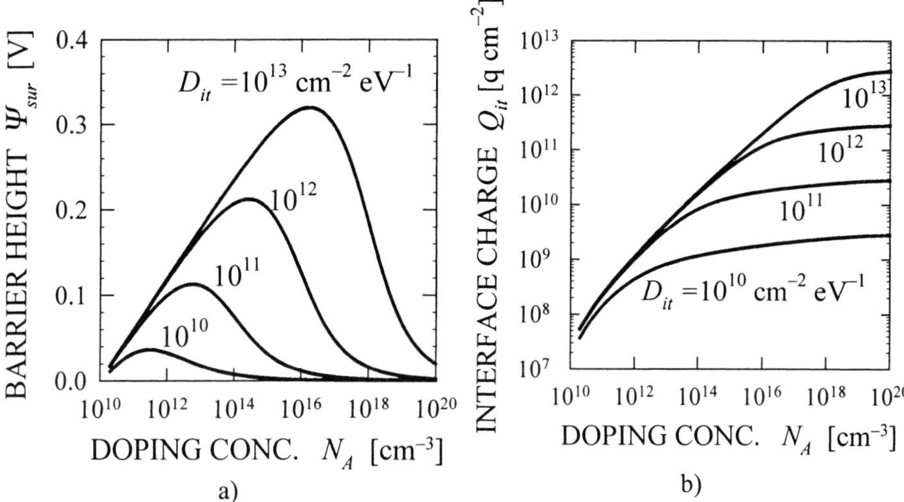

Figure B.14. a) Barrier height Ψ_{sur} at a grain boundary with symmetric interface properties as a function of Si doping and interface state density: b) Corresponding charge stored in the interface.

$$Q_{sc} = q2N_A W_{scr} = \sqrt{0.8N_A \varepsilon_o \varepsilon_s E_g} \tag{B.39}$$

where the relation $\Psi_{sur} = qN_A W_{scr}^2/(\varepsilon_o\ \varepsilon_s)$ applies. These equations define a critical doping concentration N_{Ac} at which the typical charge in the interface equals the charge in the semiconductor. For an interface state density of $D_{it} = 10^{12}$ cm^{-2} eV^{-1}, we calculate $N_{Ac} = 2\times10^{17}$ cm^{-3}. Figure B.14b shows the simulated charge at the interface. For $D_{it} = 10^{12}$ cm^{-2} eV^{-1}, the critical doping $N_{Ac} = 2 \times 10^{17}$ cm^{-3} corresponds to the onset of interface charge saturation. The charge in the interface cannot become larger than $D_{it}\,E_g$, and therefore has to saturate.

Now let us consider the case of a doping $N_A << N_{Ac}$. In this case, the semiconductor does not provide sufficient charges for the Fermi level to move away from the neutrality level Φ_0. The Fermi level is pinned to the neutrality level. Hence the band bending Ψ_{sur} increases linearly with $\Phi_n = kT/q\ \ln(N_V/N_A)$. See Figure 2.32 on p. 49 for the definition of Φ_n. This is the logarithmic increase observed in Figure B.14a.

For large doping concentrations $N_A >> N_{Ac}$, the interface charge saturates as shown in Figure B.14b. Consequently, the width of the space charge region and the band bending necessary to accommodate the almost constant charge at the interface decreases again. The relation between surface potential Ψ_{sur}, semiconductor charge Q_{sc}, and doping N_A is $\Psi_{sur} = Q_{sc}^2/(2\ \varepsilon_o\ \varepsilon_s\ q^2\ N_A)$, thus explaining the decrease of Ψ_{sur} for large N_A values.

The impact of the band bending on the grain boundary recombination rate and the grain boundary surface recombination is discussed along with Figure 2.33a on p. 50.

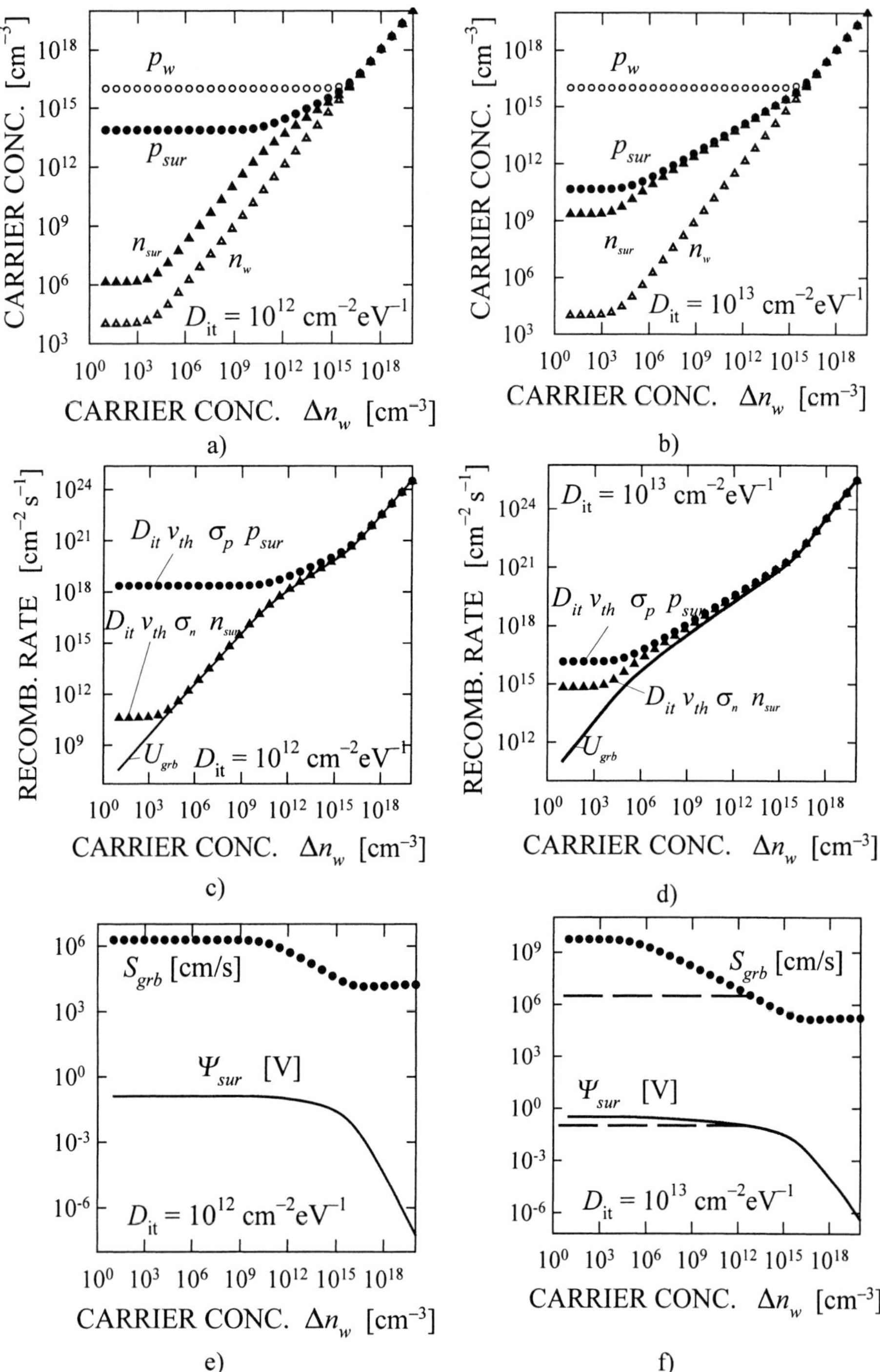

Figure B.15. a) and e): Carrier concentrations n_w, p_w at the edge of the space charge region and n_{sur}, p_{sur} at the grain boundary p-type Si having an acceptor concentration $N_A = 10^{16}$ cm^{-3} for two different interface state densities D_{it}; c) and d): aapture rates for electrons and holes and total recombination rate R_{sur}; e) and f): surface recombination velocity S_{eff} and surface potential Ψ_{sur}.

Dependence on injection level

The dependence of grain boundary recombination on the injection level is similar to the case of surface recombination. We consider a semiconductor with bulk doping $N_A = 10^{16}$ cm^{-3} and two different interface state densities $D_{it} = 10^{12}$ cm^{-2} eV^{-1} and $D_{it} = 10^{13}$ cm^{-2} eV^{-1}. Figures B.15a and B.15b show the carrier concentrations at the edge of the space charge region (subscript *w*) and at the grain boundary (subscript *sur*). Band bending enhances the minority carrier concentration. The carrier injection reduces the band bending at injection levels ranging from 10^{11} cm^{-3} to 10^{15} cm^{-3} in Figure B.15a and 10^{6} cm^{-3} to 10^{15} cm^{-3} in Figure B.15b. The electron and hole concentrations are considerably different from the equilibrium case and become almost equal ($n_{sur} = p_{sur}$) at injection levels far below high injection. Our model calculations are performed for equal capture cross-sections of electrons and holes. Thus, under injection the band bending adjusts to a value that maximizes the recombination current [181]. The total surface recombination rate U_s (see Figures B.15c and B.15d) is controlled by the smaller one of the capture rates of electrons and holes, which is the capture rate for electrons in our case. The shoulder in the total recombination rate corresponds to a decrease of the grain boundary recombination velocity S_{grb}, which is shown in Figures B.15e and B.15f together with the surface potential Ψ_{sur}.

Surface recombination velocities larger than the thermal velocity of electrons are not possible if the carriers are supplied by the grain boundary by diffusion from the bulk. In that case, the surface potential will be clamped to the value indicated by the broken line (see Figure B.15f).

B.4 Transport in polycrystalline thin films

B.4.1 Transport equations

We restrict ourselves to steady-state situations. The electric field $\boldsymbol{E} = -\nabla\Phi$ is the solution of the Poisson equation

$$-\Delta\Phi = \frac{q}{\varepsilon_o \varepsilon_s}\left(p + N_D - n - N_A\right) \tag{B.40}$$

where q is the elementary charge and $\varepsilon_o \times \varepsilon_s$ is the dielectric constant of Si. The electron current density

$$\boldsymbol{j}_n = -q\mu_n n\nabla\Phi + qD_p\nabla n \tag{B.41}$$

and the hole current density

$$\boldsymbol{j}_p = -q\mu_p p\nabla\Phi - qD_p\nabla p \tag{B.42}$$

are caused by charge carrier drift in the electric field and by diffusion. We assume mobilities $\mu_{n,p} = q\ D_{n,p}/kT$ that are independent of position and independent of the carrier concentrations. Both conditions are fulfilled only approximately. In thin-film materials the material quality is often spatially inhomogeneous. The mobility μ also depends on the carrier concentrations n and p due to carrier-carrier scattering, and is constant only under low-level injection. The carrier generation rate g is the source and the recombination rate U_{rec} is the sink for electrons and holes. The continuity equations read

$$\nabla \boldsymbol{j}_n = q(-g + U_{rec}) \tag{B.43}$$

and

$$\nabla \boldsymbol{j}_p = q(g - U_{rec}) \tag{B.44}$$

The total current has a vanishing divergence $\nabla(\boldsymbol{j}_n + \boldsymbol{j}_p) = 0$, because electrons and holes are always created and destroyed in pairs. Equations (B.40) through (B.44) form a set of coupled non-linear equations. They are non-linear because the recombination rate U_{rec} depends, in general, non-linearly on the carrier concentrations. The drift part of the current contains the electric field that also depends on the carrier concentrations and is thus a second source of non-linearity. Adequate boundary conditions for Φ, $\nabla\Phi$, n, ∇n, p, and ∇p have to be used to solve these transport equations. In general the boundary conditions are complex: at the surface of the semiconductor, for example, the occupation of defect levels depends on the position of the two quasi-Fermi energies. Hence, the surface charge stored in the defects depends on the volume carrier concentrations. The solution of the transport equations requires iterations that are generally only feasible on computers. The computer code PC-1D solves the one-dimensional transport equations [403] and is particularly adapted for simulation solar cells. This code is being updated continuously to account for the latest set of Si material parameters. More subtle effects, such as injection-dependent free carrier absorption and tunneling-enhanced recombination in space charge regions, were recently included into the PC1D code [402]. However, PC1D only permits the simulation of one-dimensional transport.

Polycrystalline thin-film cells with grain boundaries require at least a two-dimensional transport model. The waffle-shaped cells discussed in Chapter 4 have a texture depth similar to the texture period and therefore also require a multi-dimensional transport model. For these cases the DESSIS code [344] is an advanced modeling tool. An advantage of computer modeling is that the investigation of the combined influence of various physical effects is possible. A disadvantage is the need for a complex computer code that is, in general, not published. Thus modifications to the code are often not possible. Multi-dimensional computer simulations are also very time-consuming.

B.4.2 Carrier concentration in the base

Fortunately, it is often possible to find approximate analytical solutions to the transport equations that give insight into the underlying physical effects. An important simplification that is often required to enable analytical treatments is the assumption of low-level injection conditions. This assumption guarantees a recombination rate $U_{rec} = \Delta n/\tau$ that is linear in the excess minority carrier concentration Δn. If quasi-neutrality holds (as in the homogeneously doped base of a Si cell), the electric field $-\nabla\Phi$ does not need to be considered for the transport of the *minority* carriers. Then the minority carrier diffusion equation

$$D_n\left(\frac{\partial^2 \Delta n}{\partial X^2}+\frac{\partial^2 \Delta n}{\partial Y^2}+\frac{\partial^2 \Delta n}{\partial Z^2}\right)-\frac{\Delta n}{\tau}=-g \tag{B.45}$$

holds [79]. Here *X*, *Y*, and *Z* denote the Cartesian coordinates.

B.4.2.1 Thin-film cells

Figure B.16 shows a columnar square grain of edge length *G* that we consider to represent all the various grains in a polycrystalline thin-film Si cell. The thickness of the cell's base is W_{bas}. The emitter is at the illuminated front surface. The space charge region (SCR) separates the emitter and the base. The interface of the base and the back metallization has a surface recombination velocity S_b. The intra-grain volume is characterized by a minority carrier diffusion length $L = (D_n\,\tau)^{0.5}$ and a minority carrier diffusion coefficient D_n. The side walls of the grains (i.e. the grain boundaries) exhibit a surface recombination velocity S_{grb}. The origin of the *X*- and *Y*-axes of the Cartesian coordinate system is located in the center of the junction plane. In this section we analyze the base and thus locate the origin of the *Z*-axis in the interface of the SCR and the base. We assume diffusive transport in the homogeneously doped p-type base part of the cell. In the dark, the excess minority carrier concentration Δn obeys the diffusion equation

$$D\left[\frac{\partial^2 \Delta n}{\partial x^2}+\frac{\partial^2 \Delta n}{\partial y^2}+\frac{\partial^2 \Delta n}{\partial z^2}\right]-\frac{\Delta n}{l^2}=0 \tag{B.46}$$

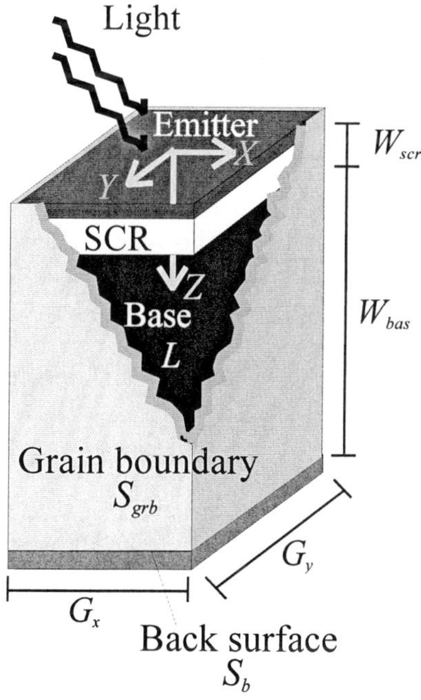

Figure B.16. Grain geometry of the transport model: square grains are considered. The base of the cell is characterized by a minority carrier diffusion length *L*. The back surface exhibits a surface recombination velocity S_b. The side walls of the grains have recombination velocity S_{grb}.

We normalize all distances by the grain size G; that is: $x = X/G$, $y = Y/G$, $z = Z/G$, and use the normalized diffusion length

$$l = L/G \tag{B.47}$$

Recombination at grain boundaries is described by the reduced recombination velocity

$$s_{grb} = S_{grb}\, G/D_n \tag{B.48}$$

The minority carrier current into the grain boundaries is

$$\left.\frac{\partial \Delta n}{\partial x}\right|_{x=\pm 1/2} = \mp\frac{s_{grb}}{2}\left.\Delta n\right|_{x=\pm 1/2} \text{ and } \left.\frac{\partial \Delta n}{\partial y}\right|_{y=\pm 1/2} = \mp\frac{s_{grb}}{2}\left.\Delta n\right|_{y=\pm 1/2} \tag{B.49}$$

in X- and Y-direction, respectively. The reduced recombination velocity $s_b = S_b\, G/D_n$ at the back surface quantifies the surface losses. The minority carrier current into the back surface is

$$\left.\frac{\partial \Delta n}{\partial z}\right|_{z=w} = s_b \left.\Delta n\right|_{z=w} \tag{B.50}$$

Here, $w_{bas} = W_{bas}/G$ denotes the reduced cell thickness. We normalize the excess minority carrier concentration Δn by the concentration at the junction; hence

$$\left.\Delta n\right|_{z=0} = 1 \tag{B.51}$$

at the junction. Note that the minority carrier concentration Δn depends on the four parameters l, s_{grb}, s_b, and w_{bas}.

Following Dugas's ansatz [34], we solve the transport Eqs. (B.46) under boundary conditions (B.49) and (B.50) by a Fourier-like decomposition of the excess minority carrier density in the X- and Y-directions. This type of transport analysis in the Fourier space was also applied to study multi-junction solar cells [404]. Using cosine functions with appropriate wave numbers c_i, c_j that fulfill

$$c_i \tan(c_i/2) = s_{grb}/2 \quad \text{with } c_i \in [\pi i, \pi(i+1)[\quad \text{for i} = 0,1,2... \tag{B.52}$$

we find

$$\Delta n(x,y,z) = \sum_{i,j=0}^{\infty} \gamma_i \gamma_j \frac{c_i \cos(c_i x) c_j \cos(c_j y)}{4\sin(c_i/2)\sin(c_j/2)} \left[\cosh(k_{ij} z) - \beta_{ij} \sinh(k_{ij} z)\right] \tag{B.53}$$

where the coefficients k_{ij} are given by

$$k_{ij}^2 = c_i^2 + c_j^2 + l^{-2} \quad \text{for } i, j = 0, 1, 2,... \tag{B.54}$$

$$\gamma_i = \frac{8\sin^2(c_i/2)}{c_i^2 + c_i \sin(c_i)}, \quad \gamma_j = \frac{8\sin^2(c_j/2)}{c_j^2 + c_j \sin(c_j)} \quad \text{for } i, j = 0, 1, 2, \ldots \tag{B.55}$$

and

$$\beta_{ij} = \frac{\sinh(k_{ij} w_{bas}) + s_b / k_{ij} \cosh(k_{ij} w_{bas})}{\cosh(k_{ij} w_{bas}) + s_b / k_{ij} \sinh(k_{ij} w_{bas})} \quad \text{for } i, j = 0, 1, 2, \ldots \tag{B.56}$$

We apply Eq. (B.53) to calculate the minority carrier profile of a polycrystalline cell with a grain size G and a thickness W_{bas} equal to 5 µm. Nine hundred Fourier components $i, j = 0 \ldots 30$ were used for the profiles displayed in Figure B.17. Although transport is a three-dimensional problem, the profile is determined in fractions of a second on a personal computer. Fast calculation is particularly helpful for fitting experimental data.

In Figure B.17a, the grain boundary recombination velocity S_{grb} and the back surface recombination velocity S_b are zero. Since the minority carrier diffusion length $L = 10$ µm is twice the base thickness W_{bas}, the excess carrier concentration is almost constant throughout the device.

In Figure B.17b, the minority carrier diffusion length is only $L = 5$ µm: the carrier density decays approximately exponentially into the depth z. The profile is still one-dimensional because $S_{grb} = 0$.

In Figure B.17c we use $S_{grb} = 10^6$ cm s^{-1} and still $S_b = 0$ as before. The carrier density sharply decreasing towards the grain boundaries at $X = \pm$ 2.5 µm causes the carriers to diffuse into the grain boundary. The largest change in carrier density relative to b) occurs at the back of the cell at $Z = 5$ µm. No carriers diffuse into the back surface, since $S_b = 0$.

In Figure B.17d, the back surface recombination velocity is $S_b = 10^6$ cm s^{-1} and consequently the carrier density is further reduced at the back of the cell.

From Eq. (B.53) we compute the average carrier concentration

$$\overline{\Delta n}(z) = \int_{x=-1/2}^{x=1/2} dx \int_{y=-1/2}^{y=1/2} dy \, \Delta n(x, y, z) = \sum_{ij} \gamma_i \gamma_j \left[\cosh(k_{ij} z) - \beta_{ij} \sinh(k_{ij} z)\right] \tag{B.57}$$

The diode saturation current density is then

$$j_0 = -q D_n \frac{n_o}{G} \left.\frac{\partial \overline{\Delta n}}{\partial z}\right|_{Z=0} = q D_n \frac{n_o}{G} \left(\sum_{i,j} \gamma_i \gamma_j \beta_{ij} k_{ij} \right) \tag{B.58}$$

Limitation of the model for large grain boundary recombination velocity S_{grb}

The model assumes that the excess carrier concentration Δn is constant in the plane that separates the quasi-neutral base and the space charge region of the injecting junction. For very large grain boundary recombination velocities S_{grb}, the recombination in the space charge region becomes large and dominant. The quasi-Fermi levels are no longer flat in the space charge region and under high forward injection the excess carrier concentration at the junction will no longer be constant. Thus the base recombination current will deviate from that calculated by Eq. (B.58).

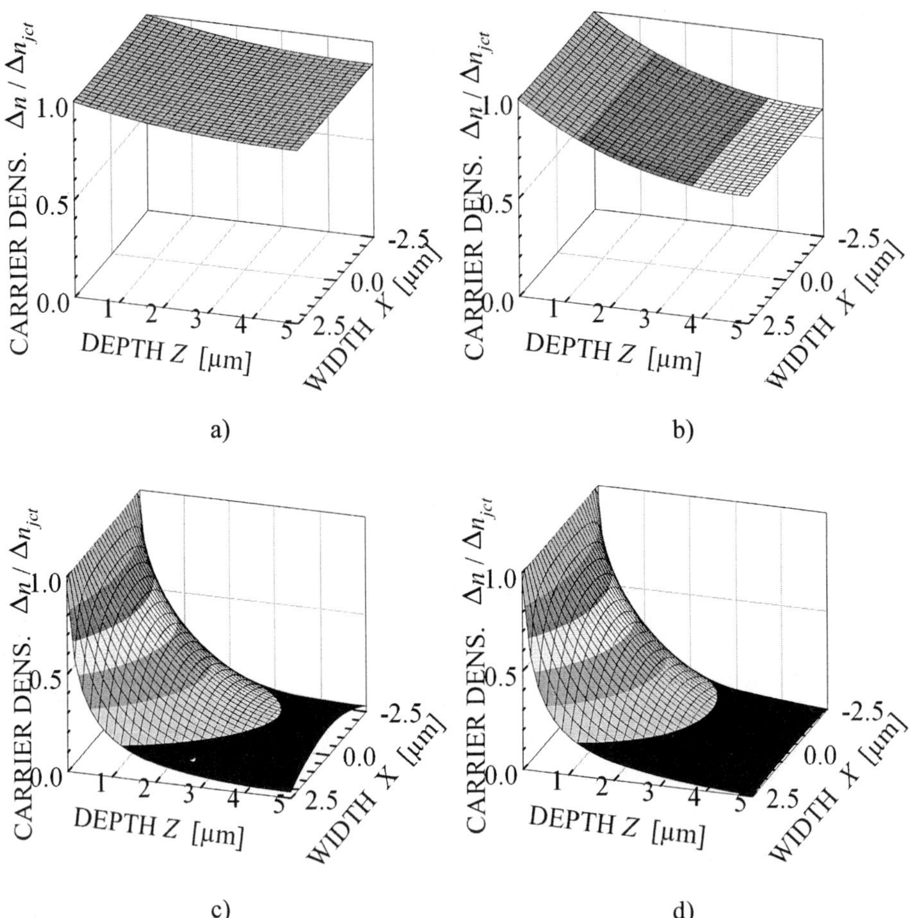

Figure B.17. Normalized excess minority carrier densities in the $Y = 0$ plane of a square silicon grain of width G = 5 μm and thickness W_{bas} = 5 μm. a) Minority carrier diffusion length L = 10 μm, grain boundary recombination velocity S_{grb} = 0, back surface recombination velocity S_b = 0; b) L = 5 μm, $S_{grb} = S_b = 0$; c) L = 5 μm, $S_{grb} = 10^6$ cm s^{-1}, and S_b = 0; d) L = 5 μm, $S_{grb} = S_b = 10^6$ cm s^{-1}.

B.4.2.2 Thick-film cells

The impact of grain boundaries on the minority carrier transport is most pronounced for small grains, i.e. where $w_{bas} = W_{bas}/G >> 1$, which is equivalent to thick cells. For thick cells, the expression for the excess minority carrier concentration

$$\Delta n(x, y, z) = \sum_{ij} \cos(c_i x)\cos(c_j y)\frac{4\sin(c_i/2)\,4\sin(c_j/2)}{c_i + \sin(c_i)\,c_j + \sin(c_j)}\exp(-k_{ij} z) \qquad \text{(B.59)}$$

from Eq. (B.53) can be simplified significantly. The carrier concentration profile depends on only the two recombination parameters, s_{grb} and l, via Eqs. (B.54) and (B.52). From Eq. (B.59), we compute the average carrier concentration

$$\overline{\Delta n}(z) := \int_{x=-1/2}^{x=1/2} \int_{y=-1/2}^{y=1/2} n(x,y,z)\,dx\,dy = \sum_{ij} \frac{8\sin^2(c_i/2)\ \ 8\sin^2(c_j/2)}{c_i(c_i+\sin(c_i))c_j(c_j+\sin(c_j))}\exp(-k_{ij}z) \quad \text{(B.60)}$$

at a depth z. We apply these expressions in Chapter 3 on p. 71 to derive bounds for the grain boundary recombination S_{grb} and intra-grain diffusion lengths L from the slope of the inverse of the measured quantum efficiency IQE^{-1} vs. the optical absorption length L_α.

B.4.3 Electrostatic potential in the space charge region

In a space charge region, the concentrations of electrons and holes varies by orders of magnitude. Since the electrons are majority carriers in the emitter and minority carriers in the base, the space charge region contains a "sweet plane" where the capture rate of electrons into the defect equals that of holes. For the one-dimensional case Choo gave an easy-to-handle analytical expression that evaluates the recombination in the space charge region of a monocrystalline asymmetric junction [347]. Rau and Werner introduced an approach that permits the analytical treatment of space charge region recombination in polycrystalline material [35]. Their trick is to assume an interface state density that is independent of energy. Then the charge at the grain boundary varies linearly with the majority carrier Fermi level and thus permits a linearization of the Poisson equation which facilitates analytical treatment.

We adopt this approach here. To calculate the carrier concentrations n and p from the electrostatic potential, we assume flat quasi-Fermi levels, which implicitly means we assume little recombination in the space charge region. The geometry we are considering here is shown in Figure B.16 on p. 233. The origin of the Z-axis is now positioned at the interface between the emitter and the space charge region. Assuming fully depleted grains the electrical potential fulfills the Poisson equation

$$-\left[\frac{\partial^2\Phi}{\partial X^2}+\frac{\partial^2\Phi}{\partial Y^2}+\frac{\partial^2\Phi}{\partial Z^2}\right] = -\frac{qN_A G^2}{\varepsilon_o\varepsilon_s} =: \beta \quad \text{(B.61)}$$

The charge on the grain boundary varies linearly with the potential

$$-\frac{\partial\phi}{\partial X} = -\frac{q^2 N_t}{2\varepsilon_o\varepsilon_s}\Phi \quad \text{(B.62)}$$

The neutrality level Φ_o off the grain boundary is at a fixed distance from band edges and thus runs parallel to the potential energy of an electron $\Phi_o(X = G/2, Y = G/2, Z) = -\Phi(X = G/2, Y = G/2, Z)$ + constant. If we chose the constant to result in a zero potential Φ at the intersection of Φ_o with the plane $(E_{Fn} + E_{Fp})/2$ we achieve zero grain boundary charge at zero potential Φ as expressed by Eq. (B.62). The electrostatic potential is constant at $Z = 0$, due to the high lateral conductivity in the n^+-type emitter. Its value

$$-q\Phi\big|_{z=0} = +qU/2 + kT\ln(N_C/N_D) - E_g + \Phi_o \quad \text{(B.63)}$$

depends on the donor concentration N_D in the emitter, the applied voltage U, and the neutrality level Φ_o. At the interface between the space charge region and the base we assume a zero electric field in the Z-direction, and hence

$$\left.\frac{\partial \Phi}{\partial Z}\right|_{Z=W_{SCR}} = 0 . \tag{B.64}$$

For the electrical potential the ansatz is

$$\Phi = \sum_{i=1}^{\infty} \sum_{j=1}^{\infty} \cos(c_i x)\cos(c_j x)\, A_{ij}\left[a_{ij} \cosh(k_{ij} z) + b_{ij} \cosh(k_{ij} z)\right] \tag{B.65}$$

as proposed by Dugas for the solution to the minority carrier diffusion problem [34]. Using cosine functions with appropriate wave numbers c_i, c_j that fulfill

$$c_i G \tan(c_i G/2) = \frac{q^2 N_t G}{2\varepsilon_o \varepsilon_s} \quad \text{with } Gc_i \in [\pi i, \pi(i+1)[\text{ for } i = 0,1,2... \tag{B.66}$$

one finds the coefficients k_{ij} are given by

$$k_{ij}^{\,2} = c_i^{\,2} + c_j^{\,2} \quad \text{for } i, j = 0, 1, 2,\ldots \tag{B.67}$$

$$A_{ij} = \beta \frac{4\sin(Gc_i/2)}{c_i^2\left(Gc_i + Gc_i \sin(Gc_i)\right)} \frac{4\sin(Gc_j/2)}{c_j^2\left(Gc_j + Gc_j \sin(Gc_j)\right)} \quad \text{for } i, j = 0, 1, 2,\ldots \tag{B.68}$$

$$a_{ij} = \frac{c_{ij}^2}{\beta} \Phi\Big|_{z=0} - 1 \quad \text{for } i, j = 0, 1, 2,\ldots \tag{B.69}$$

and

$$b_{ij} = -a_{ij} \tanh(k_{ij} W_{SCR}) \quad \text{for } i, j = 0, 1, 2,\ldots \tag{B.70}$$

Figure B.18 shows the spatial dependence of the conduction band E_C, the valence band E_V, and the equilibrium Fermi level E_F that we calculate for a p-type space charge region with an acceptor concentration $N_A = 10^{16}$ cm^{-3}.

In Figure B.18a the interface state density N_t is zero. The potential is parabolic as for the one-dimensional abrupt junction. The depth of the space charge region is 3.5 μm.

In Figure B.18b the interface state density is $N_t = 10^{12}$ cm^{-2} eV^{-1}, with the neutrality level at the midgap. The interface charge causes a bending of the bands. The band bending increases the minority carrier concentration at the grain boundary. Recombination is thus enhanced. In the middle of the grain at $X = 0$ the potential is still parabolic and undisturbed by the grain boundaries. Let us now consider a gradual reduction in grain size.

Figure B.18c shows the same situation as in Figure B.18b, but the grain size is G = 0.5 μm, which is a tenth of case b). The band bending reduces and the Fermi level E_F shifts towards midgap. The space charge in the smaller grain is now insufficient to com-

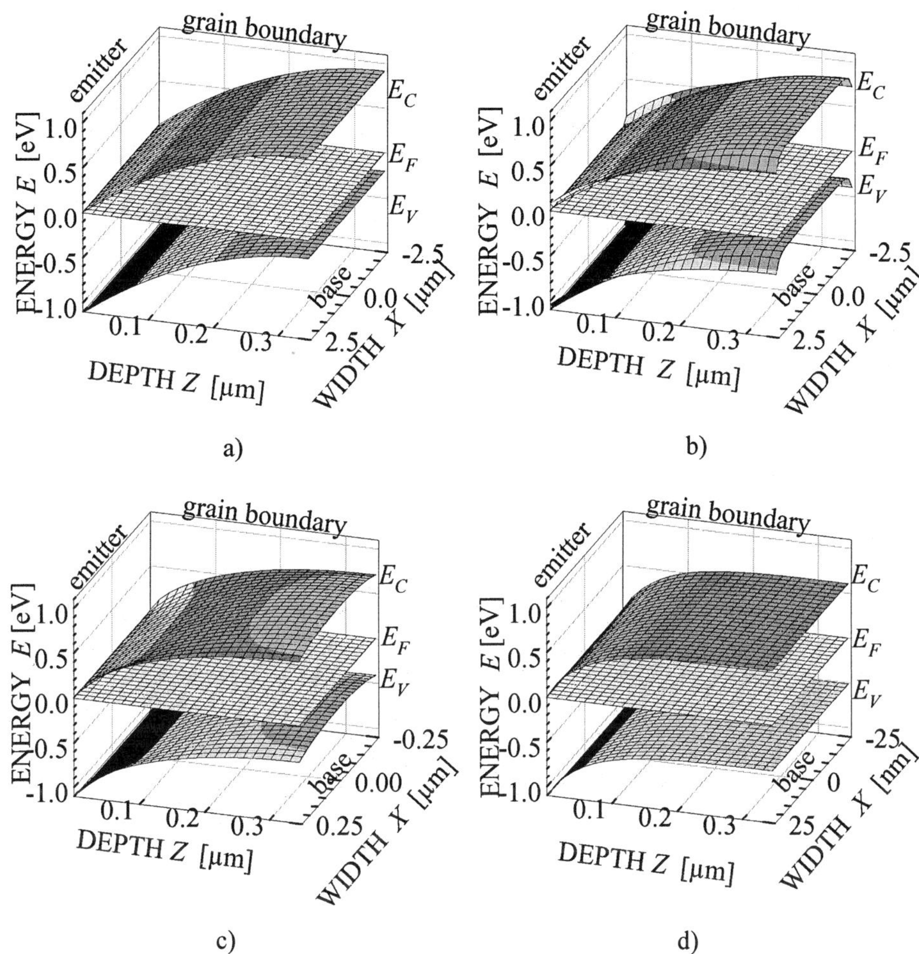

Figure B.18. Equilibrium energy-position diagrams for a fully depleted p-type semiconductor (N_A = 10^{16} cm^{-3}) with a constant interface state density N_t. The conduction band E_c, the valence band E_V and the equilibrium Fermi level E_F are shown. a) G = 5 μm, N_t = 0; b) G = 5 μm, N_t = 10^{12} cm^{-2} eV^{-1}; c) G = 0.5 μm, N_t =10^{12} cm^{-2} eV^{-1}; d) G = 50 nm, N_t =10^{12} cm^{-2} eV^{-1}.

pensate the charging of the grain boundary. Hence, the Fermi level is hardly allowed to shift away from the neutrality level.

This effect is even larger at a grain size reduced by another factor of ten to G = 50 nm, as shown in Figure B.18d. The interface state density is still $N_t = 10^{12}$ cm^{-2} eV^{-1}. No band bending prevails. On an increasing fraction of the grain boundaries the Fermi level is pinned to the neutrality level, which is at midgap in our case. Consequently, the electron concentration is identical to the hole concentration and the total recombination rate (per grain boundary area) is greater in Figure B.18d than in cases c) and b). In addition, smaller grains result in a larger grain boundary area.

APPENDIX C

C Quantum efficiency

C.1 Measurement

C.1.1 Light-biased

Quantum efficiency spectra are routinely measured under time-constant bias light in order to investigate the device under conditions similar to operation. We adjust the distance from the lamp to the cell so as to yield a short-circuit current that is 50% to 100% of the short-circuit current that we measure with an AM1.5G spectrum at 1000 W m^{-2} [13]. The measurement set-up is shown in Figure C.1. Chopped monochromatic light is directed via a beam splitter onto the cell under test, towards and onto a monitor cell. Two lock-in amplifiers measure the modulation of currents Δj_{cell} and Δj_{mon}, respectively. We record the ratio $R_{cell} = \Delta j_{cell} / \Delta j_{mon}$. In a second measurement the cell under test is replaced by a calibrated reference cell with a known external quantum efficiency EQE_{cal}. Again, Δj_{cal} and Δj_{mon} from the calibrated cell and from the monitor cell are measured and the ratio $R_{cal} = \Delta j_{cal} / \Delta j_{mon}$ is recorded. The external quantum efficiency of the cell under test is

$$EQE = EQE_{cal} \times R_{cell} / R_{cal} \tag{C.1}$$

The use of a monitor cell permits correction of low-frequency fluctuations of the lamp intensity. The monitor cell and the cell under test are placed on a sample holder that is at

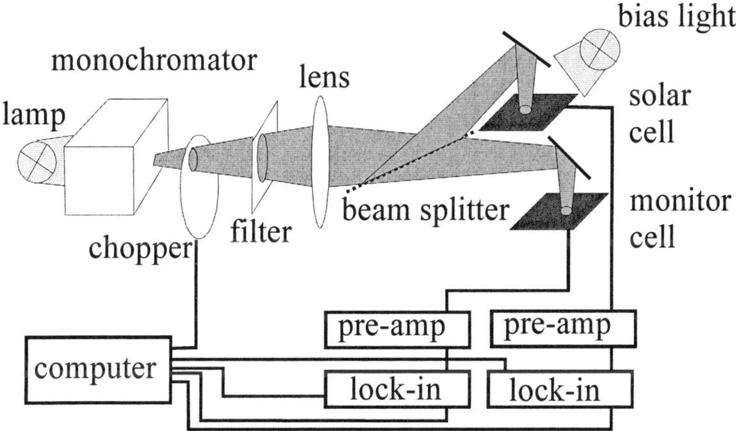

Figure C.1. Experimental set-up for quantum efficiency measurements.

a temperature of 298 K. A vacuum pump holds the cell to the chuck and guarantees sufficient thermal contact. The monitor cell is a textured high-efficiency cell with an external quantum efficiency as high as $EQE = 0.9$ at 1050 nm.

The calibration of the reference cell has a relative accuracy of 2%. The accuracy of the hemispherical reflection measurement is 1.5% between 300 nm and 1100 nm. The accuracy of the external quantum efficiency measurement is also 1.5%. Both latter measurement errors are deduced from measuring the same samples with different measurement systems. We estimate a 5% error for the internal quantum efficiency measurement.

The monochromatic light is chopped at a frequency f. The time $1/f$ has to be greater than the time constant of the fastest process in the cell. For the measurements reported in this work the frequency is $f = 30$ Hz. For the high-efficiency cells we measure an internal quantum efficiency that is independent of the modulation frequency up to 160 Hz.

C.1.2 Voltage-biased

When studying injection level-dependent recombination processes, carrier injection by a bias voltage is an attractive alternative to carrier generation by bias illumination. The determination of the injection level is easier for a voltage bias since, in contrast to bias illumination, the complex optical properties of the cell do not need to be known.

Experimental set-up

Figure C.2 shows the circuit that we use to apply a time-constant bias voltage U_b [31]. The illuminated solar cell is represented by an ideal diode, a current source, and a series resistance R_s. The current source represents only the modulated fraction of the photogeneration due to the chopped monochromatic light. The cell has a current-voltage curve $I(U)$ in the dark. The IQE signal U_{out} is generated by the modulated current I_m. The current $I_m = I_R + I_D$ splits into a current I_D through the diode and a current I_R through the series resistance R_s. The operational amplifier OP converts the current I_R to an output voltage $U_{out} = R_{OP} I_R$. The external voltage U_b biases the cell via the operational amplifier OP.

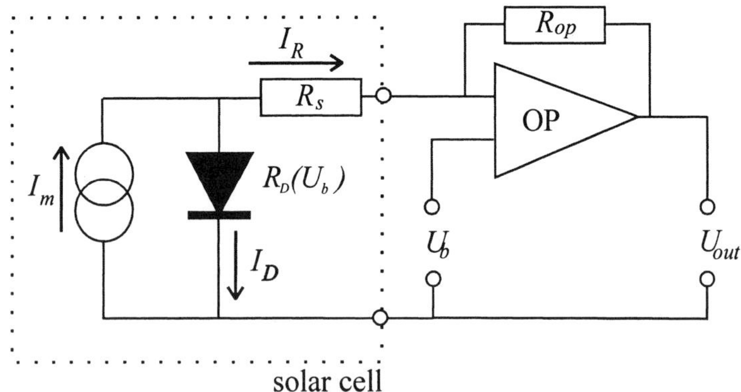

Figure C.2. Current-voltage converter for quantum efficiency measurements under forward bias voltage U_b. A fraction I_D of the light-generated current I_m flows through the diode and is thus lost for the output voltage U_{out}.

Measurement example

Figure C.3 shows the quantum efficiency of a high-efficiency thin-film Si solar cell with a thickness of 46.5 μm (see description of Cell A on p. 83 for more details) under illumination with light of wavelengths 500 and 800 nm. For both wavelengths the *IQE* data depend on the forward bias voltage U_b. At bias voltages up to 500 mV the value of *IQE*(500 nm) is constant, while *IQE*(800 nm) starts to increases at 350 mV. For voltages U_b >550 mV *IQE*(500 nm) and *IQE*(500 nm) both decrease.

Interpretation of the measurement

Light of wavelength 500 nm penetrates a distance of 0.9 μm into the cell. Hence, the light is mainly absorbed in the emitter and the quantum efficiency therefore reflects the recombination in the emitter that is known to be, in general, independent of the injection level.

In contrast, light of wavelength 800 nm penetrates 12 μm into the cell and is mainly absorbed in the base. The quantum efficiency at 800 nm is thus dominated by recombination in the base and at the back surface. Since the diffusion length exceeds the base thickness, *IQE*(800 nm) actually reflects recombination at the Si/SiO_2 back surface, which is known to decrease with the injection level. A decreasing surface recombination increases the quantum efficiency at voltages above 350 mV.

The decrease of the differential resistance $R_D(U_b)$ with increasing forward bias U_b causes the common decrease of *IQE*(500 nm) and *IQE*(800 nm) at voltages greater than 550 mV. This decrease is thus *not* due to a change in the recombination rates with the injection level [31]. The larger the R_s (see Figure C.2), the more photogenerated current is diverted through the diode, since the value of R_D decreases with increasing bias voltage U_b. The circles in Figure C.3 depict the measured ratio *IQE*(500nm, U_b)/*IQE*(500 nm, 0). In addition, this figure also shows *IQE*(500nm, U_b)/*IQE*(500 nm, 0) as measured

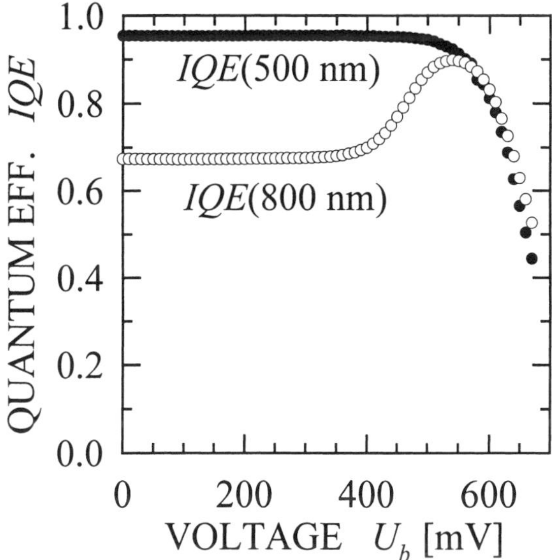

Figure C.3. Internal quantum efficiency *IQE*(500 nm) and *IQE*(800 nm) measured at two wavelengths for a thin-film high-efficiency solar cell from crystalline Si. Data from Ref. [405].

with R_s being increased by 2.2 Ω (open circles) and by 10 Ω (filled triangles) with external resistors. As expected, the voltage dependence is enhanced by increasing R_s.

Ratio $IQE(U_b)/IQE(0)$ determined from the current-voltage curve $I(U)$

If the reduction of the measurement signal is a purely resistive effect, it should be possible to calculate the reduction factor from the measured dark current-voltage curve $I(U)$ of the solar cell. With increasing forward bias U_b the small signal resistance

$$R_D(U_b) = \left(\left. \frac{\mathrm{d}I}{\mathrm{d}U} \right|_{U_b} \right)^{-1} - R_s \tag{C.2}$$

of the diode decreases. The analysis of the equivalent circuit shown in Figure C.2 yields U_b-dependent output voltage

$$U_{out}(U_b) = Q_R(U_b) R_{OP} I_m \tag{C.3}$$

that depends on a voltage-dependent resistance factor

$$Q_R = \frac{R_D(U_b)}{R_D(U_b) + R_s} \tag{C.4}$$

Using Eq. (C.2) we derive the resistance factor

$$Q_R = 1 - R_s \left. \frac{\mathrm{d}I}{\mathrm{d}U} \right|_{U_b} \tag{C.5}$$

directly from the measured current-voltage curve $I(U)$ of the solar cell. Our above interpretation of the experimental data shown in Figure C.3 is equivalent to the hypothesis

$$Q_R = \frac{IQE(500\ \mathrm{nm}, U_b)}{IQE(500\ \mathrm{nm}, 0)} \tag{C.6}$$

We test this hypothesis by comparing the quantum efficiency ratio $IQE(500\ \mathrm{nm}, U_b)\ /\ IQE(500\ \mathrm{nm}, 0)$ and the resistance factor Q_R in Figure C.4. For the calculation of the resistance factor Q_R, we choose a value for R_s that best fits the quantum efficiency ratio $IQE(500\mathrm{nm}, U_b)/IQE(500\ \mathrm{nm}, 0)$. The fits are the solid line lines in Figure C.4 and are achieved for R_s = 0.16 Ω, 2.0 Ω, and 9.6 Ω. The close agreement to the actually inserted resistor values of 0 Ω, 2.2 Ω, and 10 Ω confirms our interpretation of the decrease of IQE(500 nm) in Figure C.3.

Correction of experimental IQE spectra for resistance factor Q_R

The quantum efficiency spectrum has to be corrected for this effect, which is not related to the recombination properties, by

$$IQE^*(\lambda, U_b) = IQE(\lambda, U_b) / Q_R(U_b) \tag{C.7}$$

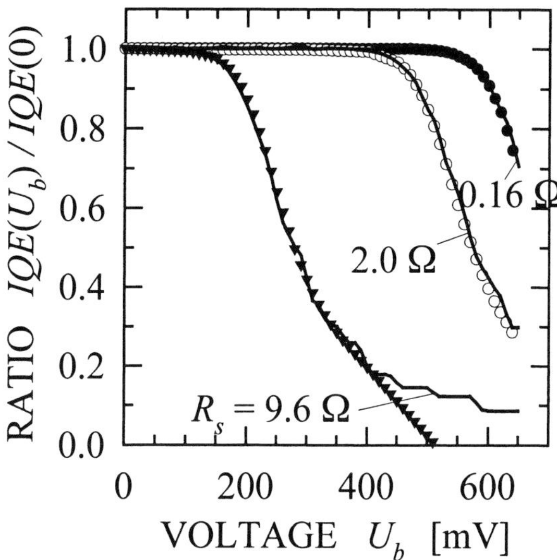

Figure C.4. Measured quantum efficiency ratio for wavelength 500 nm (symbols). Resistance factor Q_R calculated from the dark current-voltage curve for various series resistance values R_s (solid lines). Data from Ref. [405].

In the rest of this work only the corrected quantum efficiency IQE^* is used when quantum efficiency data under bias voltage are presented. We omit the asterisk to simplify the notation. The impact of a series resistance on quantum efficiency measurements is also discussed in Refs. [406, 407, 408].

C.2 Standard analysis

C.2.1 Front illumination

The standard technique for the evaluation of internal quantum efficiency spectra [36] assumes a cell that is thick, compared to the optical absorption length $L_\alpha = 1/\alpha_s$, and has a spatially homogeneous minority carrier diffusion length. Here, α_s denotes the absorption coefficient of crystalline Si. The profile of the minority carrier generation rate $g(Z)$ decays exponentially with the distance from the cell surface. When the optical absorption in the emitter of the solar cell is neglected, the inverse internal quantum efficiency

$$IQE^{-1} = 1 + L_\alpha / L_Q \tag{C.8}$$

depends linearly on the optical absorption length. Hence an effective diffusion length L_Q may be determined as the slope of the inverse internal quantum efficiency IQE^{-1}, when plotted versus the optical absorption length L_α.

Experimental example

Figure C.5 shows an example of the standard IQE evaluation technique. We measure the internal quantum efficiency of a monocrystalline Si solar cell with a base doping of 5×10^{18} cm^{-3}. The P-doped emitter is 0.5 µm thick. The cell thickness is $W_f = 525$ µm. The effective diffusion length is only $L_Q = 6.4$ µm, due to the high doping. The minority carriers do not reach the back surface of the cell, and consequently the effective diffusion length equals the diffusion length L in the base. At absorption lengths L_α that are larger than the film thickness W_f, the optical reflection at the back surface enhances the quantum efficiency IQE, and the measured IQE^{-1} data deviate from the linear relation expressed by Eq. (C.8).

C.2.2 Rear illumination

Illuminating the cell through the back achieves a higher sensitivity to the back surface recombination than illumination through the front surface. Obviously, the cell has to be bifacially sensitive to permit rear illumination. The quantum efficiency for back illumination is [409, 410]

$$IQE = \frac{1 + \frac{S_b L_\alpha}{D_n}}{\cosh(W_{bas}/L) + \frac{S_b L}{D_n}\sinh(W_{bas}/L)} \tag{C.9}$$

Here, plotting IQE versus L_α yields a linear relation. The slope and intercept provide information on S_b and L. Equation (C.9) predicts that the ratio of the slope to the intercept is S_b/D_n; here D_n is the diffusion coefficient of the minority carriers (electrons).

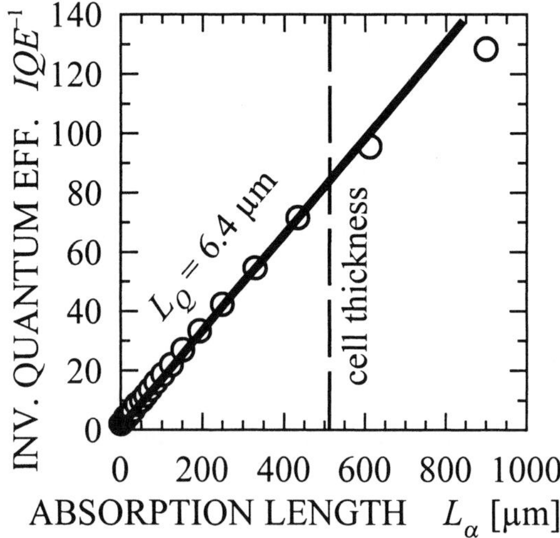

Figure C.5. The inverse internal quantum efficiency is linear in the optical absorption length. For absorption lengths greater than the cell thickness the linear relation does not hold.

C.3 Effective diffusion lengths (formulas)

As sketched in Figure 3.2 on p. 56, a quantum efficiency spectrum of a crystalline thin-film cell shows, in general, two linear regions if plotted as $IQE^{-1}(L_\alpha)$. Both regions define an effective diffusion length:

- The length L_Q that we call the quantum efficiency diffusion length and that we derive from the IQE spectrum measured with strongly absorbed light ($L_\alpha << W_{bas}$). The definition of L_Q is given on p. 55.
- The length L_C that we call the collection diffusion length and that we derive from the IQE spectrum measured with weakly absorbed light ($L_\alpha >> W_{bas}$). The definition of L_C is given on p. 56.

We give the formulas for both effective diffusion lengths L_Q and L_C in monocrystalline and polycrystalline material.

C.3.1 Monocrystalline semiconductors

Quantum efficiency diffusion length L_Q

For planar monocrystalline sheets of Si material, the effective *quantum efficiency diffusion length*

$$L_{Q,mono} = L\,\frac{S_b L \sinh(W_{bas}/L) + D_n \cosh(W_{bas}/L)}{S_b L \cosh(W_{bas}/L) + D_n \sinh(W_{bas}/L)} \qquad \text{(C.10)}$$

depends on the diffusion length L of the semiconductor, the back surface recombination velocity S_b, the base thickness W_{bas}, and the minority carrier diffusion coefficient D_n [24]. The value of $L_{Q,mono}$ equals the effective current-voltage diffusion length $L_{J,mono}$, which determines the dark saturation current density

$$j_o = \frac{q\, n_o D_n}{L_{J,mono}} \qquad \text{(C.11)}$$

of a single-sided p-n junction (Eqs. (8) and (9) of Ref. [24]). The symbol q denotes the elementary charge and n_o is the equilibrium minority carrier concentration. The quantum efficiency can therefore be used to evaluate the amount of recombination in the base of a single-crystalline solar cell [24, 177, 411].

Collection diffusion length L_C

The collection diffusion length

$$L_{C,mono} = L\,\frac{S_b L(\cosh(W_{bas}/L) - 1) + D_n \sinh(W_{bas}/L)}{S_b L \sinh(W_{bas}/L) + D_n \cosh(W_{bas}/L)} \qquad \text{(C.12)}$$

for monocrystalline cells is closely related to the collection probability for weakly absorbed light (Eq. 6 in Ref. [24]). A thin-film cell with base thickness W_{bas}, diffusion length L, SRV S_b, and negligible recombination in the emitter would generate exactly the same short-circuit current under spatially homogeneous carrier generation as a semi-infinitely thick monocrystalline cell with a base diffusion length of $L_{C,mono}$.

C.3.2 Polycrystalline semiconductors

Minority carrier recombination in polycrystalline semiconductor materials occurs in the volume of the grains, at the grain boundaries and at the surface of the cell. The quantum efficiency model has therefore to account for all three locations of recombination. Three-dimensional modeling is necessary.

Referring to the cell geometry shown in Figure 2.28 on p. 46, we calculate the quantum efficiency under monochromatic illumination with light that has the absorption coefficient α_s. The spatial carrier generation rate

$$g(x,y,z) = \alpha_s \exp(-\alpha_s z) \tag{C.13}$$

decays exponentially into the depth Z. The reduced absorption coefficient α_s is measured in units of inverse grain size G^{-1} here. The position $z = Z/G$, and similarly x and y, are also scaled by the grain size. The incident photon flux is unity. With this normalization, the quantum efficiency reads

$$IQE(\alpha_s) = \int_{x=-1/2}^{1/2} dx \int_{y=-1/2}^{1/2} dy \int_{z=0}^{w_{bas}} dz\, \eta_c(x,y,z) g(z) \tag{C.14}$$

Applying the reciprocity theorem [37] that we discuss in section 3.2 on p. 57, the collection efficiency

$$\eta_c(x,y,z) = \Delta n(x,y,z) \tag{C.15}$$

at any position (x, y, z) equals the excess minority carrier concentration $\Delta n(x,y,z)$ at this position in the dark with unity excess carrier concentration at the injecting junction. Hence, the internal quantum efficiency is

$$IQE(\alpha_s) = \int_{x=-1/2}^{1/2} dx \int_{y=-1/2}^{1/2} dy \int_{z=0}^{w_{bas}} dz\, \Delta n(x,y,z) \alpha_s \exp(-\alpha_s z) = \alpha_s \int_{z=0}^{w_{bas}} dz\, \overline{\Delta n}(z) \exp(-\alpha_s z) \tag{C.16}$$

We assume a cell with zero back surface reflectance. Integrating the carrier concentration $\overline{\Delta n}(z)$ under forward bias as given by Eq. (B.57) on p. 235, we find the internal quantum efficiency

$$IQE(\alpha_s)=\sum_{ij}\frac{\gamma_i\gamma_j}{\alpha_s^2-k_{ij}^2}\{\alpha_s^2-\beta_{ij}\alpha k_{ij}$$
$$-e^{-\alpha_s w_{bas}}\left[\alpha_s^2\cosh(k_{ij}w_{bas})+\alpha_s k_{ij}\sinh(k_{ij}w_{bas})\right]$$
$$+e^{-\alpha_s w_{bas}}\beta_{ij}\left[\alpha_s^2\sinh(k_{ij}w_{bas})+\alpha_s k_{ij}\cosh(k_{ij}w_{bas})\right]\}\qquad\text{(C.17)}$$

for which the coefficients are defined in Eqs. (B.54) through (B.56) on p. 234. This expression for the quantum efficiency of a thin-film cell applies for grains with a square cross-section and with the grain boundaries oriented perpendicularly to the junction. For the case of a thick cell ($W_{bas}/G = w_{bas} \to \infty$), we find $\beta_{ij} = 1$, and hence the expression for the quantum efficiency

$$IQE(\alpha_s)=\sum_{i,j}\frac{\gamma_i\gamma_j}{1+k_{ij}\alpha_s^{-1}}\qquad\text{(C.18)}$$

is considerably simplified.

Example

With the above result for the internal quantum efficiency of polycrystalline thin-film cells, we calculate the IQE^{-1} as a function of the optical absorption length L_α. Figure C.6 shows the inverse quantum efficiency IQE^{-1} from Eq. (C.17) for a Si thin-film cell with a thickness $W_f = 5$ µm, a diffusion length $L = 10$ µm that exceeds the thickness by a factor of two, a medium quality back surface passivation with $S_b = 10^3$ cm s^{-1}, a grain size $G =$ 5 µm that is equal to the film thickness, and a typical grain boundary recombination

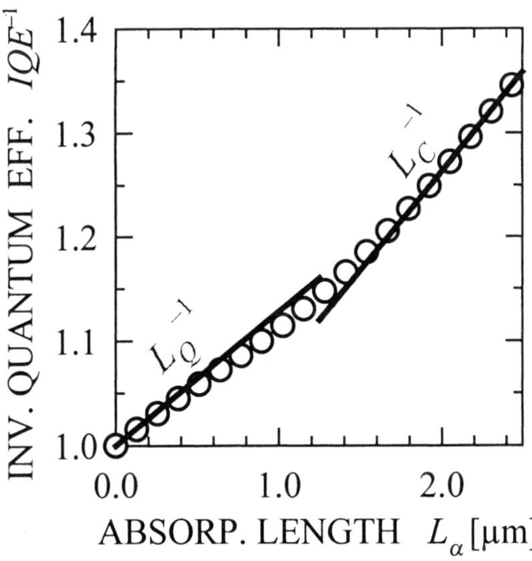

Figure C.6. Simulated inverse internal quantum efficiency of a thin-film cell of base thickness $W_{bas} = 5$ µm with grain size $G = 5$ µm, diffusion length $L = 10$ µm, back surface recombination velocity $S_b = 10^3$ cm s^{-1}, grain boundary recombination velocity $S_{grb} = 10^4$ cm s^{-1}, and diffusion coefficient $D_n = 20$ cm^2 s^{-1}.

velocity of $S_{grb} = 10^4$ cm s^{-1}. The inverse quantum efficiency IQE^{-1} is not linear in L_α. Instead, the curve $IQE^{-1}(L_\alpha)$ exhibits two linear regions, one with a smaller slope at small L_α that defines L_Q, and the other with a larger slope for large L_α that defines L_C.

For other examples the slope for large L_α may also be less than the slope at small L_α.

Quantum efficiency diffusion length L_Q for thin films

From the definition of the quantum efficiency diffusion length L_Q in Eq. (3.4) on p. 55 and the expression for the quantum efficiency (C.17), we derive

$$L_Q = \left(\sum_{i,j} \gamma_i \gamma_j \beta_{ij} k_{ij} \right)^{-1} \tag{C.19}$$

for polycrystalline thin-film cells. We proved on p. 65 that a thin-film cell with effective diffusion length L_Q has the diode saturation current $j_o = qn_oD_n / L_Q$ (recombination other than in the base being neglected). Thus, the diode saturation current of the polycrystalline thin-film cell equals that of an infinitely thick monocrystalline cell that has the base diffusion length $L = L_Q$.

Quantum efficiency diffusion length L_Q for thick films

Figure C.7 shows lines of the constant reduced effective diffusion length $l_Q = L_Q / G$ for the case of an infinitely thick cell ($W_{bas} > L$). Then, we find $\beta_{ij} = 1$, and thus

$$l_Q = \left(\sum_{i,j} \gamma_i \gamma_j k_{ij} \right)^{-1} \tag{C.20}$$

The convergence of this series is poor for large grain boundary recombination velocities in Figure C.7. Here the calculation of 2^9 terms ($i, j = 0,1,2, ...511$) is required for a precision of 1% in l_Q. Figure C.7 shows that l_Q is almost independent of the grain boundary recombination velocity S_{grb} for diffusion lengths L much smaller than grain size G, i.e. $l \ll 1$. In contrast, the effective diffusion length l_Q only depends on the grain boundary recombination velocity s_{grb} for very large intra-grain diffusion lengths $l \gg 1$. In the upper right-hand corner of Figure C.7 at $s_{grb} = l = 10^3$ we calculate an effective diffusion length $l_Q = 0.073$. The effective diffusion length L_Q is more than one order of magnitude smaller than the grain size G.

Simplified calculation of L_Q for thick cells

In this section, we derive an approximate analytical equation for l_Q since Eq. (C.20) requires some computational effort. The derivation of the approximate expression bases on the summation rule $\tau_{eff}^{-1} = \tau_{bulk}^{-1} + \tau_{grb}^{-1}$ for the effective lifetime $\tau_{eff}^{-1} := D_n / L_Q^2$, the bulk lifetime $\tau_{bulk}^{-1} = D_n / L^2$, and the grain boundary lifetime $\tau_{grb}^{-1} = D_n / L_{grb}^2$. Here, the grain boundary-controlled diffusion length L_{grb} describes the minority carrier current into the grain

$$n \frac{D_n}{L_{grb}} G^2 = n \frac{S_{grb}}{2} 4 G L_{grb} \tag{C.21}$$

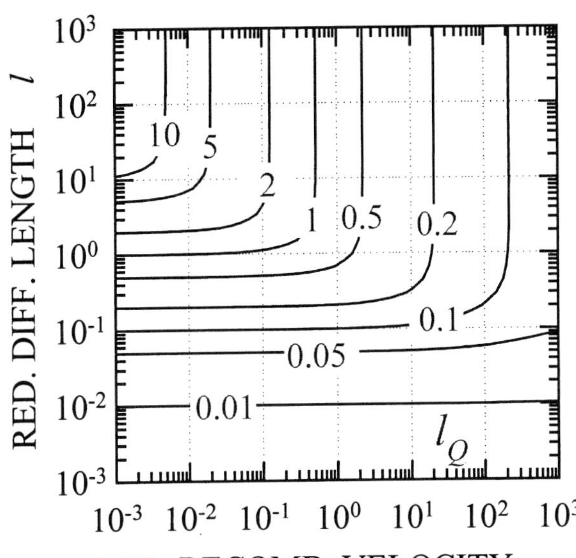

Figure C.7. Iso-lines of the effective diffusion length l_Q (from Eq. (C.20)), which depends on the grain boundary recombination velocity $s_{grb} = S_{grb}\, G/D_n$ and the intra-grain diffusion length $l = L/G$. For diffusion lengths smaller than grain size, $l < 1$, we find little influence of the recombination velocity s_{grb}.

as a recombination current into a surface of recombination velocity S_{grb} /2 and area $4\,G\,L_{grb}$. We calculate a grain boundary lifetime $\tau_{grb}^{-1} = D/L_{grb}^{2} = 2\,S_{grb}\,D_n\,/G$. With the above summation rule for the lifetimes τ_{eff}, τ_{grb}, and τ_{grb}, we find

$$\frac{1}{l_Q} = \sqrt{\left(\frac{1}{l}\right)^2 + 2s_{grb}} \tag{C.22}$$

This relation also follows immediately from Eq. (C.20) by expansion to the first order in s_{grb}. Equation (C.22) is thus accurate only for small s_{grb}. To account for the limiting behavior we found for large s_{grb}, we use the expression

$$\frac{1}{l_Q} = \sqrt{\left(\frac{1}{l}\right)^2 + \left(\frac{1}{(2s_{grb})^{-1/2} + 0.072}\right)^2} \tag{C.23}$$

Comparison of l_Q from Eqs. (C.20) and (C.23) reveals an agreement better than 30% in the range $l < 10^3$ and $s_{grb} < 10^2$.

Simplified calculation of L_Q for thin cells

Equation (C.22) holds only for thick cells. In order to find a simple analytic expression for L_Q that also holds for thin cells with the bulk diffusion length L, the grain boundary recombination velocity S_{grb}, the back surface recombination velocity S_b, the base thickness W_{bas}, and grain size G, we first assume that the grain boundary recombination velocity is $S_{grb} = 0$. This cell has a quantum efficiency diffusion length $L_{Q,mono}$ given by Eq. (C.10). Now consider an infinitely thick polycrystalline cell that has an intra-grain diffusion length equal to $L_{Q,mono}$. Let this thick cell have the grain boundary recombination velocity S_{grb} of the thin polycrystalline cell. We may then use the approximate expression (C.23). We thus find an easy-to-handle, approximate expression

$$L_Q = \left(\left(L \frac{S_b L \sinh(W_{bas}/L) + D_n \cosh(W_{bas}/L)}{S_b L \cosh(W_{bas}/L) + D_n \sinh(W_{bas}/L)} \right)^{-2} + \left(\left(\frac{G D_n}{2 S_{grb}} \right)^{1/2} + 0.072\ G \right)^{-2} \right)^{-1/2} \quad (C.24)$$

for the quantum efficiency diffusion length L_Q of polycrystalline thin-film cells. Please note that the derivation of the expression assumes an excess carrier concentration at the junction that is spatially constant. For large S_{grb} this assumption may become incorrect.

Collection diffusion length L_C for thin films

From the defining Eq. (3.5) for L_C and from the quantum efficiency expressed in Eq. (C.17), we erive the collection diffusion length

$$L_C = \sum_{i,j} \gamma_i \gamma_j \frac{1}{k_{ij}} \left[\beta_{ij} \left(1 - \cosh(k_{ij} w_{bas})\right) + \sinh(k_{ij} w_{bas}) \right] \quad (C.25)$$

for polycrystalline thin-film solar cells. This expression also holds for cells with light trapping and no parasitic absorption (absorption that does not create electron-hole pairs), since we use the *internal* quantum efficiency to define L_C. In general, solar cells exhibit parasitic absorption for very weakly absorbed light, and thus the value of L_C depends in practice on the optics of the device.

The physical interpretation of L_C is as follows: the collection diffusion length L_C is the diffusion length that a thick monocrystalline cell has to have in order to yield the same short-circuit current density as a thin-film cell with no light trapping under spatially homogeneous photogeneration. Evidence for this interpretation of L_C was first given for the special case of an infinitely thick cell with either zero or infinite grain boundary recombination by Donolato [412].

Collection diffusion length L_C for thick films

For thick cells, which means base layer thickness $W_{bas} >> L$, Eq. (C.22) simplifies to

$$l_C = \sum_{i,j} \gamma_i \gamma_i \frac{1}{k_{ij}} \quad (C.26)$$

We neglect the emitter here. This series for l_C converges much faster than the series for l_Q and 2^5 terms are sufficient for a numerical precision better than 1% in the parameter range $l < 10^3$ and $s_{grb} < 10^3$.

Iso-lines of the reduced collection diffusion length $l_C = L_C/G$ are shown in Figure C.8. The behavior of l_C is similar to that of l_Q in Figure C.7. However, in the upper right-hand corner of Figure C.8 at $s_{grb} = 10^3$ and $l = 10^3$ we find $l_C = 0.172$, a value 2.4 times greater than l_Q. For the effective diffusion length l_C we find numerically the limiting value

$$l_{C\infty} = \lim_{l, s_{grb} \to \infty} \left\{ l_C(l, s_{grb}) \right\} = 0.170 \quad (C.27)$$

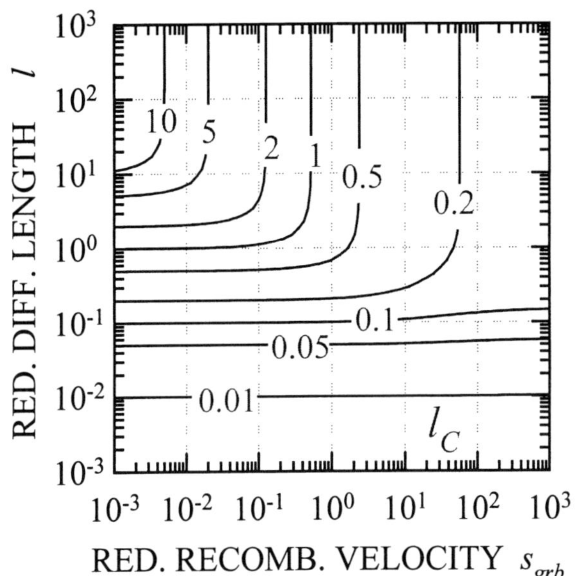

Figure C.8. Iso-lines of the effective diffusion length l_C, which depends on the grain boundary recombination velocity $s_{grb} = S_{grb}\,G/D$ and the intra-grain diffusion length $l = L/G$. The behavior is similar to that of l_C depicted in Figure C.7. However, the numerical values are different for large s_{grb} and large l.

We first reported this limiting value without a derivation in Eq. 4 of Ref. [197], a result that was confirmed by Donolato [412]. A polycrystalline material with grain size G and negligible volume recombination (that is, $l = \infty$), and a worst-case grain boundary recombination (that is, $s_{grb} = \infty$), collects current from weakly absorbed light as well as a material with a volume diffusion length $L = 0.17\ G$ and no grain boundary recombination.

Simplified calculation of L_C for thick cells

A slight modification of the empirical relation (C.22)

$$\frac{1}{l_C} = \sqrt{\left(\frac{1}{l}\right)^2 + \left(\frac{1}{\left(2s_{grb}\right)^{-1/2} + 0.17}\right)^2} \tag{C.28}$$

accounts for the finite limit $l_{C\infty} = 0.17$. Comparison of l_C from Eqs. (C.28) and (C.26) reveals a precision better than 10% in the range $l < 10^3$ and $s_{grb} < 10^3$.

Simplified calculation of L_C for thin cells

If we now substitute $l = L_{C,mono}/G$ from Eq. (C.12) on p. 247 to find an easy-to-handle expression for the effective carrier collection length

$$L_C = \left(\left(L\,\frac{S_b L\left(\cosh(W_{bas}/L)-1\right) + D_n \sinh(W_{bas}/L)}{S_b L \sinh(W_{bas}/L) + D_n \cosh(W_{bas}/L)}\right)^{-2} + \left(\left(\frac{G\,D_n}{2\,S_{grb}}\right)^{1/2} + 0.17\,G\right)^{-2}\right)^{-1/2} \tag{C.29}$$

of polycrystalline thin-film cells. As noted in the corresponding section on L_Q, the applicability of this equation may be limited to small values of S_{grb}.

C.4 Laplace transform of quantum efficiency spectra

The local carrier collection probability $\eta_c(Z)$ is defined as the probability that a carrier generated at the depth Z is collected at the junction of a short-circuited cell. In this section, we assume a planar cell of thickness W_f with one-dimensional transport and the cell surface located at $Z = 0$. In contrast to the previous section, the thickness of the emitter W_e and the space charge region W_{scr} are not neglected and $W_f = W_e + W_{scr} + W_{bas}$. For a zero back surface reflectance of a cell with thickness W_f, the internal quantum efficiency

$$IQE = \int_{Z=0}^{W_f} \eta_c(Z)\,\alpha_s \exp(-\alpha_s Z)dZ = \alpha_s \mathrm{L}\{\eta_c(Z)\} \tag{C.30}$$

is the Laplace transform $\mathrm{L}\{\eta_c(Z)\}$ of the local carrier collection efficiency $\eta_c(Z)$ [209]. Sinkkonen et al. suggested that the local carrier collection efficiency

$$\eta_c(Z) = \int_{c-i\infty}^{c+i\infty} \frac{IQE(\alpha_s)}{\alpha_s} d\alpha_s = L^{-1}\left\{\frac{IQE(\alpha_s)}{\alpha_s}\right\} \tag{C.31}$$

could be determined from the inverse Laplace transform of experimental data $IQE(\alpha_s)$ and α_s. For the inverse Laplace transformation the quantum efficiency $IQE(\alpha_s(\lambda))$ is interpreted as a function of the absorption coefficient α_s. Unfortunately, the inverse Laplace transformation is numerically unstable. Therefore, the experimental data

$$\frac{IQE(\alpha_s)}{\alpha_s} = \frac{P(\alpha_s)}{Q(\alpha_s)} \tag{C.32}$$

are first fitted by the ratio of two polynomials $P(\alpha_s)$ and $Q(\alpha_s)$ with the inverse Laplace transform

$$\eta(z) = \sum_i \frac{P(\beta_i)}{Q'(\beta_i)} \exp(\beta_i z) \tag{C.33}$$

calculated by the Heaviside expansion theorem [362, 413]. The β_i are the roots of the polynomials $Q(\alpha_s)$.

Figure C.9 shows the measured internal quantum efficiency of a crystalline Si solar cell with a film thickness $W_f = 150$ µm. The experimental quantum efficiency data (circles) as well as the fit to the experimental data with the test function $\alpha_s P(\alpha_s)/Q(\alpha_s)$ are taken from Ref. [413]. The inverse Laplace transform of the test function yields the local carrier collection efficiency that is shown in Figure C.10. The maximum of the collection efficiency $\eta_c(Z)$ indicates the depth of the junction, which is around $Z = 0.4$ µm here.

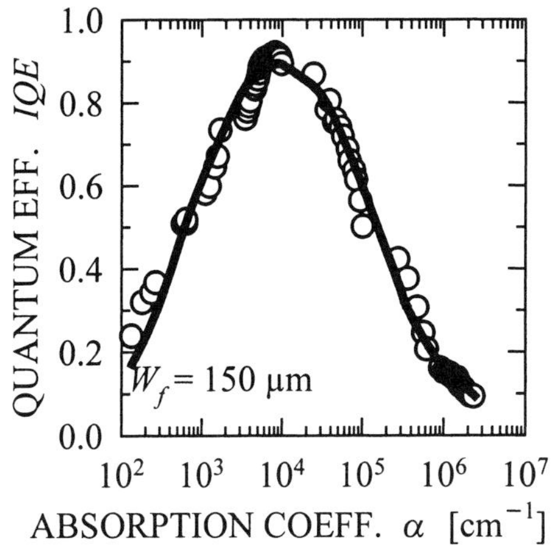

Figure C.9. Internal quantum efficiency *IQE* of a W_f = 150 μm-thick crystalline Si solar cell (circles) and the fitted function $P(\alpha)/Q(\alpha)$ (solid line). Data from Ref. [413].

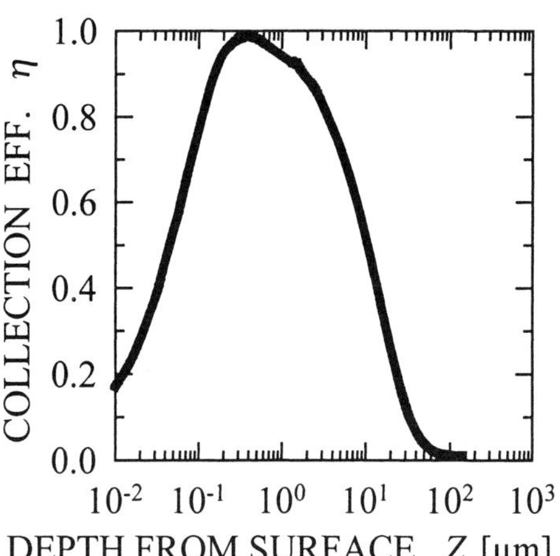

Figure C.10. Local collection efficiency determined by the inverse Laplace transform of the IQE shown in Figure C.9. Data from Ref. [413].

For the extraction of recombination parameters in the emitter, fitting of the local carrier collection efficiency in Figure C.10 was suggested with the analytical expression for the collection efficiency

$$\eta_c(Z) = \frac{K_e \exp(Z/L_e) + \exp(-Z/L_e)}{K_e \exp(W_e/L_e) + \exp(-W_e/L_e)} \quad \text{for } 0 \le Z \le W_e \tag{C.34}$$

Tha latter results from the solution of the diffusion equation in the emitter [209]. Within the space charge region of thickness W_{scr}, unity collection efficiency

$$\eta_c(Z) = 1 \quad \text{for } W_e \le Z \le W_e + W_{scr} \tag{C.35}$$

is assumed. The recombination parameters in the base are deduced from fitting the collection efficiency

$$\eta_c(Z) = \frac{K_{bas} \exp((W_f - Z)/L_{bas}) + \exp(-(W_f - Z)/L_{bas})}{K_{bas} \exp(W_{bas}/L_{bas}) + \exp(-W_{bas}/L_{bas})} \quad \text{for } W_e + W_{scr} \le Z \le W_f \tag{C.36}$$

to the data in Figure C.10. The coefficients

$$K_{e,bas} = \frac{D_{e,bas} + S_{e,bas} L_{e,bas}}{D_{e,bas} - S_{e,bas} L_{e,bas}} \tag{C.37}$$

depend on the minority carrier diffusion length $D_{e,bas}$, the surface recombination velocity $S_{e,bas}$, and the minority carrier diffusion length $L_{e,bas}$ in the emitter and the base, respectively. The result of the fit for our example is given in Ref. [413].

The authors of Refs. [209, 362, 413] claim that fitting the local carrier collection efficiency has advantages over the direct fitting of the internal quantum efficiency. Their Laplace technique does, however, include two fitting procedures: fitting the *IQE* data with a test function and fitting the theoretical collection efficiency η_c to the Laplace transform of the test function. The ambiguity in choosing an appropriate test function is fully avoided by directly fitting the *IQE* data with an analytical solution of the diffusion equation. In addition, for thin cells with light trapping, the carrier generation profile is not a single exponential function, and the basic Eq. (C.30) underlying this Laplace transform analysis is no longer valid [414].

In this work, we therefore prefer a direct fitting of the internal quantum efficiency *IQE* with appropriate analytical models to determine the recombination parameters. The local carrier collection efficiency is then given by Eqs. (C.34) to (C.36).

C.5 Effective grain size for log-normal grain size distribution

Polycrystalline cells consist of grains with different sizes. Figure 2.31 on p. 49 shows the measured log-normal grain size distribution that is well described by the function [200, 201]

$$P(G) = \frac{1}{(2\pi)^{1/2}\sigma G}\exp\left(-\frac{1}{2}\left(\frac{\ln(G/\bar{G})}{\sigma}+\frac{\sigma}{2}\right)^2\right) \tag{C.38}$$

which that describes a Gaussian distribution of the logarithm of the grain size G. The average grain size is $\bar{G}$, and σ is the width parameter of this distribution. For simulations we would not want to simulate many grains with various grain sizes. The question we address here is: what is an appropriate choice for an effective grain size G_{eff} that we choose to represent an ensemble of grains with a log-normal size distribution? To answer this question we proceed in two steps. First, we calculate the effective quantum efficiency diffusion length $L_{Q,logn}$ for a log-normal grain size distribution. We tactically assume that all grains have the same intra-grain diffusion length L, diffusion coefficient D_n, and grain boundary recombination velocity S_{grb}. In a second step, we calculate the effective grain sizes G_{eff} that yields a quantum efficiency diffusion length $L_Q(G_{eff}) = L_{Q,logn}$. As a result, we express $G_{eff}(\sigma, \bar{G})$ as a function of the parameters of the log-normal distribution.

To calculate $L_{Q,logn}$, we first study the log-normal distribution. A variable substitution for $\ln(G / \bar{G})$ yields the nth moment

$$\langle G^n \rangle = \int_0^\infty G^n P(G)\, dG = \bar{G}^n \exp\left(\frac{\sigma^2}{2}(n^2 - n)\right) \tag{C.39}$$

of the distribution $P(G)$. The symbol < > denotes an average over the log-normal distribution. The first moment is the average grain size is $\langle G \rangle = \bar{G}$. The area-weighted average grain size is

$$G_{aw} = \langle G^3 \rangle / \langle G^2 \rangle = \bar{G}\exp(2\,\sigma^2) \tag{C.40}$$

The log-normal-averaged diffusion length is

$$\frac{1}{L_{Q,logn}} = \frac{\langle G^2/L_Q(G)\rangle}{\langle G^2\rangle} \tag{C.41}$$

We average the area-weighted $1/L_{Q,logn}$, since the diode saturation current $j_o = (q\, n_o\, D_n)/L_{Q,logn}$ is also proportional to $1/L_{Q,logn}$ and since the factor G^2 ensures that the diode saturation currents from various grains add up to the total saturation current. We assume all grains to be independent of each other. Carrier injection from one grain across a grain boundary into a neighboring grain is neglected.

The effective grain size G_{eff} we are looking for fulfills the condition $L_Q(G_{eff}) = L_{Q,logn}$, with $L_Q(G_{eff})$ from Eq. (C.22) and $L_{Q,logn}$ from Eq. (C.41). In Eq. (C.22), the grain size enters via the second term s_{grb} only. We therefore first consider a grain without volume recombination ($l = \infty$). We find from Eq. (C.22)

$$\frac{1}{L_Q(G)} = \sqrt{\frac{S_{grb}}{D_n G}} \tag{C.42}$$

From Eqs. (C.41) and (C.42) we calculate the effective diffusion length

$$L_{Q,logn} = \frac{\langle G^2\rangle}{\langle G^2/L_Q\rangle} = \sqrt{\frac{D_n}{S_{grb}}}\frac{\langle G^2\rangle}{\langle G^{3/2}\rangle} = \sqrt{\frac{D_n \overline{G}\exp(5\sigma^2/4)}{S_{grb}}} \tag{C.43}$$

for a log-normal grain size distribution. Thus, the effective grain size is

$$G_{eff} = \left(\frac{\langle G^2\rangle}{\langle G^{3/2}\rangle}\right)^2 = \overline{G}\exp(5\sigma^2/4) \tag{C.44}$$

which we may also express as

$$G_{eff} = \overline{G}^{3/8}\overline{G}_{aw}^{\;5/8} \tag{C.45}$$

We can thence deduce G_{eff} from the average grain size $\overline{G}$ and the area-weighted average size G_{aw} of an experimental grain size distribution.

We checked Eq. (C.44) numerically for many cases ($\sigma \in [0.1, 2]$, $L \in [0.1\ \mu m, 10^3\ \mu m]$, $s_{grb}/D_n \in [0, 10^3\ \mu m^{-1}]$, $\overline{G} = 1\ \mu m$,) including cases of non-negligible volume recombination. In all these cases we find that a cell with periodic grain structure of grain size G_{eff} has the same diode saturation current as a cell with non-interacting grains that have a log-normal grain size distribution that is characterized by $\overline{G}$ and σ.

References

[1] J. Zhao, A. Wang, and M. A. Green, *24.5% efficiency silicon PERT cells on MCZ substrates and 24.7% efficiency PERL cells on FZ substrates*, Progress in Photovoltaics **7**, 471 (1999).

[2] P. Campbell and M. A. Green, *Light trapping properties of pyramidally textured surfaces*, J. Appl. Phys. **62**, 243 (1987).

[3] M. A. Green, J. Zhao, Q. Wang, and S. R. Wenham, *Progess and outlook for high-eficiency crystalline Si solar cells*, in *Technical Digest 11th International Photovoltaic Science and Engineering Conf.* (Int. PVSEC Publishing, Kyoto, 1999), p. 21.

[4] M. Schöfthaler, U. Rau, W. Füssel, and J. H. Werner, *Optimization of the back contact geometry for high efficiency solar cells*, in *Proc. 23rd IEEE Photovoltaic Specialists Conf.* (IEEE, New York, 1993), p. 315.

[5] M. A. Green, *Silicon Solar Cells: Advanced Principles & Practice* (Center for Photovoltaic Devices and Systems, Sydney, 1995), p. 333.

[6] G. Heiser, A. G. Aberle, S. R. Wenham, and M. A. Green, *Two-dimensional numerical simulations of high-efficiency silicon solar cells*, Microelectronics Journal **26**, 273 (1995).

[7] A. G. Aberle, P. P. Altermatt, G. Heiser, S. J. Robinson, A. Wang, J. Zhao, U. Krumbein, and M. A. Green, *Limiting loss-mechanisms in 23% efficient silicon solar cells,* J. Appl. Phys. **77**, 3491 (1995).

[8] J. P. Kalejs and W. Schmidt, *High productivity methods of preparation of EFG ribbon silicon wafers*, in *Proc. 2nd World Conf. Photovoltaic Solar Energy Conf.*, edited by J. Schmid, H. A. Ossenbrink, P. Helm, H. Ehmann, and E. D. Dunlop (Joint Research Center European Commission, Ispra, 1998), p. 1822.

[9] R. Hezel, R. Meyer, and A. Metz, *A new generation of crystalline Si solar cells: simple processing and record efficiencies for industrial-size devices*, in *Technical Digest 11th International Photovoltaic Science and Engineering Conf.* (Int. PVSEC Publishing, Kyoto, 1999), p. 913.

[10] R. Hezel, Ch. Schmiga, and A. Metz, *Next generation of industrial silicon solar cells with efficiencies above 20%*, in *Proc. 28th IEEE Photovoltaic Specialists Conf.* (IEEE, New York, 2000), p.184.

[11] A. Metz and R. Hezel, *High quality passivated rear contact structure for silicon solar cells based on simple mechanical abrasion*, in *Proc. 28th IEEE Photovoltaic Specialists Conf.* (IEEE, New York, 2000), p. 172.

[12] S. Glunz, R. Preu, S. Schaefer, E. Schneiderlöchner, W. Pfleging, R. Lüdemann, and G. Willeke, *New simplified methods for patterning the rear contact of RP-PERC high-efficiency solar cells*, in *Proc. 28th IEEE Photovoltaic Specialists Conf.* (IEEE, New York, 2000), p. 168.

[13] R. Hulstrom, R. Bird, and C. Riordan, *Spectral solar irradiance data sets for selected terrestrial conditions*, Solar Cells **15**, 365 (1985).

[14] W. Shockley and H. J. Queisser, *Detailed balance limit of efficiency of p-n junction solar cells*, J. Appl. Phys. **32,** 510 (1961).

[15] A. Martí and G. L. Araújo, *Limiting efficiency for photovoltaic energy conversion in multigap systems*, Solar Energy Materials and Solar Cells **43**, 203 (1996).

[16] J. C. Miñano, *Optical confinement in photovoltaics*, in Physical Limitations to Photovoltaic Energy Conversion, edited by A. Luque and G. L. Araùjo (Adam Hilger, Bristol, 1990), p. 50.

[17] J. H. Werner, R. Bergmann, and R. Brendel, *The challenge of crystalline thin film silicon solar cells*, in *Festkörperprobleme/Advances in Solid State Physics*, Vol. 34, edited by R. Helbig (Vieweg, Braunschweig, 1994), p. 115.

[18] R. Brendel, J. H. Werner, and H. J. Queisser, *Thermodynamic efficiency limits for semiconductor solar cells with carrier multiplication*, Solar Energy Materials and Solar Cells **41/42**, 419 (1996).

[19] T. Tiedje, E. Yablonovitch, G. D. Cody, and B. G. Brooks, *Limiting efficiency of silicon solar cells*, IEEE Trans. Electron. Devices **ED-31**, 711 (1984).

[20] M. A. Green, *Limiting efficiency of bulk and thin-film silicon solar cells in the presence of surface recombination*, Progress in Photovoltaics **7**, 327 (1999).

[21] P. P. Altermatt, J. Scmidt, G. Heiser, and A. G. Aberle, *Assessment and parameterization of Coulomb-enhanced Auger recombination coefficients in lowly injected crystalline Si*, J. Appl. Phys. **82**, 4938 (1997).

[22] M. A. Green and M. Keevers, *Optical properties of intrinsic silicon at 300 K*, Progress in Photovoltaics **3**, 189 (1995).

[23] M. A. Green, *Intrinsic concentration, effective density of states, and effective mass in silicon*, J. Appl. Phys. **67**, 2944 (1990).

[24] P. A. Basore, *Extended spectral analysis of the internal quantum efficiency*, in *Proc. 23rd IEEE Photovoltaic Specialists Conf.* (IEEE, New York, 1993), p. 147.

[25] R. Brendel, M. Hirsch, R. Plieninger, and J. H. Werner, *Quantum efficiency analysis of thin layer silicon solar cells*, IEEE Trans. Electron. Devices **ED-43**, 1104 (1996).

[26] R. Brendel, *SUNRAYS: A versatile solar cell ray tracing program for the photovoltaic community*, in *Proc. 12th European Photovoltaic and Solar Energy Conf.*, edited by R. Hill, W. Palz, and P. Helm (H. S. Stephens, Bedford, 1994), p. 1339.

[27] R. Brendel, *Optical design of crystalline thin layer silicon solar cells on glass*, in *Proc. 13th European Photovoltaic Solar Energy Conf.*, edited W. Freiesleben, W. Palz, H. A. Ossenbrink, and P. Helm (H. S. Stephens, Bedford, 1995), p. 436.

[28] R. Brendel, R. B. Bergmann, P. Lölgen, M. Wolf, and J. H. Werner, *Ultrathin crystalline silicon solar cells on glass substrates*, Appl. Phys. Lett. **70**, 390 (1997).

[29] J. Zettner, M. Thönissen, T. Hierl, R. Brendel, and M. Schulz, *Novel porous silicon backside light reflector for thin silicon solar cells*, Progress in Photovoltaics **6**, 423 (1998).

[30] R. Brendel, *Note on the interpretation of injection level dependent surface recombination velocities*, Appl. Phys. A **60**, 523 (1995).

[31] R. Brendel and M. Wolf, *Differential and actual surface recombination velocities*, in *Proc. 13th European Photovoltaic Solar Energy Conf.*, edited by W. Freiesleben, W. Palz, H. A. Ossenbrink, and P. Helm (H. S. Stephens, Bedford, 1995), p. 428.

[32] A. G. Aberle, J. Schmidt, and R. Brendel, *On the data-analysis of light-biased photo conductance decay measurements*, J. Appl. Phys. **79**, 1491 (1996).

[33] R. Brendel and U. Rau, *Injection and collection diffusion lengths of polycrystalline thin-film solar cells*, in *Polycrystalline Semiconductors V – Bulk Materials, Thin Films, and Devices*, edited by J. H. Werner, H. P. Strunk, and H. W. Schock (ScitechPubl., Uetticon am See, Switzerland, 1999), p. 81.

[34] J. Dugas, *3D modelling of a reverse cell made with improved multicrystalline silicon wafers*, Solar Energy Materials and Solar Cells **32**, 71 (1994).

[35] U. Rau and J. H. Werner, *An analytical model for rectifying contacts on polycrystalline semiconductors*, in *Polycrystalline Semiconductors V – Bulk Materials, Thin Films, and Devices*, edited by J. H. Werner, H. P. Strunk, and H. W. Schock (ScitechPubl., Uetticon am See, Switzerland, 1999), p. 553.

[36] N. D. Arora, S. G. Chamberlain, and D. J. Roulston, *Diffusion length determination in p-n junction diodes and solar cells*, Appl. Phys. Lett. **37**, 325 (1980).

[37] C. Donolato, *A reciprocity theorem for charge collection*, Appl. Phys. Lett. **46**, 270 (1985).

[38] K. Misiakos and F. A. Lindholm, *Generalized reciprocity theorem for semiconductor devices*, J. Appl. Phys. **58**, 4743 (1985)

[39] C. Donolato, *An alternative proof of the generalized reciprocity theorem for charge collection*, J. Appl. Phys. **66**, 4524 (1989).

[40] T. Markvart, *Relationship between dark carrier distribution and photogenerated carrier collection in solar cells*, IEEE Trans. Electron. Devices **ED-43**, 1034 (1996).

[41] M. A. Green, *Generalized relationship between dark carrier distribution and photocarrier collection in solar cells*, J. Appl. Phys. **81**, 269 (1997).

[42] U. Rau and R. Brendel, *The detailed balance principle and the reciprocity theorem between photocarrier collection and dark carrier distribution in solar cells*, J. Appl. Phys. **81**, 6412 (1998).

[43] W. Shockley, M. Sparks, and G. K. Teal, *p-n junction transistors*, Phys. Rev. **83**, 151 (1951).

[44] L. Onsager, *Reciprocal relations in irreversible processes I*, Phys. Rev. **37**, 405 (1931); *Reciprocal relations in irreversible processes II*, Phys. Rev. **38**, 2265 (1931).

[45] M. Schöfthaler, R. Brendel, G. Langguth, and J. H. Werner, *High-quality surface passivation by corona-charged oxides for semiconductor surface characterization*, in *Proc. 1st World Conf. Photovoltaic Energy Conversion* (IEEE, New York, 1994), p. 1509.

[46] T. Baba, M. Shima, T. Matsuyama, S. Tsuge, K. Wakisaka, and S. Tsuda, *9.2% efficiency thin-film polycrystalline silicon solar cell by a novel solid phase crystallization method*, in *Proc. 13th European Photovoltaic Solar Energy Conf.*, edited by W. Freiersleben, W. Palz, H. A. Ossenbrink, and P. Helm (H. S. Stephens, Bedford, 1995), p. 1708.

[47] R. Auer, J. Zettner, J. Krinke, G. Polisski, T. Hierl, R. Hezel, M. Schulz, H.-P. Strunk, F. Koch, D. Nikl, and H. v. Campe, *Improved performance of thin-film silicon solar cells on graphite substrates*, in *Proc. 26th IEEE Photovoltaic Specialists Conf.* (IEEE, New York, 1997), p. 739.

[48] C. Hebling, S. Reber, K. Schmidt, R. Lüdemann, and F. Lutz, *Oriented recrystallization of Si layers for silicon thin-film solar cells*, in *Proc. 26th*

IEEE Photovoltaic Specialists Conf. (IEEE, New York, 1997), p. 623.

[49] K. Yamamoto, M. Yoshimi, T. Suzuki, Y. Okamoto, Y. Tawada, and A Nakajima, *Thin film poly-Si solar cell with "Star Structure" on glass substrate fabricated at low temperature*, in *Proc. 26th IEEE Photovoltaic Specialists Conf.* (IEEE, New York, 1997), p. 575.

[50] R. W. McClelland, C. O. Bolzer, and C. J. J. Fan, *A technique for producing epitaxial films on re-usable substrates*, Appl. Phys. Lett. **37**, 560 (1980).

[51] T. Yonehara, K. Sakaguchi, and N. Sato, *Epitaxial layer transfer by bond and etch back of porous Si*, Appl. Phys. Lett. **64**, 2108 (1994).

[52] H. Tayanaka and T. Matsushita, in *Proceedings of the 6th Sony Research Forum* (Sony, 1996), p. 556 (in Japanese).

[53] R. Brendel, *A novel process for ultrathin monocrystalline silicon solar cells on glass*, in *Proc. 14th European Photovoltaic Solar Energy Conf.*, edited by H. A. Ossenbrink, P. Helm, and H. Ehmann (H. S. Stephens, Bedford, 1997), p. 1354.

[54] W. Ruppel and P. Würfel, *Upper limit for conversion of solar cells*, IEEE Trans. Electron. Devices **ED-27**, 877 (1980).

[55] P. Würfel, *The chemical potential of radiation*, J. Phys. C: Solid State Phys. **15**, 3967 (1982). K. Schick, E. Daub, S. Finkbeiner, *Verification of a generalized Planck Law for luminescence radiation from silicon solar cells*, Appl. Phys. A **54**, 109 (1992).

[56] M. A. Green, *Limits on the open circuit voltage and efficiency of solar silicon cells by intrinsic Auger process*, IEEE Trans. Electron. Devices **ED-31** (1984) 671.

[57] S. Deb and H. Saha, *Secondary ionization and its possible bearing on the performance of a solar cell*, Solid-State Electron. **15**, 1389 (1972).

[58] V. S. Vavilov and K. I. Britsyn, *Quantum yield of photoionization in silicon,* Sov. Phys. – JETP **7**, 935 (1958).

[59] S. Kolodinski, J. H. Werner, T. Wittchen, and H. J. Queisser, *Quantum efficiencies exceeding unity due to impact ionization in silicon solar cells*, Appl. Phys. Lett. **63**, 2405 (1993).

[60] P. T. Landsberg, H. Nussbaumer, and G. Willeke, *Band-band impact ionization and solar cell efficiency*, J. Appl. Phys. **74,** 1451 (1993).

[61] S. Kolodinski, J. H. Werner, and H. J. Queisser, *Quantum efficiencies exceeding unity in silicon leading to novel selection principles for solar cell materials*, Sol. En. Mat. Sol. Cells **33,** 275 (1994).

[62] H. Kiess and W. Rehwald, *On the ultimate efficiency of solar cells*, Solar Energy Mater. Solar Cells **38**, 45 (1995).

[63] J. H. Werner, S. Kolodinski, and H. J. Queisser, *Novel optimisation principles and efficiency limits for semiconductor solar cells*, Phys. Rev. Lett. **72,** 3851 (1994).

[64] W. Spirkl and H. Ries, *Luminescence and efficiency of an ideal photovoltaic cell with charge carrier multiplication*, Phys. Rev. B **52**, 11319 (1995).

[65] J. K. Liakos and P. T. Landsberg, *Auger recombination and impact ionization in heterojunction photovoltaic cells*, Semicond. Sci. Technol. **11**, 1895 (1996).

[66] J. H. Werner, R. Brendel, and H. J. Queisser, *New upper efficiency limits for semiconductor solar cells*, in *Proc. 1st World Conf. Photovoltaic Energy Conversion* (IEEE. New York, 1994), p. 1747.

[67] J. H. Werner, R. Brendel, and H. J. Queisser, *Radiative efficiency limit of*

terrestrial solar cells with internal carrier multiplication, Appl. Phys. Lett. **67**, 1028 (1995) .

[68] A. Luque and A. Marti, *Entropy production in photovoltaic conversion*, Phys. Rev. B **55**, 6994 (1997).

[69] H. Ries and W. Spirkl, *A generalized Kirchoff law for quantum absorption and luminescence*, Solar Energy Material Solar Cells **38**, 39 (1995).

[70] A. B. Sproul, M. A. Green, *Intrinsic carrier concentration and minority carrier mobility of silicon from 77 to 300 K*, J. Appl. Phys. **73**, 1214 (1993).

[71] J. Schmidt, M. Kerr, and P. P. Altermatt, *Coulomb-enhanced Auger recombination in crystalline silicon at intermediate and high-injection densities*, J. Appl. Phys. **88**, 1494 (2000).

[72] A. De Vos, *Endoreversible Thermodynamics of Solar Energy Conversion* (Oxford University Press, Oxford, 1992).

[73] G. L. Araujo, *Limis to efficiency of single and multiple bandgap solar cells*, in *Physical Limitations to Photovoltaic Energy Conversion*, edited by A. Luque and G. L. Araujo (Adam Hilger, Bristol, 1990), p.106.

[74] G. L. Araujo and A. Marti, *Absolute limiting efficiencies for photovoltaic energy conversion*, Solar Energy Materials and Solar Cells **33**, 213 (1994).

[75] A. Luque, *Solar Cells and Optics for Photovoltaic Concentration* (Adam Hilger, Bristol, 1989), p. 309.

[76] M. J. Keevers and M. A. Green, *Absorption edge of silicon from solar cell spectral response measurements*, Appl. Phys. Lett. **66**, 174 (1995).

[77] D. K. Schroder, R. N. Thomas, and J. C. Swartz, *Free carrier absorption in Silicon*, IEEE Trans. Electron. Devices **ED-25**, 254 (1978).

[78] E. Daub and P. Würfel, *Ultralow values of the absorption coefficient of Si obtained from luminescence*, Phys. Rev. Lett. **74**, 1020 (1995).

[79] S. Sze, *Physics of Semiconductor Devices*, 2nd edition (Wiley & Sons, New York, 1981), Eq. (15), p. 87 and p. 850.

[80] B. L. Sopori and R. A. Pryor, *Design of antireflection coatings for textured silicon solar cells*, Solar Cells **8**, 249 (1983).

[81] J. Zhao and M. A. Green, *Optimized antireflection coatings for high efficiency silicon solar cells*, IEEE Trans. Electron. Devices **ED-38**, 1925 (1991).

[82] G. F. Zheng, J. Zhao, M. Gross, and E. Chen, *Very low light-reflection from the surface of incidence of a silicon solar cell,* Solar Energy Materials and Solar Cells **40**, 89 (1996).

[83] L. Schirone, G. Sotgiu, and F. P. Califano, *Chemically etched porous Si as an antireflection coating for high efficiency Si solar cells*, Thin Solid Films **297**, 296 (1997).

[84] G. A. Landis, *A light trapping solar cell coverglass*, in *Proc. 21st IEEE Photovoltaic Specialists Conf.* (IEEE, New York, 1990), p. 1304.

[85] J. M. Gee, R. Gordon, and H. Liang, *Optimization of textured-dielectric coatings for crystalline-silicon solar cells*, in *Proc. 25th IEEE Photovoltaic Specialists Conf.* (IEEE, New York, 1996), p. 733.

[86] P. Campbell, S. R. Wenham, and M. A. Green, *Improved reflection and light trapping using tilted pyramids and grooves*, in *Proc. 4th Int. Photovoltaic Science Engineering Conf. in Sydney* (1989), p. 615.

[87] T. Nunoni, S. Okamoto, K. Nakajima, S. Tanaka, N. Shibuya, K. Okamoto, T. Nammori, and H. Itho, *Cast polycrystalline silicon solar cell with grooved surface,* in *Proc. 21st IEEE Photovoltaic Specialists Conf.* (IEEE, New York,

1990), p. 664.

[88] A. Scheydecker, A. Goetzberger, and V. Wittwer, *Reduction of reflection losses by structured surfaces,* in *Proc. 10th European Photovoltaic Solar Energy Conf.*, edited by A. Luque, G. Sala,W. Palz, G. D. Santos, and P. Helm (Kluwer, Dordrecht, 1992), p. 39.

[89] P. Campbell, S. R. Wenham, and M. A. Green, *Light trapping and reflection control in solar cells using tilted crystallographic surface textures*, Solar Energy Materials and Solar Cells **31**, 133 (1993).

[90] G. Willeke, H. Nussbaumer, H. Bender, and E. Bucher, *A simple and effective light trapping technique for polycrystalline silicon solar cells*, Solar Energy Materials and Solar Cells **26**, 345 (1992).

[91] H. Nussbaumer, G. Willeke, and E. Bucher, *Optical behaviour of textured silicon*, J. Appl. Phys. **75**, 2202 (1994).

[92] A. Poruba, Z. Remes, J. Springer, M. Vanecek, A. Feijfar, J. Kocka, J. Meier, P. Torres, and A. Shah, *Light scattering in microcrystalline silicon thin film cells*, in *Proc. 2nd World Conf. Photovoltaic Solar Energy Conf.*, edited by J. Schmid, H. A. Ossenbrink, P. Helm, H. Ehmann, and E. D. Dunlop (Joint Research Center European Commission, Ispra, 1998), p. 781.

[93] R. Shimokawa, T. Takahashi, H. Takato, A. Ozaki, and Y. Takano, *2 μm thin film c-Si cells on near-Lambertian Al_2O_3 substrates*, in *Technical Digest of the 11th Int. Photovoltaic Science and Engineering Conf.* (Tanaka Printing, Kyoto, 1999), p. 763.

[94] A. Goetzberger, *Optical confinement in thin Si-solar cells by diffusive back reflectors*, in *Proc. 15th IEEE Photovoltaic Specialists Conf.* (IEEE, New York, 1981), p. 867.

[95] E. Yablonovitch and G. D. Cody, *Intensity enhancement in textured optical sheets for solar cells*, IEEE Trans. Electron. Devices **ED-29**, 300 (1982).

[96] R. A. Arndt, J. F. Allison, J. G. Haynos, and A. Meulenberg Jr., *Optical propeties of the COMSAT non-reflective cell* in in *Proc. 11th IEEE Photovoltaic Specialists Conf.* (IEEE, New York, 1975), p. 40.

[97] C. R. Barona, H. W. Brandhorst, *V-groove silicon solar cells*, in *Proc. 11th IEEE Photovoltaic Specialists Conf.* (IEEE, New York, 1975), p. 44.

[98] A. W. Blakers, A. Wang, A. M. Milne, J. Zhao, and M. A. Green, *22.8% efficient silicon solar cell*, Appl. Phys. Lett. **55**, 1363 (1989).

[99] S. G. Uyehata, and E. Kolesat Jr., *Micromachined silicon surfaces for maximizing optical absorption at 632.8 nm*, J. Micromechanics and Microengineering **1**, 171 (1991).

[100] T. Machida, K. Nakajiama, Y. Takeda, S. Tanaka, N. Shibuya, K. Okamoto, T. Nammori, T. Nunoi and T. Tsuji, *Efficiency improvement in polycrystalline silicon solar cells with grooved surfaces*, in *Proc. 22nd IEEE Photovoltaic Specialists Conf.* (IEEE, New York, 1991), p. 2943.

[101] H. Bender, J. Szlufcik, H. Nussbaumer, G. Palmers, O. Evrand, J. Nijs, E. Bucher, and G. Willeke, *Polycrystalline Si solar cells with a mechanically formed texturisation*, Appl. Phys. Lett. **62**, 2941 (1993).

[102] M. A. Green, *Surface texturing and patterning in solar cells,* in *Advances in Solar Energy*, Vol. 8, edited by M. Prince (American Solar Energy Society, Boulder, 1993), p. 231.

[103] A. Luque, *Coupling light to solar cells*, in *Advances in Solar Energy*, Vol. 8., edited by M. Prince (American Solar Energy Society, Boulder, 1993), p. 161.

[104] G. A. Landis, *Cross grooved solar cell*, US Patent 4608451, August 26, 1986.

[105] D. Redfield, *Multiple pass thin-film silicon solar cell*, Appl. Phys. Lett. **25**, 647 (1974).

[106] G. Schumm, H.-D. Mohring, and G. H. Bauer, *An approach to thin stable a-Si:H solar cells with high efficiency by a novel light-trapping structure*, in *Proc. 11th European Countries Photovoltaic Solar Energy Conf.*, edited by L. Guimaraes, W. Palz, C. De Reyff, H. Kiess, and P. Helm (Harwood Academic, Chur, 1992), p. 207.

[107] D. Thorp, P. Campbell, and S. R. Wenham, *Absorption enhancement in conformally textured thin-film silicon solar cells*, in *Proc. 25th IEEE Photovoltaic Specialists Conf.* (IEEE, New York, 1996), p. 705.

[108] D. Thorp, P. Campbell, and S. R. Wenham, *Conformal films for light-trapping in thin silicon solar cells*, Progress in Photovoltaics **4**, 205 (1996).

[109] R. B. Bergmann, R Brendel, M. Wolf, P. Lölgen, J. H. Werner, J. Krinke, and H. P. Strunk, *Crystalline silicon films by chemical vapour deposition on glass for thin film solar cells*, in *Proc. 25th IEEE Photovoltaic Specialists Conf.* (IEEE, New York, 1996), p. 365.

[110] J. C. Miñano and A. Luque, *Geometrical patterns producing almost isotropical light confinement*, in *Proc. 8th European Photovoltaic Solar Energy Conf.*, edited by I. Solomon and B. Equer (Kluwer, Dordrecht, 1988), p. 1387.

[111] R. Brendel, *Coupling of light into mechanically textured silicon solar cells: a ray tracing study*, Progress in Photovoltaics **3**, 25 (1995).

[112] A Luque, *The confinement of light in solar cells*, Solar Energy Materials **23**, 152 (1991).

[113] R. Brendel, M. Hirsch, R. Plieninger, and J. H. Werner, *Experimental analysis of quantum efficiency for thin layer silicon solar cells with back surface fields and light trapping schemes*, in *Proc. 13th European Photovoltaic Solar Energy Conf.*, edited by W. Freiesleben, W. Palz, H. A. Ossenbrink, and P. Helm (H. S. Stephens, Bedford, 1995), p. 428.

[114] R. Brendel, *Quantitative infrared study of ultrathin MIS structures by grazing internal reflection*, Appl. Phys. A **50**, 587 (1990).

[115] W. P. Dumke, *Spontaneous radiative recombination in semiconductors*, Phys. Rev. **105**, 139 (1957).

[116] M. A. Green and S. R. Wenham, *Novel parallel multijunction solar cell*, Appl. Phys. Lett. **65**, 2907 (1994).

[117] S. Koc, *The quantum efficiency of the photo-electric effect in germanium for the 0.3 to 2 μm wavelength region*, Czech. J. Phys. **7**, 91 (1957).

[118] J. Tauc and A. Abraham, *The quantum efficiency of the internal photo-electric effect in indium antimonide*, Czech. J. Phys. **9**, 95 (1959).

[119] O. Christensen, *Quantum efficiency of the internal photoelectric effect in silicon and germanium*, J. Appl. Phys. **47**, 689 (1976).

[120] F. J. Wilkinson, A. J. D. Farmer, and J. Geist, *The near ultraviolet quantum yield of silicon*, J. Appl. Phys. **54**, 1172 (1983).

[121] J. Geist and C. S. Wang, *New calculations of the quantum yield of silicon in the near ultraviolet*, Phys. Rev. B **27**, 4841 (1983).

[122] M. V. Fischetti and S. E. Laux, *Understanding hot electron transport in silicon devices: Is there a short cut?* J. Appl. Phys. **78**, 1058 (1995).

[123] P. A. Wolff, *Theory of electron multiplication in silicon and germanium*, Phys. Rev. **95**, 1415 (1954).

[124] W. Shockley, *Problems related to p-n junctions in silicon*, Solid-State Electron. **2**, 35 (1961).

[125] G. A. Baraff, *Distribution functions and ionization rates for hot electrons in semiconductors*, Phys. Rev. **128**, 2507 (1962).

[126] L. V. Keldysh, *Concerning the theory of impact ionization in semiconductors*, Sov. Phys. – JETP **21**, 1135 (1965).

[127] D. J. Robbins, *Aspects of the theory of impact ionization in semiconductors*, phys. stat. sol. B **97**, 9 (1980).

[128] N. Sano and A. Yoshii, *Impact-ionization theory consistent with a realistic band structure of silicon*, Phys. Rev. B **45**, 4171 (1992).

[129] X. Wang, V. Chandramouli, C. M. Maziar, and A. F. Tasch, Jr., *Simulation program suitable for hot carrier studies: an efficient multiband Monte Carlo model using both full and analytic band structure description for silicon,* J. Appl. Phys. **73**, 3339 (1993).

[130] W. Quade, E. Schöll, and M. Rudan, *Impact ionization within the hydrodynamic approach to semiconductor transport*, Solid-State Electron. **36**, 1493 (1993).

[131] E. Cartier, M. V. Fischetti, E. A. Eklund, and F. R. McFeely, *Impact ionization in silicon*, Appl. Phys. Lett. **62**, 3339 (1993).

[132] N. Sano and A. Yoshii, *Impact ionization rate near thresholds in Si*, J. Appl. Phys. **75**, 5102 (1994).

[133] Y. Wang and K. Brennan, *K dependence of the impact ionization transition rate in bulk InAs, GaAs, and Ge*, J. Appl. Phys. **71**, 2736 (1992).

[134] J. Bude and K. Hess, *Thresholds of impact ionization in semiconductors*, J. Appl. Phys. **72**, 3554 (1992).

[135] J. Kolnik, Y. Wang, I. H. Oguzman, and K. F. Brennan, *Theoretical investigation of wave-vector-dependent analytical and numerical formulations of the interband impact-ionization transition rate for electrons in bulk silicon and GaAs*, J. Appl. Phys. **76**, 3542 (1994).

[136] M. Stobbe, R. Redmer, and W. Schattke, *Impact ionization rate in GaAs*, Phys. Rev. B **49**, 4494 (1994).

[137] N. Sano and A. Yoshii, *Impact-ionization model consistent with the band structure of semiconductors*, J. Appl. Phys. **77**, 2020 (1995).

[138] M. Reigrotzki, M. Stobbe, R. Redmer, and W. Schattke, *Impact ionization rate in ZnS*, Phys. Rev. B **52**, 1456 (1995).

[139] R. C. Alig, S. Bloom, and C. W. Struck, *Scattering by ionization and phonon emission in semiconductors,* Phys. Rev. B **22**, 5565 (1980).

[140] M. Wolf, R. Brendel, H. J. Werner, and H. J. Queisser, *Solar cell efficiency and carrier multiplication in* $Si_{1-x}Ge_x$ *alloys*, J. Appl. Phys. **83**, 4213 (1998).

[141] K. Hess, *Advanced theory of semiconductor devices* (Prentice Hall, New Jersey, 1988), p. 25.

[142] D. A. Papaconstantopoulos, *Handbook of the band structure of elemental solids* (Plenum Press, New York, 1986), p. 231.

[143] J. R. Chelikowsky and M. L. Cohen, *Nonlocal pseudopotential calculations for the electronic structure of eleven diamond and zinc-blende semiconductors*, Phys. Rev. B **14**, 556 (1976).

[144] J. C. Slater and G. F. Koster, *Simplified LC AO method for the periodic potential problem*, Phys. Rev. **94**, 1498 (1954).

[145] A. R. Beattie and P. T. Landsberg, *Auger effect in semiconductors*, Proc. Royal Soc. A **249**, 16 (1958).

[146] W. Shockley, W. W. Hooper, H. J. Queisser, and W. Schroen, *Mobile electric charges on insulating oxides with application to oxide covered silicon p-n junctions*, Surface Science **2**, 277 (1964).

[147] A. Goetzberger, *Improved properties of silicon dioxide layers grown under bias*, J. Electrochem. Soc. **113**, 138 (1966).

[148] J. A. Giacometti and O. N. Oliveira, *Corona charging of polymers*, IEEE Trans. Elect. Insul. **27**, 924 (1992).

[149] R. Williams and M. H. Woods, *High electric fields in silicon dioxide produced by corona charging*, J. Appl. Phys. **44**, 1026 (1973).

[150] M. Rennau, A. Beyer, G. Ebest, and P. Arzt, *Surface-photo-Kelvin-voltage (SPKV) method for determination of different characteristics of solar cells*, in *Proc. 12th Europ. Photovoltaic Solar Energy Conf.* (H. S. Stephens, Bedford, 1994), p. 489.

[151] M. Schöfthaler, *Transiente Mikrowellenreflexion zur kontaktlosen Lasungstraäger Lebensdauermessung in Silizium für Solarzellen*, PhD thesis, University of Stuttgart (Shaker, Aachen, 1995).

[152] G. Beck and M. Kunst, *Contactless scanner for photo active materials using laser-induced microwave absorption*, Rev. Sci. Instrum. **57**, 197 (1986).

[153] M. Kunst and G. Beck, *The study of charge carrier kinetics in semiconductors by microwave conductivity measurements*, J. Appl. Phys. **60**, 3558 (1986).

[154] P. A. Basore and B. R. Hansen, *Microwave-detected photoconductance decay*, in *Proc. 21st IEEE Photovoltaic Specialists Conf.* (IEEE, New York, 1990), p. 374.

[155] M. Schöfthaler and R. Brendel, *Sensitivity and transient response of microwave reflection measurements*, J. Appl. Phys. **77**, 3162 (1995).

[156] J. Schmidt, and A. G. Aberle, *Easy to use surface passivation technique for bulk carrier lifetime measurements on silicon wafers*, Progr. Photovoltaics **6**, 259 (1998)

[157] P. Warwer, M. Rochel, and H.-G. Wagemann, *Unambigous distinction between diffusion length and surface recombination velocity of solar cells at different excitation levels*, J. Appl. Phys. **85**, 7764 (1999).

[158] J. Dziewior and W. Schmid, *Auger coefficients for highly doped and highly excited silicon*, Appl. Phys. Lett. **31**, 346 (1977).

[159] R. Häcker and A. Hangleiter, *Intrinsic upper limits of the carrier lifetime in silicon*, J. Appl. Phys. **75**, 7570 (1994).

[160] J. Schmidt and A. G. Aberle, *Accurate method for the determination of bulk minority-carrier lifetimes of mono- and multicrystalline silicon wafers*, J. Appl. Phys. **81**, 6186 (1997).

[161] J. Beck and R. Conradt, *Auger recombination in silicon*, Solid State Commun. **13**, 93 (1973).

[162] J. del Alamo, S. Swirhun, and R. M. Swanson, *Simultaneous measurement of hole lifetime, hole mobility and bandgap narrowing in heavily doped n-type silicon*, Techn. Dig. Int. Electron. Devices Meet., 290 (1985).

[163] C. H. Wang and A. Neugroschel, *Minority carrier transport parameters in degenerate n-type silicon*, IEEE Electron. Device Lett. **11**, 576 (1990).

[164] A. Wieder, *Emitter effects in shallow bipolar devices: measurements and consequences*, IEEE Trans. Electron. Devices **27**, 1402 (1980).

[165] A. Hangleiter and R. Häcker, *Enhancement of band-to-band Auger recombination by electron-hole correlations*, Phys. Rev. Lett. **65**, 215 (1990).

[166] R. A. Sinton and R. M. Swanson, *Recombination in highly injected silicon*, IEEE Trans. Electron. Devices **ED-34**, 1380 (1987).

[167] E. Yablonovitch and T. Gmitter, *Auger recombination in silicon at low carrier densities*, Appl. Phys. Lett. **49**, 587 (1986).

[168] R. Brendel and H. J. Queisser, *On the thickness dependence of open circuit voltages of p-n junction solar cells*, Solar Energy Materials and Solar Cells **29**, 397 (1993).

[169] W. Shockley and W. T. Read, *Statistics of recombination of holes and electrons*, Phys. Rev. **87**, 835 (1952).

[170] R. N. Hall, *Electron-hole recombination in germanium*, Phys. Rev. **87**, 387 (1952).

[171] M. P. Godlewski, C. R. Barona, and H. W. Brandhorst, *Low-high junction theory applied to solar cells*, in *Proc. 10th IEEE Photovoltaic Specialists Conf.* (IEEE, New York, 1973), p. 40.

[172] N. S. Singh and P. K. Singh, *Modeling of minority carrier surface recombination velocity at low-high junction of an n^+-p-p^+ silicon diode*, IEEE Trans. Electron. Devices **ED-38**, 337 (1991).

[173] S. N. Singh and P. K. Singh, *A new expression for minority carrier surface recombination velocity at low-high junction of n^+-p-p^+ silicon diode*, Solid-State Electron. **ED-33**, 968 (1990).

[174] M. Y. Ghannam, *A new n^+pnn^+ structure with backside floating junction for high efficiency silicon solar cells*, in *Proc. 2nd IEEE Photovoltaic Specialists Conf.* (IEEE, New York, 1991), p.284.

[175] M. Ghannam, E. Demesmaecker, J, Nijs, R. Mertens, and R. Ocerstraten, *Two-dimensional study of alternative back surface passivation methods for high efficiency solar cells*, in *Proc. 11th European Countries Photovoltaic Solar Energy Conf.*, edited by L. Guimaraes, W. Palz, C. De Reyff, H. Kiess, and P. Helm (Harwood Academic, Chur, 1992), p. 45.

[176] A. Aberle, S. Glunz, and W. Warta, *Impact of illumination level and oxide parameters on Shockley Read Hall recombination at the Si-SiO_2 interface*, J. Appl. Phys. **71**, 4422 (1992).

[177] R. Brendel, M. Hirsch, M. Stemmer, U. Rau, and J. H. Werner, *Internal quantum efficiency analysis of high efficiency thin film silicon solar cells*, Appl. Phys. Lett. **66**, 1261 (1995).

[178] H. Nomura, T. Saitho, T. Uematsu, and T. Warabisko, *Measurement and analysis of surface recombination velocity at the Si-SiO_2 interfaces under high level injection*, in *Technical Digest of the 7th International Photovoltaic Science Engineering Conf.* (Nagoya Institute Technology, Nagoya, 1993), p. 223.

[179] S. W. Glunz, A. B. Sproul, W. Warta, and W. Wettling, *Injection-level-dependent recombination velocities at the Si-SiO_2 interface for various dopant concentrations*, J. Appl. Phys.**75**, 1611 (1994).

[180] M. Hirsch, U. Rau, and J. H. Werner, *Admittance of minority carriers in p-n junction silicon solar cells, in Proc. 12th European Photovoltaic Solar Energy Conf.*, edited by R. Hill, W. Palz, and P. Helm (H. S. Stephens, Bedford 1994), p. 537.

[181] H. C. Card and E. S. Yang, *Electronic processes at grain boundaries in polycrystalline semiconductors under optical illumination*, IEEE Trans. Electron. Devices **ED-24**, 397 (1977).

[182] T. M. Buck and F. S. McKim, *Effects of certain chemical treatments and ambient atmospheres on surface properties of silicon,* J. Electrochem. Soc. **105**, 709 (1958).

[183] W. D. Eades and R. M. Swanson, *Calculation of surface generation and recombination velocities at the Si-SiO_2 interface*, J. Appl Phys. **58**, 4267 (1985).

[184] R. B. M. Girisch, R. P. Mertens, and R. F. De Keersmaecker, *Determination of Si-SiO_2 interface recombination parameters using a gate controlled point junction diode under illumination*, IEEE Trans. Electron. Devices **ED-35**, 203 (1988).

[185] A. W. Stephens, A. G. Aberle, and M. A. Green, *Surface recombination velocity measurements at the silicon-silicon dioxide interface by microwave-detected photoconductance decay*, J. Appl. Phys. **76**, 363 (1994).

[186] J. Schmidt, F. M. Schurmanns, W. C. Sinke, S. W. Glunz, and A. G. Aberle, *Observation of multiple defect states at silicon-silicon nitride interfaces fabricated by low-frequency plasma-enhanced chemical vapour deposition*, Appl. Phys. Lett. **71**, 252 (1997).

[187] J. Schmidt and A. G. Aberle, *Carrier recombination at silicon-silicon nitride interfaces fabricated by plasma-enhanced chemical vapor deposition*, J. Appl. Phys. **85**, 3626 (1999).

[188] S. J. Robinson, S. R. Wenham, and M. A. Green, *Recombination rate saturation mechanism at oxidized surfaces of high-efficiency silicon solar cells*, J. Appl. Phys. **78**, 4740 (1995).

[189] A. G. Aberle, T. Lauinger, J. Schmidt, and R. Hezel, *Injection-level dependent surface recombination velocities at the silicon-plasma silicon nitride interface*, Appl. Phys. Lett. **66**, 2828 (1995).

[190] T. Lauinger, J. Schmidt, A. G. Aberle, and R. Hezel, *Record low surface recombination velocities on 1 Ωcm p-silicon using remote plasma nitride passivation*, Appl. Phys. Lett. **68**, 1232 (1996).

[191] J. Schmidt, *Untersuchung zur Ladungsträgerrekombination an den Oberflächen und im Volumen von kristallinen Silicium-Solarzellen*, Dissertation Universität Hannover (Shaker, Aachen, 1998).

[192] S. Dauwe, A. Metz, and R. Hezel, *A novel mask-free low-temperature rear surface passivation scheme based on PECVD silicon nitride for high-efficiency silicon solar cells*, in *Proc. 16th European Photovoltaic Solar Energy Conf.*, edited by H. Scheer, B. McNelis, W. Palz, H. A. Ossenbrink, and P. Helm (James & James, London, 2000), p. 1747.

[193] J. R. Ellminger and M. Kunst, *Investigation of charge carrier injection in silicon/silicon nitride junctions*, Appl. Phys. Lett. **69**, 517 (1996).

[194] A. G. Aberle, *Crystalline silicon solar cells: Advanced surface passivation and analysis* (University of New South Wales, Sydney, 1999).

[195] A. Cuevas, M. Stuckings, J. Lau, and M. Petravic, *The recombination velocity of boron doped diffused silicon surfaces*, in *Proc. 14th Europ. Photovoltaic Solar Energy Conf.*, edited by H. A. Ossenbrink, P. Helm, and H. Ehmann (H. S. Stephens, Bedford, 1997), p. 2416.

[196] M. A. Green, K. Emery, K. Bücher, D. L. King, and S. Ihari, *Solar cell efficiency tables (version 14)*, Progress in Photovoltaics **7**, 321 (1999).

[197] R. Brendel, R. B. Bergmann, B. Fischer, J. Krinke, R. Plieninger, U. Rau, J. Reiß, H. P. Strunk, H. Wanka, and J. H. Werner, *Transport analysis for polycrystalline silicon solar cells on glass substrates*, in *Proc. 26th IEEE Photovoltaic Specialists Conf.* (IEEE, New York, 1997), p.635.

[198] C. Donolato, *Theory of beam induced current characterization of grain boundaries in polycrystalline solar cells*, J. Appl. Phys. **54**, 1314 (1983).

[199] J. Zettner, *Eigenschaften kristalliner Silizium-Dünnschicht Solarzellen auf SiC/Graphit Substraten*, Dissertation University of Erlangen-Nürnberg (University of Erlangen-Nürnberg, Erlangen, 2000).

[200] R. B. Bergmann and J. Krinke, *Large grained polycrystalline silicon films by solid phase crystallization of phosphorus doped amorphous silicon*, J. Crystal Growth **177**, 191 (1997).

[201] J. Aitchison and J. A. C. Brown, *The Lognormal Distribution* (Cambridge University Press, London, 1969).

[202] G. A. M. Hurkx, D. B. M. Klaassen, and M. P. G. Knuvers, *A new recombination model for device simulation including tunneling*, IEEE Transactions Electron. Devices **39**, 331 (1992).

[203] S. A. Edminston, G. Heiser, A. B. Sproul, and M. A. Green, *Improved modeling of grain boundary recombination in bulk and p-n junction regions of polycrystalline silicon solar cells*, J. Appl. Phys. **80**, 678 (1996).

[204] R. Brendel and U. Rau, *Effective diffusion lengths for minority carriers in solar cells as determined from internal quantum efficiency*, J. Appl. Phys. **85**, 3634 (1999).

[205] A. Kay and M. Grätzel, *Low-cost photovoltaic modules based on dye sensitized nanocrystalline titanium dioxide and carbon powder*, Solar Energy Materials and Solar Cells **44**, 99 (1996).

[206] J. A. del Alamo and R. M. Swanson, *The physics and modelling of heavily doped emitters*, IEEE Trans. Electron. Devices **ED-31**, 1878 (1984).

[207] G. Güttler and H. J. Queisser, *Impurity photovoltaic effect in silicon*, Energy Conversion **10**, 51 (1970).

[208] M. J. Keevers and M. A. Green, *Efficiency improvements of silicon solar cells by the impurity photovoltaic effect*, J. Appl. Phys. **75**, 4022 (1994).

[209] J. Sinkkonen, J. Roukolainen, P. Uotila, and A. Hovinen, *Spatial collection efficiency of a solar cell*, Appl. Phys. Lett. **66**, 206 (1995).

[210] M. R. Spiegel, *Theory and Problems of Laplace Transforms* (Mc Graw-Hill, New York, 1983).

[211] A.-A. S. Al-Omar and M. Y. Ghannam, *Direct calculation of two-dimensional collection probability in pn junction solar cells, and study of grain-boundary recombination in polycrystalline silicon cells*, J. Appl. Phys. **79**, 2103 (1996).

[212] M. Hirsch, U. Rau, and J. H. Werner, *Analysis of internal quantum efficiency and a new graphical evaluation scheme*, Solid-State Electron. **38**, 1009 (1995).

[213] A. W. Smith, A. Rohatgi, and S. C. Neel, *Texture: a ray tracing program for the photovoltaic community*, in *Proc. 21st IEEE Photovoltaic Specialists Conf.* (IEEE, New York, 1990), p. 426.

[214] R. Brendel, *SUNRAYS user's manual*, distributed by Garching Innovation GmbH, Königinstr. 19, D-80539 München, Germany, Fax +49-89-21081593.

[215] J. O. Schumacher, S. Sterk, B. Wagner, and W. Warta, *Quantum efficiency*

analysis of high-efficiency solar cells with textured surfaces, in *Proc. 13th European Photovoltaic Solar Energy Conf.*, edited by W. Freiersleben, W. Palz, H. A. Ossenbrink, and P. Helm (H. S. Stephens, Bedford, 1995), p. 96.

[216] C. Zechner, P. Fath, G. Wileke, and E. Bucher, *Two- and three-dimensional optical carrier generation determination in crystalline silicon solar cells*, Solar Energy Materials and Solar Cells **51**, 255 (1998).

[217] A. S. Glassner (editor), *An introduction to ray-tracing* (Academic Press, Cambridge, 1997).

[218] B. Harbecke, *Coherent and incoherent reflection and transmission of multilayer structures*, Appl. Phys. B **39**, 165 (1986).

[219] E. D. Palik (editor), *Handbook of Optical Constants of Solids* (Academic Press, Boston, 1985), p. 396, 753.

[220] J. M. Rodriguez, I. Tobias, and A. Luque, *Random pyramidal texture modelling*, Solar Energy Materials and Solar Cells **45**, 241 (1997).

[221] T. Uematsu, M. Ida, K. Hane, Y. Hayashi, and T. Saitoh, *A new light trapping structure for very thin, high efficiency silicon solar cells*, in *Proc. 20th IEEE Photovoltaic Specialists Conf.* (IEEE, New York, 1988), p. 792.

[222] A. W. Smith and A. Rohatgi, *Ray tracing analysis of the inverted pyramid texturing geometry for high efficiency silicon solar cells*, Solar Energy Materials and Solar Cells **29**, 37 (1993).

[223] M. Goldbach, Th. Meyer, R. Brendel, and U. Rau, *Two-dimensional simulation of silicon thin-film solar cells with innovative device structures*, Progress in Photovoltaics **7**, 85 (1999).

[224] R. Häcker and A. Hangleiter, *Intrinsic upper limits of the carrier lifetime in silicon*, J. Appl. Phys. **75**, 7570 (1994).

[225] J. Schmidt, *Measurement of differential and actual recombination parameters on crystalline silicon wafers*, IEEE Trans. Electron. Devices **ED-46**, 2018 (1999).

[226] R. A. Sinton and A. Cuevas, *Contactless determination of current-voltage characteristics and minority-carrier lifetimes in semiconductors from quasi-steady-state photoconductance data*, Appl. Phys. Lett. **69**, 2510 (1996).

[227] J. Zhao, A. Wang, S. . Wenham, and A. M. Green, *21.5% efficient thin-layer silicon solar cell*, in *Proc. 13th European Photovoltaic Solar Energy Conf.*, edited by W. Freiersleben, W. Palz, H. A. Ossenbrink, and P. Helm (H. S. Stephens, Bedford, 1995), p. 1566.

[228] M. A. Green, A. Wang, J. Zhao, S. R. Wenham, P. Campbell, and D. Thorp, *Enhanced light-trapping in 21.5% efficient thin silicon solar cells,* in *Proc. 13th European Photovoltaic Solar Energy Conf.*, edited by W. Freiesleben, W. Palz, H. A. Ossenbrink, and P. Helm (H. S. Stephens, Bedford, 1995), p. 13.

[229] R. B. Bergmann, *Crystalline Si thin-film solar cells: a review*, Appl. Phys. A **69**, 187 (1999).

[230] K. R. Catchpole, M. J. McCann, K. J. Weber, and A. W. Blakers, *A review of thin-film crystalline silicon for solar cell applications. Part 2: Foreign substrates,* Solar Energy Materials and Solar Cells **68**, 173 (2001).

[231] M. J. McCann, K. R. Catchpole, K. J. Weber, and A. W. Blakers, *A review of thin-film crystalline silicon for solar cell applications. Part 1: Native substrates,* Solar Energy Materials and Solar Cells **68**, 135 (2001).

[232] R. Brendel, *Perspectives of crystalline thin-film Si solar cells*, in *Proc. International Workshop on Photovoltaic Technologies towards the 21st century*, edited by T. Fujuki (NAIST, Nara, 1999), p. 6.

[233] R. Brendel, *Review of layer transfer processes for crystalline thin-film Si solar cells*, Jpn. J. Appl. Phys. **40**, 4431 (2001).

[234] R. Brendel, R. Auer, and H. Artmann, *Textured monocrystalline thin-film Si solar cells from the porous Si (PSI) process*, Progress in Photovoltaics **9**, 217 (2001).

[235] W. Zimmermann, S. Bau, F. Haas, K. Schmidt, and A. Eyer, *Silicon sheets from powder as low cost substrates for crystalline silicon thin film solar cells*, in *Proc. 2nd World Conf. Photovoltaic Solar Energy Conf.*, edited by J. Schmid, H. A. Ossenbrink, P. Helm, H. Ehmann, and E. D. Dunlop (Joint Research Center European Commission, Ispra, 1998), p. 1790.

[236] R. Lüdemann, S. Schaefer, C. Schüle, and C. Hebling, *Dry processing of mc-silicon thin film solar cells*, in *Proc. 26th IEEE Photovoltaic Specialists Conf.* (IEEE, New York, 1997), p. 159.

[237] S. Bourdais, S. Reber, H. Lautenschlager, A. Slaoui, G. Fanozzi, and A. Hurrle, *Recrystallized silicon thin-film solar cells on mullite ceramic substrates*, in *Proc. 16th European Photovoltaic Solar Energy Conf.*, edited by H. Scheer, B. McNelis, W. Palz, H. A. Ossenbrink, and P. Helm (James & James, London, 2000), p. 1493.

[238] J. G. Darrant, R. B. Bergmann, and J. H. Werner, *Glass or glass ceramic substrates*, UK patent application no. GB9603028.3, filed Feb. 14, 1996, published Aug. 20, 1997.

[239] J. Dietl, D. Helmreich, and E. Sirtl, *Solar silicon*, in *Crystals: Growth, Properties and Applications,* vol. 5 (Springer, Berlin, 1981), p. 43.

[240] S. Reber, *Electrical confinement for crystalline silicon thin-film solar cell on foreign substrate*, PhD-thesis, Physics Department University of Mainz in Germany (University of Mainz, Mainz, 2000), p. 111.

[241] S. Reber, A. Eyer, D. Oßwald, H.-J. Pohlmann, C.Lutz, and A. Rosen, *Simultaneous infiltration and recrystallization of SiSiC ceramics for crystalline silicon thin-film solar cells*, in *Proc. 20th IEEE Photovoltaic Specialists Conf.* (IEEE, New York, 1988), p. 146.

[242] A. Eyer, N. Schillinger, S. Schelb, and A. Räuber, *Silicon sheets grown from powder layers by a zone melting process*, J. Crystal Growth **82**, 151 (1987).

[243] S. Bourdais, G. Beaucarne, F. Mazel, A. Slaoui, J. Poortmans, and F. Fantozzi *Polycrystalline silicon thin films deposited onto mullite substrates: from material preparation to photovoltaic devices*, in *Proc. 16th European Photovoltaic Solar Energy Conf.*, edited by H. Scheer, B. McNelis, W. Palz, H. A. Ossenbrink, and P. Helm (James & James, London, 2000), p. 1496.

[244] A. J. van Zutphen, M. Zeman, and J. W. Metselaar, *Film silicon on ceramic substrates for solar cells*, in *Proc. 16th European Photovoltaic Solar Energy Conf.*, edited by H. Scheer, B. McNelis, W. Palz, H. A. Ossenbrink, and P. Helm (James & James, London, 2000), p. 1412.

[245] C. Hebling, *Die kristalline Silicium Dünnschichtsolarzelle auf isolierenden Substraten*, Dissertation University of Freiburg 1998 (Hartung Gorre, Konstanz, 1999).

[246] H. Morikawa, Y. Kawama, Y. Matsuno, S. Hamamoto, K. Imida, T. Ishihara, K. Kojima, and T. Ogama, *Development of high efficiency thin film Si solar cells using zone melting recrystallization*, in *Technical Digest of the 11th Int. Photovoltaic Science and Engineering Conf.* (Tanaka Printing, Kyoto, 1999), p. 529.

[247] M. Schulz, D. Karg, and H. v. Campe, *Abschlußbericht FORSOL: Entwicklung von Dünnschicht-Solarmodulen* (Arbeitsgemeinschaft der Bayerischen Forschungsverbünde, München, 1999).

[248] K. Heidler (Red.), *Fraunhofer Institut Solare Energiesysteme, Leistungen und Ergebnisse Jahresbericht 1997* (FhG-ISE, Freiburg, 1997), p. 80.

[249] R. Lüdemann, S. Schaefer, and J. Reiß, *Dry solar cell processing for low-cost and high-efficiency concepts*, in *Proc. 2nd World Conf. Photovoltaic Solar Energy Conf.*, edited by J. Schmid, H. A. Ossenbrink, P. Helm, H. Ehmann, and E. D. Dunlop (Joint Research Center European Commission, Ispra, 1998), p. 1499.

[250] V. Gazuz, K. Feldrapp, R. Auer, R. Brendel, and M. Schulz, *Dry processing of silicon solar cells in a large-area microwave plasma reactor*, Solar Energy Materials & Solar Cells **72**, 277 (2002).

[251] R. Auer, M. Steinhof, and R. Hezel, *Entwicklung von MIS-Inversionsschicht- und pn-Solarzellen auf multiikristallinen Dünnschichtsystemen*, in *Abschlußbericht Bayerischer Forschungsverbund Solarenergie FORSOL: Entwicklung von Dünnschicht-Solarmodulen* (Bayerischer Forschugsverbund Solarenergie, Schwandorf, 1998), p. 126.

[252] O. Vetterl, P. Hapke, O. Kluth, A. Lambertz, S. Wieder, B. Rech, F. Finger, and H. Wagner, *Intrinsic microcrystalline silicon for solar cells*, Solid State Phenomena **67-68**, 101 (1999).

[253] J. Meier, H. Keppner, S. Dubail, Y. Ziegler, L. Feitknecht, P. Torres, Ch. Hof, U. Kroll, D. Fischer, J. Cuperus, J. A. Selvan, and A. Shah, *Microcrystalline and micromorph thin-film silicon solar cells*, in *Proc. 2nd World Conf. Photovoltaic Solar Energy Conf.*, edited by J. Schmid, H. A. Ossenbrink, P. Helm, H. Ehmann, and E. D. Dunlop (Joint Research Center European Commission, Ispra, 1998), p. 375.

[254] K. Yamamoto, M. Yoshimi, T. Suzuki, Y. Tawada, Y. Okamoto, A. Nakajima, S. Igari, *Thin-film poly-Si solar cells on glass substrates fabricated at low temperature*, Appl. Phys. A **69**, 179 (1999).

[255] T. Matsuyama, M. Tanaka, S. Tsuda, S. Nakano, and Y. Kuwano, *Improvement of n-type poly-Si film properties by solid phase crystallization*, Jpn. J. Appl. Phys. **32**, 3720 (1993).

[256] B. Rech and H. Wagner, *Potential of amorphous silicon for solar cells*, Appl. Phys. A **69**, 155 (1999).

[257] O. Kluth, A. Löffl, S. Wieder, C. Beneking, W. Appenzeller, L. Houben, B. Rech, H. Wagner, S. Hoffmann, R. Waser, J. A. A. Selvan, and H. Keppner, *Texture etched Al-doped ZnO: a new material for enhanced light trapping in thin film solar cells*, in *Proc. 26th IEEE Photovoltaic Specialists Conf.* (IEEE, New York, 1997), p. 715.

[258] A. Löffl, S. Wieder, B. Rech, O. Kluth, C. Benekng, and H. Wagner, *Al-doped ZnO films for thin-film solar cells with very low sheet resistance and controlled texture*, in *Proc. 14th Europ. Photovoltaic Solar Energy Conf.*, edited by H. A. Ossenbrink, P. Helm, and H. Ehmann (H. S. Stephens, Bedford, 1997), p.

2089.

[259] N. Wyrsch, P. Torres, M. Goetz, S. Dubail, L. Feitknecht, J. Kuperus, A. Shah, B. Rech, O. Kluth, S. Wieder, O. Vetterl, H. Stiebig, C. Beneking, and H. Wagner, *Development of inverted micromorph solar cells*, in *Proc. 2nd World Conf. Photovoltaic Solar Energy Conf.*, edited by J. Schmid, H. A. Ossenbrink, P. Helm, H. Ehmann, and E. D. Dunlop (Joint Research Center European Commission, Ispra, 1998), p. 467.

[260] T. Matsuyama, K. Wakisaka, M. Kameda, M. Kanaka, T. Matsuoka, S. Tsuda, S. Nakano, Y. Kishi, and Y. Kuwano, *Preparation of high-quality n-type poly-Si films by solid phase crystallization*, Jpn. J. Appl. Phys. **29**, 2372 (1990).

[261] R. Flückinger, J. Meier, H. Keppner, M. Götz, and A. Shah, *Preparation of undoped and doped microcrystalline silicon (μc-Si:H) by VHF-GD for p-I-N solar cells*, in *Proc. 23rd IEEE Photovoltaic Specialists Conf.* (IEEE, New York, 1993), p. 839.

[262] C. C. Tsai, R. Thompson, C. Donald, F. A. Ponce, G. B. Anderson, and B. Wacker, *Transition from amorphous to crystalline Si: effect of hydrogen on film growth*, in *Amorphous Silicon Technology*, edited by A. Madan, M. J. Thompson, P. C. Taylor, P. G. LeCombert, and Y. Hamakawa, Materials Research Symp. Proc. **118,** 49 (1988).

[263] A. Matsuda, *Formation kinetics and control of microcrystallite in μc-Si:H from glow discharge plasma*, J. Non-Cryst. Solids **59-60**, 767 (1983).

[264] R. E. I. Schropp and M. Zeman, *Amorphous and Microcrystalline Silicon Solar Cells* (Kluwer Academic Publishers, Boston, 1998), p. 21.

[265] U. Kroll, J. Meier, P. Torres, J. Pohl, and A. Shah, *From amorphous to microcrystalline silicon films prepared by hydrogen dilution using VHF (70 MHz) GD technique*, J. Non-Crystalline Solids **227-240**, 68 (1998).

[266] J. Dutta, U. Kroll, P. Chabloz, A. Shah, A. A. Hwling, J.-L. Dorier, and C. Hollenstein, *Dependence of intrinsic stress in hydrogenated amorphous silicon on excitation frequency in a plasma-enhanced chemical vapor deposition process*, J. Appl. Phys. **72**, 3220 (1992).

[267] S. Vepreck, F. A. Sarrott, S. Rambert, and E. Taglauer, *Surface hydrogen content and passivation of silicon deposition by plasma induced chemical vapor deposition from silane and the implications for the reaction mechanisms*, J. Vac. Sci. Technol. A **7**, 2614 (1989).

[268] J. Meier, S. Dubail, R. Flückiger, D. Fischer, H. Keppner, and A. Shah, *Intrinsic microcrystalline silicon (μc-Si;H) – a promising new thin-film solar cell material*, in *Proc. 1st World Conf. on Photovoltaic Energy Conversion* (IEEE, New York, 1995), p. 409.

[269] N. Wyrsch, P. Torres, M. Gerlitzer, E. Vallat, U. Kroll, A. Shah, A. Poruba, and M. Vanecek, *Hydrogenated microcrystalline silicon for photovoltaic applications*, Solid State Phenomena **67-68**, 89 (1999).

[270] M. Vanecek, A. Poruba, Z. Remes, N. Beck, N. Nesladek, *Optical properties of microcrystalline materials*, J. Non-Crystalline Solids **227-230**, 967 (1998).

[271] M. Goerlitzer, P. Torres, C. Droz, and A. Shah, *Extension of the a-Si:H electronic transport model to μc-Si:H: use of the $\mu^o\tau^o$ product to correlate electronic transport properties and solar cell performance*, Solar Energy Materials and Solar Cells **60**, 195 (2000).

[272] K. Yamamoto, M. Yoshimi, T. Suzuki, Y. Tawada, Y. Okamoto, and A. Nakajima, *Below 5 μm thin film poly-Si solar cell on glass substrate fabricated at low temperature*, in *Proc. 2nd World Conf. Photovoltaic Solar Energy Conf.*, edited by J. Schmid, H. A. Ossenbrink, P. Helm, H. Ehmann, and E. D. Dunlop (Joint Research Center European Commission, Ispra, 1998), p. 1284.

[273] K. Yamamoto, T. Suzuki, K. Kondo, T. Okamoto, M. Yamaguchi, M. Izumina, and Y. Tawada, *Low temperature crystal growth by alternating deposition and hydrogen etching sequences and its application to p-layer of a-Si:H solar cells*, Solar Energy Materials and Solar Cells **34**, 501 (1994).

[274] K. Yamamoto, A. Nakashima, T. Suzuki, M. Yoshimi, H. Nishio, and M. Izumina, *Thin film polycrystalline Si solar cell on glass substrate by a novel low temperature process*, Jpn. J. Appl. Phys. **33**, L1751 (1994).

[275] J. H. Werner, R. Dassow, T. J. Rinke, J. R. Köhler, and R. Bergmann, *From polycrystalline to single-crystalline Si on glas*,Thin Solid Films **383**, 95 (2001).

[276] D. A. Smith, C.S. Nichols, *Polycrystalline semiconductors: structure-property relationships*, Solid State Phenom. **51-52**, 105 (1996).

[277] T. Baba, T. Matsuyama, T. Sawada, T. Takahama, K. Wakisaka, S. Tsuda, and S. Nakano, *Polycrystalline silicon thin-film solar cell prepared by the solid phase crystallization (SPC) method,* in *Proc. 1st World Conf. on Photovoltaic Energy Conversion* (IEEE, New York, 1994), p. 1315.

[278] K. Yamamoto, T. Suzuki, M. Yoshimi, and A Nakajima, *Optical confinement effect for below 5 μm thin film poly-Si solar cell on glass substrate*, Jpn. J. Appl. Phys. **36**, L569 (1997).

[279] M. Hack and M. Shur, *Physics of amorphous silicon alloy p-i-n solar cells*, J. Appl. Phys. **58**, 997 (1985).

[280] K. J. Weber, K. Catchpole, M. Stocks, and A. W. Blakers, *Lift-off of silicon epitaxial layers for solar cell applications*, in *Proc. 26th IEEE Photovoltaic Specialists Conf.* (IEEE, New York, 1997), p. 107.

[281] M. Bruel, *Silicon on insulator material technology*, Electronics Lett. **31**, 1201 (1995).

[282] C. O. Bozler, R. W. McClelland, and J. C. C. Fan, *Ultrathin high-efficiency solar cells made from GaAs films prepared by the CLEFT process*, IEEE Electron. Device Lett. **EDL-2**, 203 (1981).

[283] G. A. Landis, *A processing sequence for manufacture of ultra-thin light-trapping silicon solar cells,* Solar Cells **29**, 257 (1990).

[284] M. Deguchi, Y. Kawama, Y. Matsuno, H. Morikawa, S. Arimoto, H. Kumabe, T. Murotani, and S. Mitsui, in *Technical Digest of the 7th International Photovoltaic Science Engineering Conf.* (Nagoya Institute Technology, Nagoya, 1993), p. 243.

[285] M. Deguchi, Y. Kamama, Y. Matsuno, Y. Nishimoto, H. Morikawa, S. Arimoto, H. Kumabe, and T. Murotani, *Prospect of the high efficiency for the VEST (via hole etching for the separation of thin films) cell*, in *Proc. 1st World Conf. On Photovoltaic Energy Conversion* (IEEE, New York, 1994), p. 1287.

[286] S. Arimoto, H. Morikawa, M. Deguchi, Y. Kawama, Y. Matsuno, T. Ishira, H. Kumabe, and T. Murotani, *High-efficiency operation of large-area (100 cm^2) thin-film polycrystalline silicon solar cell based on SOI structure*, Solar Energy Materials and Solar Cells **34**, 257 (1994).

[287] J. M. Gee, W. K. Schubert, and P. A. Basore, *Emitter-wrap-through-solar-cell*, in *Proc. 23rd IEEE Photovoltaic Specialists Conf.* (IEEE, New York, 1993), p. 265.

[288] R. Plieninger, *Rekristallisierte Siliciumschichten für Solarzellenanwendungen*, PhD thesis, University of Stuttgart 1997 (Cuvillier Verlag, Göttingen, 1998).

[289] T. Ishihara, S. Arimoto, H. Morikawa, Y. Nishimoto, Y. Kawama, A. Takami, S. Hamamoto, H. Naomoto, and K. Namba, *Development of high efficiency thin film polycrystalline Si cells using the VEST process*, in *Proc. Mat. Res. Soc. Symp.* **485** (1998), p. 3.

[290] N. Sato, K. Sakaguchi, K. Yamagata, Y. Fujiyama, and T. Yonehara, *Epitaxial growth on porous Si for a new bond and etch back silicon-on-insulator*, J. Electrochem. Soc. **142**, 3116 (1995).

[291] K. Sakaguchi, N. Sato, K. Yamagata, Y. Fujiyama, and T. Yonehara, *Extremely high selective etching of porous Si for single etch-stop bond-and-etch-back silicon-on-insulator*, Jpn. J. Appl. Phys. **34**, 842 (1995).

[292] K. Sakaguchi, T. Yonehara, *SOI wafers based on epitaxial technology*, Solid State Technology **June**, 88 (2000).

[293] T: Yonehara, *Eltran, SOI-Epi and SCLIPS by epitaxial layer transfer from porous Si*, in *Extended abstracts Second International Conf. Porous Semiconductors – Science and Technology*, edited by V. Parthulik and L. Canham (Tecnical University of Valencia, 2000), p. 14.

[294] C. Ascheron, H. Bartsch, A. Setzer, A. Schindler, and P. Paufler, *The effect of hydrogen implantation induced stress on GaP single crystals*, Nucl. Instr. Meth. Phys. Res. **B28**, 350 (1087).

[295] M. Bruel, *Application of hydrogen ion beams to silicon on insulator material technology*, Nuclear Instruments and Methods in Physics Research B **108**, 313 (1996).

[296] H. Tayanaka, K. Yamauchi, T. Matssushita, *Thin-film crystalline silicon solar cells obtained by separation of a porous silicon sacrificial layer*, in *Proc. 2nd World Conf. Photovoltaic Solar Energy Conf.*, edited by J. Schmid, H. A. Ossenbrink, P. Helm, H. Ehmann, and E. D. Dunlop (Joint Research Center European Commission, Ispra, 1998), p. 1272.

[297] R. B. Bergmann, *Thin film solar cells on glass by transfer of monocrystalline Si films*, Int. J. Photoenergy **1**, 83 (1999).

[298] G. Müller, R. Brendel, and M. Schulz, *Light diffusing broad-band Bragg reflectors for thin film silicon solar cells*, in *Proc. 16th European Photovoltaic Solar Energy Conf.*, edited by H. Scheer, B. McNelis, W. Palz, H. A. Ossenbrink, and P. Helm (James & James, London, 2000), p. 1699.

[299] H. Tayanaka, K. Yamauchi, and T. Matssushita, *High minority carrier lifetime in single-crystal silicon thin films on porous silicon sacrificial layer*, in *Technical Digest of the 11th Int. Photovoltaic Science and Engineering Conf.* (Tanaka Printing, Kyoto, 1999), p. 543.

[300] T. J. Rinke, R. B. Bergmann, and J. H. Werner, *Efficient thin film solar cells by layer transfer of monocrystalline Si layers*, in *Proc. 16th European Photovoltaic Solar Energy Conf.*, edited by H. Scheer, B. McNelis, W. Palz, H. A. Ossenbrink, and P. Helm (James & James, London, 2000), p. 1128.

[301] K. J. Weber, K. Catchpole, and A. W. Blakers, *Epitaxial lateral overgrowth of Si on (100) Si substrates by liquid-phase epitaxy*, J. Crystal Growth **186**, 369 (1998).

[302] K. J. Weber, A. W. Blakers, and K. R. Catchpole, *The Epilift technique for solar cells*, Appl. Phys. A **69**, 195 (1999).

[303] K. R. Catchpole, K. J. Weber, A. B. Sproul, and A. W. Blakers, *Characterization of silicon epitaxial layers for solar cell applications*, in *Proc. 2nd World Conf. Photovoltaic Solar Energy Conf.*, edited by J. Schmid, H. A. Ossenbrink, P. Helm, H. Ehmann, and E. D. Dunlop (Joint Research Center European Commission, Ispra, 1998), p. 1336.

[304] T. J. Rinke, R. B. Bergmann, R. Brüggemann, and J. H. Werner, *Ultrathin quasi-monocrystalline silicon films for electronic devices*, Solid State Phenomena **67-68**, 229 (1999).

[305] R. B. Bergmann, T. J. H. Rinke, L. Oberbeck, and R. Dassow, *Low-temperature processing of crystalline Si films on glass for electronic applications*, in *Perspectives Science and Technologies for Novel Silicon on Insulator Devices*, edited by P. L. F. Hemment, V. S. Lysenko, and A. N. Nazarov, NATO Science Series 3. High Technology, Vol. 73 (Kluwer Academic Publishers, Dordrecht, 2000), p. 109.

[306] S. Nishida, K. Nakagawa, M. Iwane, Y. Iwasaki, N. Ukijo, M. Mitzutani, and T. Shoji, *Si film growth using liquid phase epitaxy method and its application to thin-film crystalline Si solar cell*, in *Technical Digest 11th International Photovoltaic Science and Engineering Conf.* (Int. PVSEC Publishing, Kyoto, 1999), p. 537.

[307] J. Kühnle, R. B. Bergmann, J. Krinke, and J. H. Werner, *Comparison of vapor phase and liquid phase epitaxy for deposition of crystalline Si on glass*, Mat. Res. Symp. Proc. **426**, 111 (1996).

[308] K. Sakaguchi, K. Yanagita, H. Kuris, H.Suzuki, K. Ohmi, and T. Yonehara, *Eltran by splitting porous Si layers*, in *Proc. Int. Symp. Silicon on Insulator Technology and Devices* (The Electrochemical Society, Seattle, 1999), p. 117.

[309] A. Fave, B. Semmache, . Berger, P. Kleinmann, F. Mazel, Ch. Dubois, J. M. Olchowik, and A. Laugier, *Direct LPE growth of textured silicon thin layers on c-Si etched-grid fixed on ceramic substrate*, in *Proc. 16th European Photovoltaic Solar Energy Conf.*, edited by H. Scheer, B. McNelis, W. Palz, H. A. Ossenbrink, and P. Helm (James & James, London, 2000), p. 1140.

[310] R. R. Bilyalov, C. S. Solanki, J. Poortsmans, and J. Nijs, *Thin silicon films for solar cells based on porous silicon*, in *Proc. 16th European Photovoltaic Solar Energy Conf.*, edited by H. Scheer, B. McNelis, W. Palz, H. A. Ossenbrink, and P. Helm (James & James, London, 2000), p. 1536.

[311] F. R. Faller and A. Hurrle, *High-temperature CVD for crystalline-silicon thin-film solar cells*, IEEE Transactions on Electron. Devices **46**, 2048 (1999).

[312] R. Brendel, *Verfahren zur Herstellung von schichtartigen Gebilden auf einem Substrat, Substrat sowie mittels des Verfahrens hergestellte Halbleiterbauelemente*, German patent application no. DE 197 27 791.8, date of filing June 30, 1997.

[313] R. Brendel, *Layer transfer: new perspectives for crystalline thin-film Si solar cells*, in *Technical Digest of the 12th International Photovoltaic Science and Engineering Conf.* (Korean Institute of Energy Research, Dejeon, 2001), p. 549.

[314] P. Allongue, V. Kieling, and H. Gerischer, *Etching mechanism and atomic structure of H-Si (111) surfaces prepared in NH_4F*, Electrochim. Acta **40**, 1353 (1995).

[315] P. Alongue, *Porous silicon formation mechanism*, in *Properties of Porous Silicon*, edited by L. Canham (IEE, London 1997), p. 3.

[316] H. Föll, J. Carstensen, M. Christophersen, and G. Hasse, *A new view of silicon electrochemistry*, phys. stat. sol. (a) **182**, 7 (2000).

[317] A. Halimaoui, *Porous Si formation by anodisation*, in *Properties of Porous Silicon*, edited by L. Canham (IEE, London 1997), p. 12.

[318] C. Oules, A. Halimaoui, J. L. Regolini, R. Herino, A. Perio, D. Bensahel, and G. Bomchil, *Epitaxial Si growth on porous Si by reduced pressure low temperature chemical vapour deposition*, Materials Science and Engineering **B4**, 435 (1989).

[319] G. Müller and R. Brendel, *Simulated annealing of porous Si*, phys. stat. sol. (a) **182**, 313 (2000).

[320] G. Kuchler, D. Scholten, G. Müller, J. Krinke, R. Auer, and R. Brendel, *Fabrication of textured monocrystalline Si-films using the porous silicon (PSI) process*, in *Proc. 16th European Photovoltaic Solar Energy Conf.*, edited by H. Scheer, B. McNelis, W. Palz, H. A. Ossenbrink, and P. Helm (James & James, London, 2000), p. 1695.

[321] S. Oelting, D. Martini, and D. Bonnet, *Ion assisted Si deposition of Si films for solar cells*, in *Proc. 11th European Countries Photovoltaic Solar Energy Conf.*, edited by L. Guimaraes, W. Palz, C. De Reyff, H. Kiess, and P. Helm (Harwood Academic, Chur, 1992), p. 491.

[322] Y. Lifshitz, S. R. Kasi, and J. W. Rabalais, *Subplantation model for film growth from hyperthermal species: application to diamond*, Phys. Rev. Lett. **62**, 1290 (1989).

[323] J. W. Rabalais, A. H. Albayati, K. J. Boyd, D. Marton, J. Kulik, Z. Zhang, and W. K. Chu, *Ion-energy effects in silicon ion-beam epitaxy*, Phys. Rev. B **53**, 10781 (1996).

[324] S. Oelting, D. Martini, and D. Bonnet, *Crystalline thin film silicon solar cells by ion-assisted deposition*, in *Proc. 12th European Photovoltaic Solar Energy Conf.*, edited by R. Hill, W. Palz, and P. Helm (H. S. Stephens, Bedford, 1994), p. 1815.

[325] W. Kern and D. A. Poutinen, RCA Rev. **6**, 187 (1979).

[326] H. Jorke, H.-J. Herzog, H. Kibbel, *Secondary implantation of Sb into Si molecular beam epitaxy layers*, Appl. Phys. Lett. **47**, 511 (1985).

[327] S. Oelting, D. Martini, H. Koeppen, and D. Bonnet, *Ion-assisted deposition of crystalline thin film silicon solar cells*, in *Proc. 13th European Photovoltaic Solar Energy Conf.*, edited by W. Freiersleben, W. Palz, H. A. Ossenbrink, and P. Helm (H. S. Stephens, Bedford, 1995), p. 1681.

[328] R. B. Bergmann, R. M. Haussner, N. Jensen, M. Grauvogl, L. Oberbeck, T. Rinke, M. B. Schubert, C. Zacek, R. Dassow, J. R. Köhler, U. Rau, S. Oelting, J. Krinke, H. P. Strunk, and J. H. Werner, *High-rate, low temperature deposition of crystalline silicon films for thin film solar cells on glass,* in *Proc. 2nd World Conf. Photovolt. Energy Conversion*, edited by J. Schmidt, H. A. Ossenbrink, P. Helm, H. Ehmann, and E. D. Dunlop (Joint Research Center Euoropean Commission, Ispra,1998), p. 1260.

[329] R. B. Bergmann, C. Zaczek, N. Jensen, S. Oelting, and J. H. Werner, *Low-temperature Si epitaxy with high deposition rate using ion-assisted deposition,* Appl. Phys. Lett. **72**, 2996 (1998).

[330] L. Oberbeck and R. B. Bergmann, N. Jensen, S. Oleting, and J. H. Werner, *Low-temperature silicon epitaxy by ion-assisted deposition*, Solid State Phenomena **67-68**, 459 (1999).

[331] D. B. M. Klaassen, J. W. Slotboom, and H. C. Graaff, *Unified apparent bandgap narrowing in n- and p-type silicon*, Solid State Electron. **35**, 125 (1992).

[332] L. Oberbeck and R. B. Bergmann, *Electronic properties of silicon epitaxial layers deposited by ion-assisted deposition at low temperatures*, J. Appl. Phys. **88**, 3015 (2000).

[333] L. J. van der Pauw, *Messung des spezifischen Widerstandes und des Hall – Koefizienten an Scheibchen beliebiger Form*, Phillips Technische Rundschau **8**, 230 (1958).

[334] J. C. Irvin, *Resistivity of bulk silicon and of diffused layers in silicon*, Bell System Tech. J. **41**, 387 (1962).

[335] J. Krinke, G. Kuchler, R. Brendel, H. Artmann, W. Frey, S. Oelting, M. Schulz, and H. P. Strunk, *Microstructure and electrical properties of epitaxial layers deposited on silicon by ion assisted deposition*, in *Technical Digest 11th International Photovoltaic Science and Engineering Conf.* (Int. PVSEC Publishing, Kyoto, 1999), p.757.

[336] R. Brendel, D. Scholten, M. Schulz, S. Oelting, H. Artmann, W. Frey, J. Krinke, H. P. Strunk, and J. H. Werner, *Waffle cells fabricated by the Perforated Silicon (Ψ) Process*, in *Proc. 2nd World Conf. Photovoltaic Solar Energy Conf.*, edited by J. Schmid, H. A. Ossenbrink, P. Helm, H. Ehmann, and E. D. Dunlop (Joint Research Center European Commission, Ispra, 1998), p. 1242.

[337] M. L. Hitchman and K. F. Jensen (editors), *Chemical Vapor Deposition* (Academic Press, London, 1993).

[338] S. Jin, H. Bender, L. Stalsmans, R. Bilyalov, J. Poortmans, R. Loo, and M. Caymax, *Transmission electron microscopy investigation of the crystallographic quality of silicon films grown epitaxially on porous Si*, J. Crystal Growth **212**, 119 (2000).

[339] H. v. Campe, B. Cembolista, H. Ebinger, W. Hoffmann, U. Huth, W. Warzawa, and W. Warta, *Multicrystalline Si layers for photovoltaic applications grown by a modified CVD process*, in *Proc. 11th European Community Photovoltaic Solar Energy Conf.*, edited by L. Guimares, W. Palz, C de Reyff, H. Kiess, and P. Helm (Harwood Academic, Chur, 1992), p. 1066.

[340] R. Brendel and S. Oelting, *A novel process for the integrated series connection of crystalline thin-film silicon solar cells*, in *Technical Digest 11th International Photovoltaic Science and Engineering Conf.* (Tanaka Printing, Kyoto, 1999), p. 545.

[341] R. Brendel and R. Auer, *Photovoltaic mini-modules from layer transfer using the porous Si (PSI) process*, Progress in Photovoltaics **9**, 439 (2001).

[342] R. Brendel and D. Scholten, *Modeling of light trapping and transport in thin waffle-shaped Si solar cells*, Appl. Phys. A **69**, 201 (1999).

[343] M. J. Stocks, A. Cuevas, and A. W. Blakers, *Theoretical comparison of conventional and multilayer thin silicon solar cells*, Progress in Photovoltaics **4**, 35 (1996).

[344] TCAD pacage version 4.1. including DESSIS, Integrated Semiconductor Engineering AG, ise@ise.ch

[345] R. Brendel, A. Gier, M. Mennig, H. Schmidt, and J. H. Werner, *Sol-gel coatings for light trapping in thin film solar cells*, J. Non-Cryst. Solids **218**, 391 (1997).

[346] H. J. Hovel, in *Semiconductors and Semimetals*, Vol. 11: *Solar Cells*, edited by R. K. Willardson and A. C. Beer (Academic Press, New York, 1975), pp. 50 and 128.

[347] S. C. Choo, *On space-charge recombination in pn junctions*, Solid-St. Electron. **39**, 308 (1996), Eq. 3.

[348] T. Markvart, *Radiation damage in solar cells*, Journal of Material Science – Materials in Electronics **1,** 1 (1990).

[349] M. Yamaguchi, S. J. Taylor, M. Yang, S. Matsuda, O. Kawasaki, and T. Hisamatsu, *Analysis of radiation damage to Si solar cells under high-fluence electron irradiation*, Jpn. J. Appl. Phys. **35**, 3918 (1996).

[350] A. S. Al-Omar and M. Y. Ghannam, *Optimum two-dimensional short circuit collection efficiency in thin multicrystalline silicon solar cells with optical confinement*, Solar Energy Materials and Solar Cells **52**, 107 (1998).

[351] A. Wang, J. Zhao, S. R. Wenham, and M. A. Green, *21.5% efficient thin silicon solar cell*, Progress in Photovoltiacs **4**, 55 (1996).

[352] A. K. Ghosh, C Fishman, and T. Feng, *Theory of the electrical and photovoltaic properties of plycrystalline silicon*, J. Appl. Phys. **51**, 446 (1980).

[353] A. Takami, S. Arimoto, H. Mrikawa, S. Hamamoto, T. Ishihara, H. Kumabe, and T. Murotani, *High efficiency (16.45%) thin-film silicon solar cells prepared by zone-melting recrystallization*, in *Proc. 12th European Photovoltaic Solar Energy Conf.*, edited by R. Hill, W. Palz, and P. Helm (H. S. Stephens, Bedford 1994), p. 59.

[354] H. Keppner, J. Meier, P. Torres, D. Fischer, and A. Shah, *Microcrystalline silicon and micromorph tandem solar cells*, Appl. Phys. A **69**, 169 (1999).

[355] Y. Bai, D. H. Ford, J. A. Rand, R. B. Hall, and A. M. Barnett, *16.6% efficient silicon thin-film polycrystalline silicon solar cells*, in *Proc. 26th IEEE Photovoltaic Specialists Conf.* (IEEE, New York, 1997), p. 35.

[356] R. Brendel, K. Feldrapp, and S. Oelting, *Shadow-epitaxy for flexible thin-film Si solar cell design*, Solar Energy Materials and Solar Cells **64**, 251 (2000).

[357] C. Berge, R. B. Bergmann, T. Rinke, and J. H. Werner, in *Proc. 17th European Photovoltaic Solar Energy Conf.* (WIP-Renewable Energies, Munich, 2002), p. 1277.

[358] R. Brendel, R. Auer, R. Horbelt, and K. H. Feldrapp, *15.4%-efficient thin-film crystalline Si solar cells from layer transfer using porous Si*, Phys. Stat. Sol (a) (2002), in press.

[359] K. Feldrapp, R. Horbelt, R. Auer, and R. Brendel, *Porous silicon (PSI) process for industrial fabrication of thin film mini-modules?* in *Proc. 17th European Photovoltaic Solar Energy Conf.* (WIP-Renewable Energies, Munich, 2002), p. 1458

[360] A. Luque, *Solar Cells and Optics for Photovoltaic Concentration* (Adam Hilger, Bristol, 1989), p. 508, Eq. (14.20).

[361] W. Gautschi and W. F. Cahill, in *Handbook of Mathematical Functions*, edited by M. Abramowitz and I. A. Stegun (Dover Publications, New York, 1970), p. 231, Eqs. (5.1.53) and (5.1.54).

[362] J. Sinkkonen, A. Hovivnen, T. Siirtola, E. Tuominen, and M. Acerbis, *Interpretation of the spectral response of a solar cell in terms of the spatial collection efficiency*, in *Proc. 25th IEEE Photovoltaic Specialists Conf.* (IEEE, New York, 1996), p. 561.

[363] M. Born and E. Wolf, *Principles of optics*, 3rd edition (Pergamon Press, New York, 1964).

[364] D. Marcuse, *Light transmission optics*, in Bell Laboratories Series (Van Nostrand Reinhold Company, New York, 1972), p. 82.

[365] A. A. Abouelsaood, M. Y. Ghannam, and A. S. Al Omar, *Limitation of ray tracing techniques in optical modeling of silicon solar cells and photodiodes*, J. Appl. Phys. **84**, 5795 (1998).

[366] Y. Hishikawa, H. Tari, and S. Kiyama, *Numerical analysis on the optical confinement and optical loss in high-efficiency a-Si cells with textured surfaces*, in *Technical Digest of the 11th Int. Photovoltaic Science and Engineering Conf.* (Tanaka Printing, Kyoto, 1999), p. 219.

[367] Y. Hishikawa, *Three-dimensional optical simulation for high-efficiency a-Si solar cells*, Oyo Buturi **69** , 844 (2000), in Japanese.

[368] E. Lorenzo, *The solar radiation*, in *Solar Cells and Optics for Photovoltaic Concentration*, edited by A. Luque (Adam Hilger, Bristol, 1989), p. 268.

[369] R. Brendel, *Simple prism pyramids: a new light trapping texture for silicon solar cells*, in *Proc. 23rd IEEE Photovoltaic Specialists Conf.* (IEEE, New York, 1993), p. 252.

[370] P. Campbell and M. A. Green, *High performance light trapping textures for monocrystalline silicon solar cells*, Solar Energy Materials and Solar Cells **65**, 369 (2001)

[371] B. Dale and H. G. Rudenberg, *Photovoltaic conversion 1: high efficiency silicon solar cells*, in *Proc. 14th Annual Power Sources Conf.*, U. S. Army Signal Research and Development Lab., Ft. Monmouth, New Jersey, May 1960, p. 22. Ref. from [102].

[372] P. Campbell and M. Keevers, *Light trapping and reflection control for silicon thin films deposited on glass substrates by embossing*, in *Proc. 28th IEEE Photovoltaic Specialists Conf.* (IEEE, New York, 2000), p. 355.

[373] R. H. Morf and H. Kieß, *Submicron gratings for light trapping in silicon solar cells: a theoreticel study*, in *Proc. 9th European Countries Photovoltaic Solar Energy Conf.*, edited by W. Palz, G. T. Wrixon, and P. Helm (Kluwer, Dordrecht, 1989), p. 313.

[374] H. A. McLeod, *Thin Film Optical Filters* (A. Hilger, Bristol, 1986).

[375] D. Y. Smith, *The optical properties of metallic aluminum*, in *Handbook of Optical Constants of Solids*, edited by E. D. Palik (Academic, Boston, 1985), p. 396.

[376] J. H. Werner, J. K. Arch, R. Brendel, G. Langguth, M. Konoma, E. Bauser, *Crystalline thin film silicon solar cells*, in *Proc. 12th European Photovoltaic Solar Energy Conf.*, edited by R. Hill, W. Palz, and P. Helm (H. S. Stephens, Bedford 1994), p. 1823.

[377] A. W. Blakers, K. J. Weber, M. F. Stuckings, S. Armand, G. Matlakowski, A. J. Carr, M. J. Stokcs, A. Cuevas, and T. Brammer, *17% efficient thin-film silicon solar cell by liquid phase epitaxy*, Progress in Photovoltaics **3**, 193 (1995).

[378] H. v. Campe, D. Nikl, W. Schmidt, and F. Schomann, *Crystalline silicon thin film solar cells*, in *Proc. 13th European Photovoltaic Solar Energy Conf.*, edited by W. Freiesleben, W. Palz, H. A. Ossenbrink, and P. Helm (H. S. Stephens, Bedford, 1995), p.1489.

[379] J. E. Cotter, A. M. Barnett, D. H. Ford, R. B. Hall, A. E. Ingram, J. A. Rand, T. R. Ruffins, K. P. Shreve, and C. J. Thomas, *Polycrystalline Silicon-Film^{TM} thin-film solar cells: advanced products*, Progress in Photovoltaics **3**, 351 (1995).

[380] G. F. Zheng, W. Zhang, Z. Shi, D. Thorp, and M. A. Green, *High-efficiency drift-field thin-film silicon solar cells by liquid-phase epitaxy and substrate thinning*, in *Proc. 25th IEEE Photovoltaic Specialists Conf.* (IEEE, New York, 1996), p. 693.

[381] J. Arch, R. Brendel, and J. H. Werner, *Contribution of silicon substrates to the efficiencies of silicon thin film solar cells*, Solar Energy Materials and Solar Cells **45**, 309 (1997).

[382] M. G. Mauk, P. A. Burch, S. W. Johnson, T. A. Goodwin, and A. M. Barnett, *"Buried" metal/dielectric/semiconductor reflector for light trapping in epitaxial thin-film silicon solar cells*, in *Proc. 25th IEEE Photovoltaic Specialists Conf.* (IEEE, New York, 1996), p. 147.

[383] W. Theiss, *Optical Properties of Porous Silicon*, Habilitation thesis, Aachen University of Technology (RWTH, Aachen, 1995).

[384] H. Looyenga, *Dielectric constants of heterogeneous mixtures*, Physica **31**, 401 (1965).

[385] M. G. Berger, M. Thönissen, M. Krüger, R. Arens-Fischer, and H. Lüth, *Interferenzfilter auf der Basis von porösem Silicium*, German patent aplication, No. DE 196 22 748.8-51, date of filing: July 5, 1996.

[386] A. B. Sproul and M. A. Green, *Improved value for the silicon intrinsic carrier concentration 275 to 375 K*, J. Appl. Phys. **70**, 846 (1991).

[387] K. Misiakos and D. Tsamakis, *Accurate measurements of the silicon intrinsic carrier density from 78 to 340 K*, J. Appl. Phys. **74**, 3293 (1993).

[388] A. Schenk, *Finite-temperature full random-phase approximation model of bandgap narrowing for silicon device simulation*, J. Appl. Phys. **84**, 3684 (1998).

[389] P. P. Altermatt, A. Schenk, G. Heiser, and M. A. Green, *The influence of a new bandgap narrowing model on measurements of the intrinsic carrier density in crystalline Si*, in *Technical Digest of the 11th Int. Photovoltaic Science and Engineering Conf.* (Tanaka Printing, Kyoto, 1999), p. 719.

[390] R. J. Nelson, R. G. Sobers, *Minority carrier lifetime and internal quantum efficiency of surface-free GaAs*, J. Appl. Phys. **49**, 6103 (1978).

[391] M. Ruff, M. Fick, R. Lindner, U. Rössler, and R. Helbig, *The spectral distribution of the intrinsic radiative recombination in silicon*, J. Appl. Phys. **74**, 267 (1993).

[392] J. D. Cuthbert, *Recombination kinetics of excitonic molecules and free excitons in intrinsic silicon*, Phys. Rev. B **1**, 1552 (1970).

[393] W. Gerlach, H. Schlangenotto, and H. Maeder, *On the radiative recombination rate in silicon*, phys. stat. sol. (a) **13**, 277 (1972).

[394] H. Schlangenotto, H. Maeder, and W. Gerlach, *Temperature dependence of the radiative recombination coefficient in silicon*, phys. stat. sol. (a) **21**, 357 (1974).

[395] P. Jonsson, B. Bleichner, M. Isberg, and E. Nordlander, *The ambipolar Auger coefficient: measured temperature dependence in electron irradiated and highly injected n-type Si*, J. Appl. Phys. **81**, 2256 (1981).

[396] S. D. Brotherton, P. Bradley, and A. Gill, *Iron and the iron-boron complex in silicon*, J. Appl. Phys. **57**, 1941 (1985).

[397] Y. Hayamizu, T. Hamaguchi, and S. Ushio, *Temperature dependence of of minority carrier lifetime in iron-diffused p-type silicon wafers*, J. Appl. Phys. **69**, 3077 (1991).

[398] M. Schulz, *Impurity levels (Si)*, in *Landolt-Börnstein Zahlenwerte und Funktionen aus Naturwissenschaft und Technik*, Neue Serie III/Band 22 Halbleiter, Teilband b, Herausgeber O. Madelung und M. Schulz (Springer, Berlin, 1989), p. 270.

[399] J. G. Simmons and G. W. Taylor, *Nonequilibrium steady-state statistics and associated effects for insulators and semiconductors containing an arbitrary distribution of traps*, Phys. Rev. B **4**, 502 (1971).

[400] W. Füssel, M. Schmidt, H. Angermann, G. Mende, and H. Flietner, *Defects at the Si/SiO$_2$ interface: their nature and behavior in technological processes and stress*, Nucl. Instrum. Methods Phys. Res. A **377**, 177 (1996).

[401] H. Flietner, *Passivity and electronic properties of the silicon/silicon dioxide interfuse*, Materials Science Forum **185-188**, 73 (1995).

[402] D. A. Clugston and P. A. Basore, *PC1D Version 5: 32-bit solar cell modeling on personal computers*, in *Proc. 26th IEEE Photovoltaic Specialists Conf.* (IEEE, New York, 1997), p. 207.

[403] P. A. Basore, *Numeric modeling of textured Si solar cells using PC-1D*, IEEE Trans. Elecron. Devices **ED-37**, 337 (1990).

[404] U. Rau and M. Goldbach, *Modeling of the electronic transport in multijunction solar cells*, in *Proc. 1st World Conf. on Photovoltaic Energy Conversion* (IEEE, New York, 1995), p. 1421.

[405] M. Wolf, *Quantenausbeute von Si$_{1-x}$,Ge$_x$-Solarzellen*, Dissertation University of Suttgart (Cuvillier Verlag, Göttingen, 1998).

[406] K. Bücher, J. Bruns, M. Cokgüngör, and H. G. Wagemann, *Temperature and bias-light dependent recombination parameters in Si-solar cells explaining nonlinear response under concentration,* in *Proc. 9th European Photovoltaic Solar Energy Conf.*, edited by W. Palz, G. T. Wrixon, and P. Helm (Kluwer, Dordrecht, 1989), p. 429.

[407] J. Metzdorf, *Calibration of solar cells 1: the differential spectral responsivity method*, Appl. Opt. **26**, 1701 (1987).

[408] A. Schönecker, A. Zastrow, and K. Bücher, *Accurate spectral response measurements of non-linear high-efficiency cells*, in *Proc. 12th European Photovoltaic Solar Energy Conf.*, edited by R. Hill, W. Palz, and P. Helm (H. S. Stephens, Bedford 1994), p. 500.

[409] J. C. Jimeno, A. Cuevas, and A. Luque, *Determination of minority-carrier diffusion lengths and surface recombination velocity from the internal quantum efficiency of bifacial cells under posterior illumination*, in *Proc. 18th IEEE Photovoltaic Specialists Conf.* (IEEE, New York, 1985), p. 726.

[410] J. M. Gee, *A simple procedure to analyze rear-surface internal quantum efficiency*, in *Proc. 25th IEEE Photovoltaic Specialists Conf.* (IEEE, New York, 1996), p. 557.

[411] J. H. Werner, S. Kolodinski, U. Rau, J. K. Arch, and E. Bauser, *Silicon solar cells of 16.8 μm thickness and 14.7% efficiency*, Appl. Phys. Lett. **62**, 2998 (1993).

[412] C. Donolato, *Effective diffusion length of multicrystalline solar cells*, Semicond. Sci. Technol. **13**, 781 (1998).

[413] E. Tuominen, M. Acerbis, A. Hovonen, T. Siirtola, and J. Sinkkonen, *A method of extracting solar cell parameters from spectral response by inverse Laplace transform*, Physica Scripta **T69**, 306 (1997).

[414] J. Härkönen, M. Yli-Koshi, E. Tuominen, J. Sinkkonen, and T. Marjamäki, *Extraction of solar cell parameters from the spectral response by inverse Laplace method*, in *Proc. 14th European Photovoltaic Solar Energy Conf.*, edited by H. A. Ossenbrink, P. Hielm, H. Ehmann (H. S. Stephens & Assoc., Bedford, UK, 1997), p. 2431.

[415] A. Luque, *The requirements of high efficiency solar cells*, in *Physical Limitations to Photovoltaic Energy Conversion*, edited by A. Luque and G. L. Araùjo (Adam Hilger, Bristol, 1990), p. 1 and p. 28.

[416] K. Feldrapp, D. Scholten, S. Oelting, H. Nagel, M. Steinhof, R. Auer, and R. Brendel, *Thin monocrystalline Si solar cells fabricated by the pororous Si process using ion-assisted deposition*, in *Proc. 16th European Photovoltaic Solar Energy Conf.*, edited by H. Scheer, B. McNelis, W. Palz, H. A. Ossenbrink, and P. Helm (James & James, London, 2000), p. 1703.

[417] D. H. Ford, J. A. Rand, E. J. Delledonne, A. E. Ingram, J. C. Bisaillon, B. W. Feyock, M. G. Mauk, R. B. Hall, and A. M. Barnett, *High current, thin silicon-on-ceramic solar cells*, IEEE Trans. Electron. Devices **46**, 2162 (1999).

[418] T. Mishima, S. Itho, G. Matuda, K. Yamamoto, H. Kiyama, and T. Yokoyama, *Polycrystalline silicon solar cells using the plasma spray method*, in *Technical Digest of the 9th Int. Photovoltaic Science and Engineering Conf.* (Arisumi Printing, Tokyo, 1996), p. 243.

[419] H. S. Reehal, M. J. Twaites, , T. M. Bruton, *Thin film polycrystalline silicon solar cells prepared by plasma CVD*, phys. stat. sol. (a) **145**, 623 (1996).

[420] R. Shimokawa, K. Ishii, H. Nishikawa, T. Takahashi, Y. Hayashi, I. Saito, F. Nagamina, and S. Igari, *Sub-5 μm thin film c-Si solar cell and optical confinement by diffuse reflective substrate*, Solar Energy Materials & Solar Cells **34**, 277 (1994).

Index